Theoretical and Computational Chemistry

PROPERTIES AND FUNCTIONALIZATION OF GRAPHENE

A Computational Chemistry Approach

VOLUME 21

THEORETICAL AND COMPUTATIONAL CHEMISTRY

SERIES EDITORS

VOLUME 1
Quantitative Treatments of Solute/Solvent Interactions
P. Politzer and J.S. Murray (Editors)

VOLUME 2
Modern Density Functional Theory: A Tool for Chemistry
J.M. Seminario and P. Politzer (Editors)

VOLUME 3
Molecular Electrostatic Potentials: Concepts and Applications
J.S. Murray and K. Sen (Editors)

VOLUME 4
Recent Developments and Applications of Modern Density Functional Theory
J.M. Seminario (Editor)

VOLUME 5
Theoretical Organic Chemistry
C. Párkányi (Editor)

VOLUME 6
Pauling's Legacy: Modern Modelling of the Chemical Bond
Z.B. Maksic and W.J. Orville-Thomas (Editors)

VOLUME 7
Molecular Dynamics: From Classical to Quantum Methods
P.B. Balbuena and J.M. Seminario (Editors)

VOLUME 8
Computational Molecular Biology
J. Leszczynski (Editor)

VOLUME 9
Theoretical Biochemistry: Processes and Properties of Biological Systems
L.A. Eriksson (Editor)

VOLUME 10
Valence Bond Theory
D.L. Cooper (Editor)

VOLUME 11
Relativistic Electronic Structure Theory, Part 1. Fundamentals
P. Schwerdtfeger (Editor)

VOLUME 12
Energetic Materials, Part 1. Decomposition, Crystal and Molecular Properties
P. Politzer and J.S. Murray (Editors)

VOLUME 13
Energetic Materials, Part 2. Detonation, Combustion
P. Politzer and J.S. Murray (Editors)

VOLUME 14
Relativistic Electronic Structure Theory, Part 2. Applications
P. Schwerdtfeger (Editor)

VOLUME 15
Computational Materials Science
J. Leszczynski (Editor)

VOLUME 16
Computational Photochemistry
M. Olivucci (Editor)

VOLUME 17
Molecular and Nano Electronics: Analysis, Design and Simulation
J.M. Seminario (Editor)

VOLUME 18
Nanomaterials: Design and Simulation
P.B. Balbuena and J.M. Seminario (Editors)

VOLUME 19
Theoretical Aspects of Chemical Reactivity
A. Toro-Labbe (Editor)

VOLUME 20
The Crystalline States of Organic Compounds
A. Gavezzotti

Theoretical and Computational Chemistry

PROPERTIES AND FUNCTIONALIZATION OF GRAPHENE

A Computational Chemistry Approach

VOLUME 21

Edited by

TANDABANY DINADAYALANE
Associate Professor, Clark Atlanta University, Atlanta, Georgia, United States

FRANK HAGELBERG
Professor, East Tennessee State University, Johnson City, Tennessee, United States

Elsevier
Radarweg 29, PO Box 211, 1000 AE Amsterdam, Netherlands
The Boulevard, Langford Lane, Kidlington, Oxford OX5 1GB, United Kingdom
50 Hampshire Street, 5th Floor, Cambridge, MA 02139, United States

ISBN: 978-0-12-819514-7
ISSN: 1380-7323

For information on all Elsevier publications
visit our website at https://www.elsevier.com/books-and-journals

Publisher: Candice Janco
Acquisitions Editor: Kathryn Eryilmaz
Editorial Project Manager: Emerald Li
Production Project Manager: Kumar Anbazhagan
Cover Designer: Matthew Limbert

Typeset by STRAIVE, India

Contents

7. Reversible and irreversible functionalization of graphene

Y. Bhargav Kumar, Ravindra K. Rawal, Ashutosh Thakur, and G. Narahari Sastry

8. Interaction of amino acids, peptides, and proteins with two-dimensional carbon materials

Kanagasabai Balamurugan and Venkatesan Subramanian

9. Structures, properties, and applications of nitrogen-doped graphene

Tandabany Dinadayalane, Jovian Lazare, Nada F. Alzaaqi, Dinushka Herath, Brittany Hill, and Allea E. Campbell

10. Toward graphene-based devices for nanospintronics

Macon Magno and Frank Hagelberg

Contributors

Naheem Olakunle Adesina Division of Electrical and Computer Engineering, Louisiana State University, Baton Rouge, LA, United States

Nada F. Alzaaqi Department of Chemistry, Clark Atlanta University, Atlanta, GA, United States

Asanga B. Arampath Department of Physics and Center for Functional Nanoscale Materials, Clark Atlanta University, Atlanta, GA, United States

Kanagasabai Balamurugan Centre for High Computing, CSIR—Central Leather Research Institute, Chennai, India

Allea E. Campbell Department of Chemistry, Clark Atlanta University, Atlanta, GA, United States

Tandabany Dinadayalane Department of Chemistry, Clark Atlanta University, Atlanta, GA, United States

Connor W. Frye Department of Chemistry and Physics, University of Tennessee at Chattanooga, Chattanooga, TN, United States

Panchapakesan Ganesh Center for Nanophase Materials Sciences, Oak Ridge National Laboratory, Oak Ridge, TN, United States

Sophya Garashchuk Department of Chemistry & Biochemistry, University of South Carolina, Columbia, SC, United States

Frank Hagelberg Department of Physics and Astronomy, East Tennessee State University, Johnson City, TN, United States

Dinushka Herath Department of Chemistry, Clark Atlanta University, Atlanta, GA, United States

Brittany Hill Department of Chemistry, Clark Atlanta University, Atlanta, GA, United States

Ronald S. Holt Department of Chemistry and Physics, University of Tennessee at Chattanooga, Chattanooga, TN, United States

Jingsong Huang Center for Nanophase Materials Sciences, Oak Ridge National Laboratory, Oak Ridge, TN, United States

Jacek Jakowski Center for Nanophase Materials Sciences; Computational Sciences and Engineering Division, Oak Ridge National Laboratory, Oak Ridge, TN, United States

Y. Bhargav Kumar CSIR-North East Institute of Science and Technology, Jorhat; Academy of Scientific and Innovative Research (AcSIR), Ghaziabad, India

Jovian Lazare Department of Chemistry, Clark Atlanta University, Atlanta, GA, United States

David Lingerfelt Center for Nanophase Materials Sciences, Oak Ridge National Laboratory, Oak Ridge, TN, United States

Macon Magno Department of Physics and Astronomy, East Tennessee State University, Johnson City, TN, United States

Tharanga R. Nanayakkara Department of Physics and Center for Functional Nanoscale Materials, Clark Atlanta University; Department of Physics and Astronomy, Georgia State University, Atlanta, GA, United States

Yuriy V. Pershin Department of Physics and Astronomy, University of South Carolina, Columbia, SC, United States

Natarajan Ravi Department of Physics, Spelman College, Atlanta, GA, United States

Ravindra K. Rawal CSIR-North East Institute of Science and Technology, Jorhat, India

Thomas R. Rybolt Department of Chemistry and Physics, University of Tennessee at Chattanooga, Chattanooga, TN, United States

G. Narahari Sastry CSIR-North East Institute of Science and Technology, Jorhat; Academy of Scientific and Innovative Research (AcSIR), Ghaziabad, India

Jae H. Son Department of Chemistry and Physics, University of Tennessee at Chattanooga, Chattanooga, TN, United States

Ashok Srivastava Division of Electrical and Computer Engineering, Louisiana State University, Baton Rouge, LA, United States

Venkatesan Subramanian Centre for High Computing, CSIR—Central Leather Research Institute; Academy of Scientific and Innovative Research (AcSIR), CSIR-CLRI Campus, Chennai, India

Kelvin Suggs Department of Physics and Center for Functional Nanoscale Materials, Clark Atlanta University, Atlanta, GA, United States

Bobby G. Sumpter Center for Nanophase Materials Sciences; Oak Ridge National Laboratory, Oak Ridge, TN, United States

Ashutosh Thakur CSIR-North East Institute of Science and Technology, Jorhat, India

Xiao-Qian Wang Department of Physics and Center for Functional Nanoscale Materials, Clark Atlanta University, Atlanta, GA, United States

U. Kushan Wijewardena Department of Physics and Center for Functional Nanoscale Materials, Clark Atlanta University; Department of Physics and Astronomy, Georgia State University, Atlanta, GA, United States

Ruslan D. Yamaletdinov Nikolaev Institute of Inorganic Chemistry SB RAS; Boreskov Institute of Catalysis SB RAS, Novosibirsk, Russia

Preface

Following the first fabrication of an isolated graphene sheet in the laboratory, graphene was quickly recognized as a material with unique mechanical, electronic, and optical properties. Among several materials, graphene has an extremely high carrier mobility, which is one hundred times higher than that of silicon [1]. Thus, graphene is a candidate as a medium for future ultrafast electronics. The mechanical properties of graphene turned out to be equally astounding. While being one thousand times lighter than paper, graphene sheets proved to be one hundred times stronger than steel [2]. A single graphene sheet displays a light transparency of close to 98% [3], among other remarkable optical properties. Thus, graphene exhibits high efficiency in converting light into electricity [4]. It has been reported to exceed the thermal conductivity of pyrolytic graphite by a large margin [5]. New discoveries continue to add to the already substantial list of graphene superlatives [6].

Pristine graphene is not operational in digital applications due to its semi-metallic nature. As the valence and conduction bands of its π electron system coincide at the corners of its hexagonal Brillouin zone, the band energy gap of graphene vanishes, precluding its use as transistor material for post-silicon electronics. This obstacle is overcome by deriving novel nanomaterials from pure graphene. Various avenues may be followed to open a band gap within the graphene electronic system. One possibility is dimensional manipulation: graphene nanoribbons (GNRs), which are one-dimensional structures generated by truncating the graphene sheet, have finite energy gaps. Another possibility is multilayer graphene, a stacked assembly of several sheets. Further options involve graphene chemistry. For instance, attaching hydrogen atoms to the graphene sheet opens a tunable band gap that increases with increasing coverage [7]. A maximum gap of 3.9 eV was reported for graphane, the fully hydrogenated analogue of graphene [8].

Controlled manipulation is needed to turn pristine graphene into a material suitable for electronics applications. A parallel case can be made for spintronics. The low atomic number of carbon implies weak spin-orbit and hyperfine coupling, and thus long spin relaxation times. This feature in conjunction with the high carrier mobility of graphene suggests excellent spin transport properties. Unprocessed graphene, however, is nonmagnetic and therefore does not qualify as a spintronics medium. As in the electronic case, graphene derivatives that display the desired trait may be designed and realized. Again, this may be achieved by altering the dimension of the graphene sheet, or by functionalizing graphene. The first of these alternatives can be realized by reducing graphene to GNRs of the zigzag type [9], the second by importing magnetism into the intact sheet. This may be done by embedding magnetic units, such as transition metal atoms, into the graphene fabric [10].

Turning graphene, a nonmagnetic semimetal, into a material useful for electronics or for spintronics—these two priorities of recent graphene research and development exemplify what this volume is about: strategies for generating new systems that maintain some of the outstanding properties of graphene while exhibiting desired novel properties. This motif can be traced through all chapters assembled in this volume. In the following, we present a brief summary of these chapters, leading from pure yet dimensionally modified graphene to graphene combined with a rich variety of adsorbate species.

The initial sections of Chapter 1 (*Graphene—Technology and integration with semiconductor electronics*) provide a condensed characterization of pristine graphene. The chapter also focuses on recent efforts to employ graphene-based units as elements of nanoelectronic systems. These are tailored by dimensional manipulation, i.e., dimensional reduction (GNRs) and augmentation (bilayer graphene). Emphasis is placed on the use of GNRs as transistors and interconnects for integrated circuits. Thus, graphene-based field effect transistors (GFETs) have been proposed [11]. As transistors shrink, interconnects have to keep up. Traditionally, these units consist of copper, which responds to the reduction of the device dimensions from the micro- to the nano-regime with a distinct resistivity increase, resulting in enhanced power consumption. Replacing the present copper interconnects with graphene elements, and exploiting the experimentally detected ballistic conduction property of epitaxially grown GNRs [12], could offer a way out of this predicament. Surveying recent progress on GNRs as electronic nanodevices, Chapter 1 provides a condensed overview of the state of the art in establishing graphene as a basic material of integrated circuit nanotechnology.

Altering the properties of pure graphene can also proceed through structural deformation. This mode of inquiry is followed in Chapter 2 (*Kinks in buckled graphene stressed and unstressed in the longitudinal direction*). Specifically, the authors simulate GNRs traversed by travelling wrinkles, which may be understood as topological excitations (kinks) in buckled graphene membranes. The emphasis of the chapter is on the energetics and dynamics associated with kinks in the graphene-based host material. Molecular dynamics (MD) simulations applied to a GNR reveal several characteristic features of kinks in these elementary systems. In particular, the kinetic energy of moving kinks turns out to be relativistic. Remarkable effects associated with kink dynamics include negative radiation pressure (NPR), which can appear when allowance is made for kinks interacting with radiation that causes phonon excitation in the graphene network. Kink propagation may proceed away from or toward the radiation source, corresponding to positive or negative radiation pressure. The latter is understood in terms of resonance between the kink and the incoming radiation. With respect to methodology, this chapter demonstrates that some fundamental properties of graphene are accessible to modeling in a classical framework.

The following chapter (Chapter 3: *From classical to quantum dynamics of atomic and ionic species interacting with graphene and its analogue*) focuses on irradiation of graphene flakes with atomic or ionic beams, placing emphasis on a multiscale approach to materials modeling. Classical and quantum methods of modeling nuclear dynamics on the potential energy surface of the electronic ground state, described by density functional tight binding (DFTB) theory, are compared. Nuclear quantum effects, such as tunneling and zero-point corrections, become important for light beam particles. They are taken into account through quantum trajectories, calculated on the basis of Bohmian mechanics [13]. This approach is shown to capture various key phenomena associated with interactions between atoms or ion beams and fragments of graphene, among them scattering and transmission as well as reactions between the beam species and the graphene target. Simulations of Ar cluster ions, H/D, and H+/D+ projectiles interacting with graphene, provide detailed information about fundamental atomic-scale processes involved in these interactions, such as forming and breaking covalent bonds, isotopic substitution phenomena, and nuclear quantum effects for light atomic or ionic projectiles.

Chapter 4 (*From ground to excited electronic state dynamics of electron and ion irradiated graphene*) continues the thread of graphene engineering by beam irradiation, dealing with electronic excitations of graphene and graphene derivatives. More specifically, it focuses on present computational efforts to study the dynamics of excited electronic states in these media, induced by interaction with beams of charged particles. The highly local information achievable by these techniques allows to probe the electronic structure of graphene-based nanomaterials to a degree of spatial resolution far beyond optical procedures. Further, irradiation methods may be employed to modify these materials at an atomic level of precision and thus could be of high interest as nanotechnological tools. This is demonstrated by the example of defects in graphene quantum dots (GQDs) with different edge morphologies, armchair versus zigzag. Controlled electronic excitation by beam irradiation techniques is shown to provide a sensitive instrument for altering the stability and local geometry of substitutional atoms in the GDQs. These concepts are discussed in the framework of two complementary approaches to modeling the interaction between a graphene surface and a charged projectile. In particular, a linear-response model involving time-dependent density functional theory (TD-DFT) in the frequency domain is compared with direct simulation in the time domain. Transition rates computed with the former and excited states populations obtained from the latter provide evidence for the consistency of the two approaches.

Chapters 5–10 highlight various aspects of functionalized graphene and related systems. Chapter 5 (*Molecule-graphene and molecule-carbon surface binding energies from molecular mechanics*) considers adsorption to graphene, multilayer graphene, carbon nanotubes, and surface modified graphene in the framework of classical molecular mechanics. The authors assess the capacity of MM2 and MM3 force field procedures to account for adsorption energies of various molecules on carbon nanostructure surfaces. By the standard of experimental values, adsorption energies calculated with these force field methods turn out to compare very favorably with DFT results uncorrected for dispersion forces. Molecular mechanics is shown in this chapter to perform well when noncovalent interactions between gas species and graphene or graphene-derived substrates are involved. Therefore, this approach can open functionalized graphene systems consisting of thousands of atoms to quantitative analysis, while DFT computations, if properly adjusted for van der Waals interactions, might be prohibitively time-consuming in this size regime.

Covalent functionalization of pristine graphene, treated with a multiplicity of quantum-theory-based methods, is at the center of Chapter 6 (*Structural and electronic properties of covalently functionalized graphene*). Accentuating the importance of turning graphene into an electronic material, the authors discuss covalent functionalization as a way to endow graphene with a band gap. A variety of systems are analyzed, including nitrophenyl functionalized graphene, nitrogen-seeded twisted bilayer, cycloaddition of fluorinated olefins on graphene, and graphene oxide. In particular, functionalization with nitrophenyl is found to alter the π conjugation of graphene near the Fermi level substantially. As a result of this modification, a band gap opens. The size of this gap is shown to vary with the degree of nitrophenyl coverage. A diversity of computational approaches are used to study the structures discussed in this chapter, reaching from DFT to the many-body Bethe-Salpeter equation. The latter approach, allowing for an approximate treatment of electron-hole and electron-electron quasiparticle excitations, proves to be appropriate for investigating the band structure of a twisted graphene bilayer seeded with nitrogen atoms.

A survey of reversible versus irreversible modes of graphene functionalization is presented in Chapter 7 (*Reversible and irreversible functionalization of graphene*), where reversible and irreversible manipulation of graphene is correlated with noncovalent and covalent bonding, respectively. Reversible mechanisms are distinguished by their special type of linkage between an external chemical agent and the graphene π subsystem. Thus, a large set of noncovalent interactions that typically can be reversed is specified, among them cation-π, anion-π, π-π, π_{cation}-π, or lone pair-π. The authors underscore the importance of combining graphene with biomolecules, exemplified by nucleobases and amino acids, to assemble biologically active complexes. Interactions with these adsorbates are noncovalent and fall under the cation-π and π-π interactions. Functionalizing graphene by the above procedures is contrasted with various covalent, and irreversible, procedures. Covalent connections between graphene and single atoms are discussed hydrogen, oxygen, and various halogen species as examples. Further, graphene, as an inherently π electron-rich material, undergoes a wide range of cycloaddition reactions associated with the formation of two new σ-bonds from the reactant π electrons, giving rise to a cyclic product. These irreversible modifications make it possible to develop the graphene sheet into novel materials. Reversible functionalization, on the other hand, permits to accommodate the properties of graphene to a large scale of possible applications.

Expanding one of the motifs addressed in Chapter 7, namely graphene functionalization by combination with biomolecules, the following chapter (Chapter 8: *Interaction of amino acids, peptides, and proteins with two-dimensional carbon materials*) focuses on proteins attaching to graphene or graphene derivatives. This topic is of major importance for the use of graphene in biological or medical contexts. Among the numerous applications of biological nanoscience involving graphene is biomolecular and cellular imaging as well as gene and drug delivery. Pristine graphene as well as graphene oxide, however, have proved to be toxic [14]. The therapeutic use of pure graphene templates thus requires functionalizing them with biocompatible polymers or other organic species [15,16]. The biomolecular adsorbates covered in this survey are proteins, amino acids, and oligopeptides. The substrate carbon materials include graphene, graphene oxide, and carbon nanotubes. The large size variation of biologically relevant molecules requires adjusting the computational method to the demands of the system of interest. While quantum mechanical procedures are viable for describing a graphene-amino acid complex, modeling mesoscale systems that involve large proteins may necessitate using coarse-grained schemes based on classical mechanics.

Chapter 9 (*Structures, properties and applications of nitrogen-doped graphene*) presents a survey on graphene with nitrogen impurities. Beyond computational modeling, the chapter covers methods of fabricating

N-doped graphene, as well as characterization techniques for these nanomaterials. A wide variety of systems can be derived from three basic morphologies, associated with N atoms in *graphitic*, *pyridinic*, and *pyrrolic* configurations. Graphitic N atoms occupy substitutional positions, replacing C atoms within the hexagonal graphene network. They may also be located at the edges of graphene fragments. Pyridinic N bonds to two C atoms and resides adjacent to a vacancy in the graphene fabric or, alternatively, at the edge of a truncated graphene sheet. Pyrrolic N is located in a five-membered ring and donates two p electrons to the graphene π system, in contrast to pyridinic N, which donates a single p electron. As outlined in Chapter 9, these fundamental modes of combining graphene sheets with N impurities have been extensively investigated by both experiment and computation. From the literature review in this chapter, wide perspectives for future applications of graphene with added N atoms have emerged. These encompass areas as diverse as Li-ion batteries, fuel cells, field-effect transistors (FETs), ultra- and super-capacitors, photocatalysis, and sensors.

The final chapter (Chapter 10: *Toward graphene-based devices for nano-spintronics*) in this collection deals with graphene as a potential material for elements of spintronics, or spin electronics, where the spin of the electron, rather than its charge, is employed as the basic physical quantity for processing information. Spin devices include spin valves, spin transistors, and spin filters [17]. A common feature of these designs is that they subject charge carriers to magnetic interactions in order to generate or manipulate spin currents. As pointed out previously, both theory and experiment have identified graphene as a highly suitable medium for spin currents, which is rationalized by its extreme mechanical and thermal stability, its high charge carrier mobility, and, most importantly, its long spin relaxation times [18]. Naturally, research on transmission elements for spintronics applications has focused on zigzag graphene nanoribbons (zGNRs), as these are intrinsically magnetic. At the same time, the zigzag edges are highly reactive and thus difficult to stabilize. Armchair graphene nanoribbons (aGNRs) are chemically more robust but nonmagnetic. Magnetically active aGNRs, on the other hand, may be generated by functionalizing the ribbon with magnetic agents. The work summarized in this chapter explores the impact of one or two substitutional Fe atoms on the spin transport characteristics of aGNRs. In particular, the question of optimizing the spin-filtering efficiency of these composites is addressed. Sizeable degrees of current spin polarization were recorded, amounting to more than 90% in selected ranges of the bias across the device. This approaches the limit of half-metallicity, where conduction is restricted to one well-defined spin orientation only.

All chapters in this volume review current initiatives of modeling pathways toward graphene-based systems with optimized properties for a wide spectrum of potential applications. Together, they present a picture of an ongoing computational effort that, we hope, will lead to experimental examination of the models proposed here and eventually to developing unprecedented materials that combine some of the unique properties of graphene with custom-tailored novel traits.

Tandabany Dinadayalane
Department of Chemistry, Clark Atlanta University, Atlanta, GA, United States

Frank Hagelberg
Department of Physics and Astronomy, East Tennessee State University, Johnson City, TN, United States

References

[1] H.H. Radamson, Graphene, in: Springer Handbook of Electronic and Photonic Materials, Springer, 2017.

[2] ACS Material. https://www.acsmaterial.com/graphene-facts. n.d.

[3] R. Binder (Ed.), Optical Properties of Graphene, World Scientific Publishing, Co, 2016.

[4] K.-T. Lin, H. Lin, T. Yang, B. Jia, Structured graphene metamaterial selective absorbers for high efficiency and omnidirectional solar thermal energy conversion, Nat. Commun. 11 (2020) 1389.

[5] A.A. Baladin, Thermal properties of graphene and nanostructured carbon materials, Nat. Mater. (2011) 569–581.

[6] S.K. Tiwari, S. Sahoo, N. Wang, A. Huczko, Graphene research and their outputs: status and Prospect, J. Sci. Adv. Mater. Dev. 5 (2020) 10.

[7] K.A. Whitener, Hydrogenated graphene: a user's guide, J. Vac. Sci. Technol. A 36 (2018), 05G401.

[8] J. Son, S. Lee, S.J. Kim, B.C. Park, H.K. Lee, S. Kim, J.H. Kim, B.H. Hong, J. Hong, Hydrogenated monolayer graphene with reversible and tunable wide band gap and its field effect transistor, Nat. Commun. 7 (2016) 13261.

[9] K. Wakabayashi, in: T. Makarova, F. Palacio (Eds.), Carbon-Based Magnetism, Elsevier, Amsterdam, 2006, p. 279.

[10] F. Hagelberg, A. Kaiser, I. Sukuba, M. Probst, Spin filter properties of armchair graphene nanoribbons with substitutional Fe atoms, Mol. Phys. 115 (2017) 2231–2241.

[11] K.S. Novoselov, A.K. Geim, S.V. Morozov, D. Jiang, Y. Zhang, S.V. Dubonos, I.V. Grigorieva, A.A. Firsov, Electric field effect in atomically thin carbon films, Science 306 (2004) 666.

[12] J. Baringhaus, M. Ruan, F. Edler, A. Tejeda, M. Sicot, A. Taleb-Ibrahimi, A. Li, Z. Jiang, E. Conrad, C. Berger, C. Tegenkamp, W. de Heer, Exceptional ballistic transport in epitaxial graphene nanoribbons, Nature 506 (2014) 349.

[13] R.E. Wyatt, Quantum Dynamics with Trajectories, Springer, 2005.

[14] Y. Li, Y. Liu, F. Yujian, W. Taotao, L.L. Guyadera, G. Gao, L. Ru-Shi, C. Yan-Zhong, C. Chen, The triggering of apoptosis in macrophages by pristine graphene through the MAPK and TGF-beta signaling pathways, Biomaterials 33 (2012) 402.

[15] K. Yang, S. Zhang, G. Zhang, X. Sun, S.T. Lee, Z. Liu, Graphene in mice: ultrahigh *in vivo* tumor uptake and efficient photothermal therapy, Nano Lett. 10 (2010) 3318–3323.

[16] A. Sahu, W.I. Choi, G. Tae, A stimuli-sensitive injectable graphene oxide composite hydrogel, Chem. Commun. 48 (2012) 5820–5822.

[17] F. Hagelberg, Magnetism in Carbon Nanostructures, Cambridge University Press, Cambridge, 2017.

[18] T. Maassen, J. Jasper van den Berg, N. IJbema, F. Fromm, T. Seyller, R. Yakimova, B. van Wees, Long spin relaxation times in wafer scale epitaxial graphene on SiC(0001), Nano Lett. 12 (2012) 1498.

CHAPTER

1

Graphene—Technology and integration with semiconductor electronics

Ashok Srivastava and Naheem Olakunle Adesina

Division of Electrical and Computer Engineering, Louisiana State University, Baton Rouge, LA, United States

1 Introduction

Carbon is one of the most abundantly found elements in the earth's crust [1]. It is also known that carbon can exist in nature basically in two forms: crystalline and amorphous. Carbon is one of the tetravalent elements in the periodic table. It belongs to the family of semiconductors, which have their conductivities in between that of a conductor and an insulator. It has four outermost electrons, so it usually participates in covalent type of bonding. One of the crystalline allotropes of carbon is graphite. Graphite is considered as a stacked-layer structure with its atoms arranged in hexagonal structure. The layers of graphite are bonded with a weaker van der Waal's forces, which then allow free movement of electrons. Unlike diamond, which is bonded by very strong covalent bonds, graphite is a good conductor of heat and electricity. One of the most uses of graphite is lead in pencil or as a lubricant. Fig. 1.1 shows crystal structures of different allotropes of carbon [2]. Graphene is a monolayer of carbon atoms packed into a dense hexagonal honeycomb crystal structure as shown in Fig. 1.1A, which can be separated and viewed as an individual atomic plane extracted from graphite as shown in Fig. 1.1B or as an unrolled single-walled carbon nanotube as shown in Fig. 1.1C or as a giant flat fullerene molecule or buckyball as shown in Fig. 1.1D. Single layer of graphite or graphene was presumed not to exist in free stable form until 2004 when Novoselov et al. [3] experimentally first isolated single-layer graphene by micromechanical cleavage technique, peeling off repeatedly from graphite crystal using adhesive scotch tape, and reported their seminal work on the field-effect study of such atomically thin carbon film. They were

Theoretical and Computational Chemistry, Volume 21
https://doi.org/10.1016/B978-0-12-819514-7.00006-3

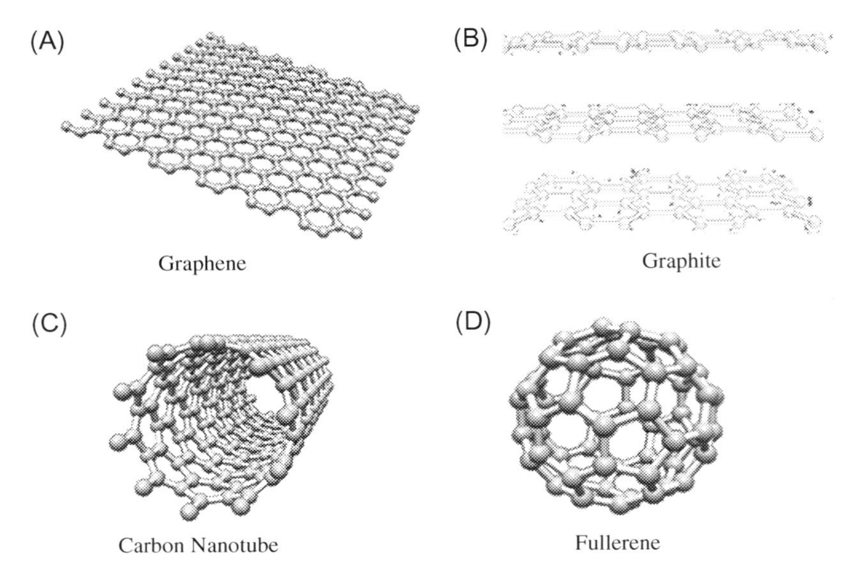

FIG. 1.1 Different allotropes of carbon in different dimensions: (A) two-dimensional (2D) atomically thick graphene, (B) three-dimensional (3D) graphite, (C) one-dimensional (1D) carbon nanotube, and (D) zero-dimensional (0D) fullerene. *Reproduced with permission from A. Srivastava, Graphene and other graphene materials technology and beyond, in: A. Srivastava, S. Mohanty (Eds.), Advanced Technologies for Next Generation Integrated Circuits, IET Press, 2020.*

able to isolate graphene with no heteroatomic contamination using a simple "Scotch Tape Method" [4]. This discovery has engendered a lot of interests in graphene research and revolutionized electronics.

Graphene is a single atom-thick sheet of hexagonally arranged carbon atoms, which uses one s-orbital and two of its p-orbitals to form bond formation (sp^2 hybridization). It has a lateral dimension, which may vary from several nanometers to microscale. Monolayer (single layer) is the purest form known and is useful for high-frequency electronics [5]. Although it can exist as two layers (bilayer) and three layers (trilayer), graphene exhibits different properties with varying number of layers. Graphene is a wonderful material with many superlatives to its name. It is the thinnest known material in the universe and the strongest ever measured [6]. From 2004, research on graphene accelerated exponentially considering graphene as an exciting condensed matter physics problem. Novoselov et al. [7] found that the electron transport in graphene is governed by relativistic Dirac equation where the charge carriers resemble Dirac fermions, relativistic particles with zero rest mass (massless particle) with an effective speed in the range of light [2]. Katsnelson et al. [8] reported that by using electrostatic barriers in a single-layer and a bilayer graphene, the massless Dirac fermions in graphene demonstrate Klein tunneling, which is the unhindered penetration of relativistic particles through a wide potential barrier. The quantized quantum Hall conductance, which is generally observed at low temperature and strong magnetic field, was also observed in graphene at room temperature [9]. Moreover, Bolotin et al. [10] found that the low-temperature carrier mobility is three times that of the best semiconductor. Its charge carriers exhibit giant intrinsic mobility, have zero effective mass, and can travel for micrometers without scattering at room temperature [6]. Thermal conductivity of graphene is also reported to be at least twice as large as that of copper for similar geometry [11]. The electron mobility in suspended graphene is found as 200,000 cm^2/V s, which is 143 times greater than that of Si (1400 cm^2/V s at 300 K) [12,13]. Therefore, graphene has very high electrical conductivity when compared with any semiconductor material.

Prior to 2004, precisely between the late 1970s to the early 1990s, major attention was focused to fullerenes (buckyballs) and carbon nanotubes, which were discovered in 1985 [14] and 1991 [15], respectively. However, some key features of currently known graphene were reported during that period. Semenoff [16] found in 1984 that the wave functions of graphene are similar to the solutions of relativistic Dirac equation. In 1987, Mouras et al. [17] coined the term "graphene" for single crystalline 2D carbon allotrope, before which graphene was commonly termed as "thin graphite lamellae." Surprisingly, even before the experimental observation of two different types of edge states, zigzag and armchair in graphene nanoribbon (GNR) [2], a nanometer dimensional form of infinite graphene sheet, Nakada et al. [18] in 1996 extensively and accurately predicted their edge states with corresponding energy band structure.

2 Properties of graphene

Many researchers have reported in their respective works that graphene exhibits remarkable properties, which are quite different from what are obtainable in other types of materials. This makes graphene a material of great utility and has wider area of applications. In this section, we will introduce the fundamental properties of graphene and their respective merits.

2.1 Band structure

Graphene is a two-dimensional (2D) material made of carbon atoms in a honeycomb-like hexagonal lattice, as shown inside Fig. 1.2A. In 1947, Wallace also considered graphene as a single graphite layer in order to estimate its band structure [19]. The carbon atoms form strong σ covalent bonds by three in-plane sp^2-hybridized orbitals, whereas the fourth bond

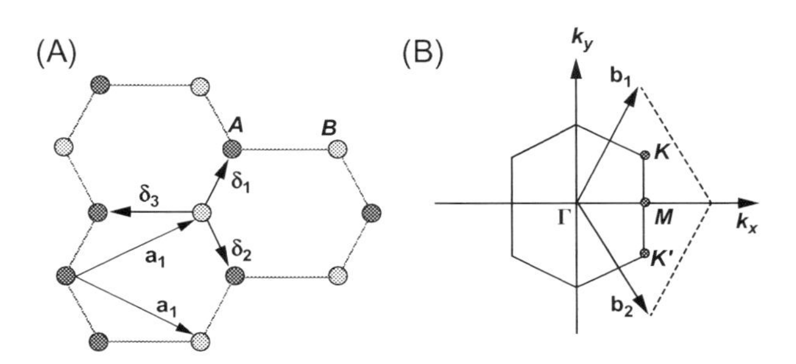

FIG. 1.2 (A) Hexagonal lattice structure of graphene consisting of two atoms A and B in a unit cell. a_1 and a_2 show the direction of the lattice vectors in the primitive unit cell and (B) reciprocal lattice vectors b_1 and b_2 in the first Brillouin zone [22].

is a π bond in z-direction [20]. The electron in this bond can move freely in the delocalized π-electronic system referred as the π-band and π^*-bands [21]. Based on the hexagonal lattice structure of graphene, we obtain a primitive unit cell, which is formed from two interpenetrating triangular lattices. The two-dimensional direct lattice vectors can be written as follows:

$$a_1 = \frac{a}{2}\left(3, \sqrt{3}\right), \quad a_2 = \frac{a}{2}\left(3, -\sqrt{3}\right) \tag{1.1}$$

where $a = 1.42$ Å is the carbon–carbon distance commonly referred to as lattice parameter. Since electronic transport can be two-dimensional in a graphene lattice, the dispersion relation for graphene also has two dimensions as shown in Fig. 1.2B. The reciprocal lattice vectors can be obtained as follows:

$$b_1 = \frac{2\pi}{3a}\left(1, \sqrt{3}\right), \quad b_2 = \frac{2\pi}{3a}\left(1, -\sqrt{3}\right) \tag{1.2}$$

Owing to honeycomb lattice structure, there are two sets of three cone-like points K and K' on the edge of the Brillouin zone, which leads to valley degeneracy of $g_v = 2$. These are named Dirac points, where the conduction and valence bands meet each other in momentum space [20] as follows:

$$K = \left(\frac{2\pi}{3a}, \frac{2\pi}{3\sqrt{3}a}\right), \quad K' = \left(\frac{2\pi}{3a}, -\frac{2\pi}{3\sqrt{3}a}\right) \tag{1.3}$$

Theoretically, the conduction π^*-band is fully filled, while the valence π-band is completely empty. This causes the Fermi energy located at the Dirac point and results in semi-metallic behavior for graphene. In graphene, the conduction band touches the valence band with no energy bandgap in between. Graphene, being semi-metallic, is universally called zero-bandgap material. The behavior of charge carriers near Dirac points resembles the Dirac spectrum for massless fermions [23] and can be described by the linear dispersion relation as follows:

$$E\left(\vec{k'}\right) = \pm\hbar\, v_F \left|\vec{k'}\right| \tag{1.4}$$

where k' is the momentum near the Dirac point, $\hbar$ is the reduced Plank constant, and v_F is the Fermi velocity. The term $\hbar v_F$ represents the gradient of dispersion. The linear dispersion relation is contrary to most of materials as the solution of Schrodinger equation has second order in space and first order in time, leading to quadratic dispersion. Charge carriers near Dirac points behave like relativistic particles ideally transporting with Fermi velocity, which is theoretically $^1/_{300}$ of the speed of light [23]. Thus, the Hamiltonian for electrons near Dirac points in graphene can be calculated by Dirac equation with zero mass [24] as follows:

$$H = \hbar\, v_F \begin{pmatrix} 0 & k_x - ik_y \\ k_x + ik_y & 0 \end{pmatrix} = \hbar\, v_F\, \vec{\sigma} \,.\, \vec{k} \tag{1.5}$$

where $\vec{\sigma} = \left(\sigma_x, \sigma_y\right)$ is the 2D vector of the Pauli matrices and $\vec{k}$ is the momentum of the quasi-particles in graphene. The term "graphene pseudospin" originated when the massless chiral Dirac equation was applied by Semenoff [16] in order to explain the low-energy band structure of graphene.

Assuming the first nearest-neighbor interaction, the close form of dispersion relation near Dirac points can be obtained [23] as follows:

$$E\left(\vec{k}\right) = \pm t\sqrt{1 + 4\cos\frac{\sqrt{3}k_x a}{2}\cos\frac{k_y a}{2} + 4\cos^2\frac{k_y a}{2}} \tag{1.6}$$

Here, t is the nearest-neighbor hopping energy (hopping between different sublattices). The plus and minus signs correspond to conduction and valence bands, respectively. It can be inferred from Eq. (1.6) that the spectrum is symmetric around zero energy where both the conduction and valence bands touch each other.

The full band structure of graphene first Brillouin zone is shown in Fig. 1.3 [25]. It can be seen that the energy dispersion around the band edges of graphene is linear. The plots also depict electron energy dispersion for π and π^*-bands in the first Brillouin zone as contour plots at equidistant energies and as pseudo-3D representations for the 2D structures. Furthermore, the linear dispersion relation established that the conduction and the valence bands touch each other at the charge neutrality point known as the Dirac point as shown by the arrow in Fig. 1.3. The Dirac points are located at the symmetric K and K' points which have been plotted in Wolfram computational dynamic player tool. Fig. 1.3 shows that graphene has zero bandgap; hence it is referred to as semi-metal.

The density of state per unit cell of graphene can be approximated by the linear dispersion relation near the Dirac point as follows:

$$D(E) = \frac{g_v g_s}{2\pi}\frac{|E|}{(\hbar v_F)^2} \tag{1.7}$$

where $g_v = 2$ is the valley degeneracy, $g_s = 2$ is the valley degeneracy corresponding to the K and K' points, and $v_F = \sqrt{3}ta/2\hbar$ is the Fermi velocity. The Fermi velocity can be as high as

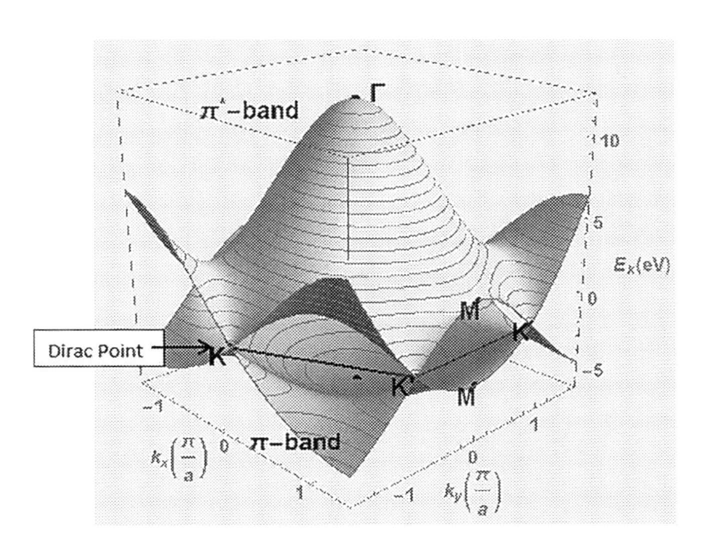

FIG. 1.3 Dispersion relation of graphene first Brillouin zone shown in reciprocal lattice space (k-space) with both x- and y-axes normalized with π/a. K and K' are the symmetric points [26].

$3\times10^6\,\mathrm{m/s}$ in suspended graphene [27], while it can be as low as $0.85\times10^6\,\mathrm{m/s}$ when electron–electron interactions are weak [28]. One of the ways to alter Fermi velocity is by changing the dielectric constant of the embedding environment. This is because the self-generating energy of the electron varies inversely with dielectric screening [29,30].

2.2 Carrier density

The carrier density of 2D electron gas sheet in graphene can be calculated from [31]:

$$n=\int_0^\infty D(E)f(E)dE \tag{1.8}$$

where $f(E)=(1+\exp[(E-E_f)/k_BT])^{-1}$ is the Fermi-Dirac distribution function and $D(E)$ is the density of states in Eq. (1.8), k_BT is the thermal energy, and E_f is the average Fermi level. The dominant carrier contribution in graphene carrier density induced by the gate voltage V_G is as follows:

$$n_G=p-n=-C_G(V_G-V_{Dirac})/q \tag{1.9}$$

where n_G is the induced carrier in graphene due to the gate voltage, C_G is the effective gate capacitance per unit area, and q is the electron charge. Since the gate-induced carriers are negligible near Dirac point and the carrier density is determined by the electron and hole puddles carriers (n^*), the formula connecting the thermally generated carriers (n_{th}) and the carrier density in graphene at Dirac point, n_{Dirac}, is derived as follows:

$$n_{Dirac}\approx\left[(n^*/2)^2+n_{th}^2\right]^{1/2} \tag{1.10}$$

Similarly, the thermally generated carriers in 2D graphene is given as [31]:

$$n_{th}=\frac{\pi}{6}\left(\frac{k_BT}{\hbar v_F}\right)^2 \tag{1.11}$$

where k_B is the Boltzmann constant and T is the absolute temperature on graphene. By assuming that the spatial electrostatic potential is a periodic step function with equal size and amplitude $\pm\Delta$, the residual charge puddle density n^* in graphene [32] is modeled as given in Eq. (1.12):

$$n^*=\int_{-\Delta}^\infty D(E+\Delta)f(E)dE+\int_{\Delta}^\infty D(E-\Delta)f(E)dE \tag{1.12}$$

By averaging the $\pm\Delta$ regions in the limit of $\Delta/k_BT>>1$, the equation can be simplified to $n^*\approx\Delta^2/\pi\hbar^2v_F^2$ [33]. Scanning tunneling microscopy [34] has been used to measure Δ for some materials. In case of graphene on SiO_2, $\Delta\approx59\,\mathrm{meV}$, so the charge puddle density is calculated as $n^*\approx2.6\times10^{11}\,\mathrm{cm}^{-2}$ at Dirac voltage of 3.66 V. Therefore, the total concentration of the electron and hole can be calculated as follows [34]:

$$n,p \approx \frac{1}{2}\left[\pm n_{g} + \sqrt{n_{g}^{2} + 4n_{\text{Dirac}}^{2}}\right] \tag{1.13}$$

where upper and lower signs correspond to the electron and hole carriers, respectively. Due to thermally generated carriers, the plot of carrier density versus gate voltage near Dirac points changes to nonlinear and its range increases with temperature. Hence, the electron and hole puddles become less important ($k_B T >> \Delta$). The carrier density increases and mobility decreases with the temperature due to scattering mechanisms, which lead to the decrease in temperature dependence of conductivity ($\sigma(E_F) = en(E_F)\mu(E_F)$) near Dirac point [34].

2.3 Conductivity

The experimental characteristic of graphene sheet shows a great deviation from an ideal theoretical graphene because of many sources of disorders such as lattice imperfections [35], impurities [36], anharmonic effects, and phonons [37]. These defects can manipulate the carrier transport in graphene by increasing the scattering, such that the carrier mobility is reduced by two orders of magnitude from ~1,000,000 cm^2/V s [38,39] to ~10,000 cm^2/V s [2]. The transport regime is determined based on the comparison between the graphene length L and the carrier mean free path [40], which scales the strength of scattering mechanisms. In ballistic transport, Landauer formalism describes the transport since mean free path is larger than the graphene length [41]. In this regime, the carrier is free of scattering and can travel at Fermi velocity (v_F). The conductivity can then be calculated as [23]:

$$\sigma_{\text{bal}} = \frac{L}{W}\frac{4e^2}{h}\sum_{n=1}^{\infty} T_n \tag{1.14}$$

In Eq. (1.14), the summation is taken over all available longitudinal transport modes and T_n is the transmission probability of mode n. In a diffusive transport, the conductivity can be described as a random walk in two-dimensions provided the graphene length is considered larger than the mean free path of carriers [31]. The carriers undergo elastic and inelastic collisions (scattering) and cause transport incoherence. The majority of literature discusses short-range scattering (defects, adsorbates), long-range scattering (Coulomb scattering by charged impurities), and electron–phonon scattering. This is because of their far-reaching effects when compared with other types of scattering. The semi-classical Boltzmann transport theory treats the scattering mechanisms using their scattering time τ to calculate the conductivity as follows [42]:

$$\sigma_{\text{sc}} = \frac{e^2}{2}\int D(\epsilon)v_k^2\,\tau(\epsilon)\left(-\frac{\partial f}{\partial \epsilon}\right)d\epsilon \tag{1.15}$$

where $v_{k'}$ and f are the carrier velocity and Fermi distribution function, respectively. At low temperature, the equation can be approximated as follows:

$$\sigma_{\text{sc}} = \frac{e^2 v_F^2}{2} D(E_F)\,\tau(E_F) \tag{1.16}$$

where $D(E_F)$ is the density of states and $\tau(E_F)$ is the scattering relaxation time at Fermi energy. By substituting Eq. (1.7) and Fermi energy with reference to K point as $E_F \approx \hbar v_F k_F \approx \hbar v_F \sqrt{\pi n}$ into Eq. (1.16), the conductivity can also be expressed in terms of carrier density as follows:

$$\sigma_{sc}(n) = \frac{e^2 v_F \tau}{\hbar} \sqrt{\frac{n}{\pi}} \tag{1.17}$$

Thus, the temperature and carrier density dependence of scattering mechanisms can be investigated to reveal the dominant scattering mechanism. A typical experimental description of graphene conductivity is presented in Fig. 1.4 [43].

Fig. 1.4A shows the raw data on a measurement of the resistivity, $\rho_{measured}$, of an exfoliated graphene sheet, while Fig. 1.4B is the fit of the conductivity, $\sigma_{sub} = 1/\rho_{measured}$ using Eq. (1.18). R is the range, and its values are constrained to be of the order of a, and n_i is the only fitting parameter.

$$\sigma_{dc} = \frac{4e^2}{h}\frac{k_f^2}{2\pi^2} \cdot \ln^2(k_f R) \tag{1.18}$$

2.4 Ambipolar field effect

By applying gate voltage (V_G) to intrinsic graphene, we can induce a surface charge density and accordingly tune the overall Fermi level [31]. Increasing (decreasing) the gate voltage increases the electron (hole) carriers and correspondingly shifts the Fermi energy toward the conduction (valence) band as shown in Fig. 1.5A [44]. The graphene is in electron and hole regimes far from Dirac point. Carrier mobilities can be extracted from Drude model $\sigma(E_F) = en(E_F)\mu(E_F)$, where μ is the mobility and n is the carrier conductivity [3,45]. Fig. 1.5B shows the calculated carrier density as a function of gate voltage, which clearly illustrates the fact that charge density is well controlled by the gate away from the Dirac point.

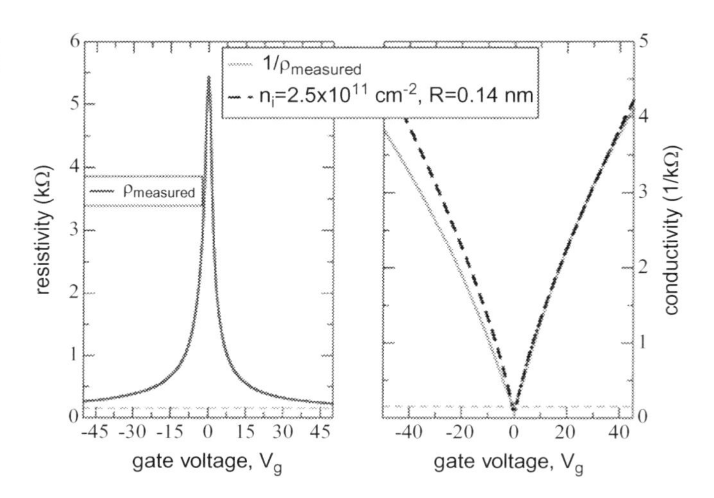

FIG. 1.4 Experimental data on graphene's conductivity: (A) resistivity and (B) conductivity [43].

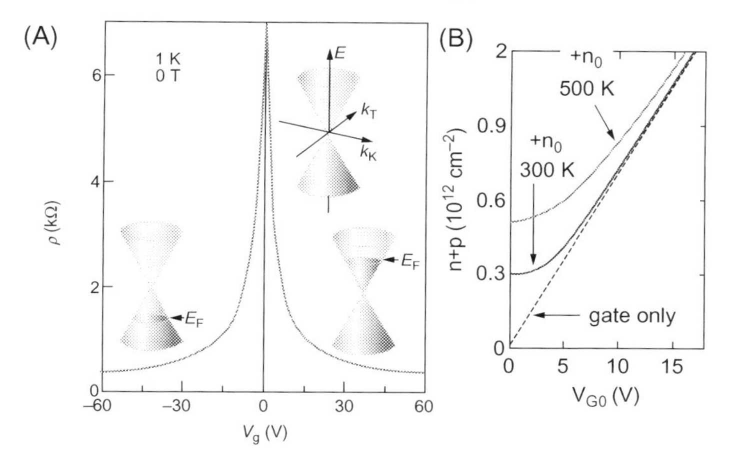

FIG. 1.5 Ambipolar electric field effect in graphene: (A) changes in the position of the Fermi level E_f as a function of gate voltage [44] and (B) calculated charge density versus gate voltage at 300 and 500 K. *Reproduced with permission from V.E. Dorgan, M.H. Bae, E. Pop, Mobility and saturation velocity in graphene on SiO_2, Appl. Phys. Lett. 97(8) (2010) 082112 with the permission of AIP Publishing.*

Since graphene is a semiconductor with no bandgap, the gate voltage can tune the charge carriers continuously between electrons and holes. Thus, the conductivity of graphene is due to either the electron transport for $V_G > V_{Dirac}$ or the hole transport regime for $V_G < V_{Dirac}$. In other words, the graphene exhibits ambipolar electric field effect when crossing the Dirac point. The graphene conductivity increases linearly by increasing $|V_G|$ away from Dirac voltage [7]. Graphene demonstrates anomalous nonzero minimum conductivity even when its carrier density vanishes at the Dirac point [46] due to the presence of inevitable disorders in graphene flakes [47,48]. The residual conductivity can be diminished for improved samples moving closer to the quantum-limited unit $4e^2/h$, which is an intrinsic property of 2D Dirac fermions [44].

2.5 High-field transport

Applying a low electric field (E) across the graphene flake can possibly make a carrier achieve a velocity υ given by $\upsilon = \mu \times E$, where μ is the carrier mobility. In scaled FETs, the definition of constant mobility cannot describe the speed of carriers due to the existence of high field across the channel region. The velocity of carriers of gapless large area graphene [49,50] saturates to $\sim 6 \times 10^7$ cm/s at high field. This value is approximately 6 times higher than the saturation velocity in conventional semiconductors. The saturation in a field-effect transistor occurs when the carrier density decreases, the induced high field maximizes the carriers' velocity to saturation velocity, and the voltage drop in a region under the channel increases. In a conventional FET, the saturation region (pinch-off) is induced in the drain side of channel, which continuously increases by increasing the drain-source voltage. In case of a graphene FET, we can achieve the minimum carrier density and maximum field at Dirac point. By increasing the drain-source voltage, the carrier distribution toward the drain region decreases and finally sets the Dirac point at the drain side of the channel. Since the saturation region does not extend with an increase in drain-source voltage, the Dirac point moves toward the source side of the channel. The carriers type between the Dirac point and drain changes to electrons, leading to an increase in the carrier density again. Thus, the slope of current (I)–voltage (V) curve of graphene decreases when the drain-source voltage becomes equal to the Dirac point, which eventually corresponds to its appearance at the drain side of the channel. Similarly, the I–V plot of the graphene FET enters the second linear regime (hole regime) when the drain-source voltage becomes greater than Dirac voltage. The optical phonon scattering contributes an order of magnitude higher than the acoustic phonons scattering at high-energy carrier transport [51]. At high field, the v_{sat} can be expressed by inelastic emission of optical phonons as follows:

$$v_{sat}(n, T) = \frac{2}{\pi}\frac{\omega_{OP}}{\sqrt{\pi n}}\sqrt{1 - \frac{\omega_{OP}^2}{4\pi n v_f^2}\frac{1}{1+N_{OP}}} \quad (1.19)$$

where ω_{op} is the effective frequency of the phonon responsible for the current saturation, and $N_{OP} = 1/[\exp(\hbar\omega_{OP}/K_B T) - 1]$ is the phonon occupation, which applies the temperature dependence of the generated optical phonon scattering. Eq. (1.19) can be simplified to $v_{sat}(n) = (2/\pi)\omega_{OP}/(\pi n)^{1/2}$ at high carrier density with the assumption that there is only the contribution of carriers in the energy window $E_F \pm \hbar\omega_{OP}/2$ [52]. At low carrier density and low temperature, the saturation velocity is maximized, $\nu_{max} = (2/\pi)\nu_F \approx 6.3 \times 10^7$ cm/s. For SiO_2 substrate, optical phonon and graphene zero-edge phonons have the energies of 55 and 160 meV, respectively. The optimized equivalent energy of the optical phonons is equal to 81 meV [34], which indicates the importance of substrate polar phonons in calculating the velocity saturation in graphene.

2.6 Contact resistance

In graphene transistors, contact resistance limits the performance as the graphene conductivity is much higher than in silicon MOSFET, and thereby, lower contact resistance is required for the realization of viable graphene material. For MOSFET, the contact resistance needs to contribute less than 10% of on-state resistance V_{DD}/I_{ON} [53]. For a junction, the screening length is given by:

$$\chi^{-1} = \sqrt{4\pi N(E_f)} \quad (1.20)$$

where $N(E_f)$ is the density of states at Fermi level. In a metal–metal junction, there is an abrupt change in the vacuum level without potential barrier at interface, and thereby, χ is very small. Because graphene is a two-dimensional material and has a low density of states, the Fermi level and screening length can be significantly changed by a small charge transfer, resulting in Fermi-level pinning and/or the dipole formation at the interface [54]. It can lead to the formation of p–n junction for positive gate voltages, resulting in asymmetric transfer characteristics of graphene field-effect transistors [55]. The built-in potential barrier at the metal-graphene contact is presented by the photocurrent distribution [56]. According to Ref. [57], the current crowding at the contact interface depends on the type of metal and contact length. Because the gate voltage can tune the DOS of graphene at the interface with contacts, the contact resistance has gate voltage dependence as well, such that it has been measured 300–500 Ω μm for Ti contacts at a charge density of 3×10^{12} cm^{-2} [58]. The charge transfer of carriers from nickel contact to graphene is large due to the large work function difference between graphene and Ni, which leads to a contact resistivity range of 400–2000 Ω μm. It has been shown that the contact resistance can reduce to 100 Ω μm because nickel contact can make strong chemical bonds with zigzag-terminated graphene [59]. In addition, Grassi et al. [60] observed negative differential resistance due to the effect of contact-induced energy broadening and Dirac point in the source and drain regions [31].

2.7 Quantum capacitance

In low-dimensional devices such as graphene, the quantum capacitance is an important quantity in their operations. It describes the response of the channel charge to the movement of the conduction and valence bands due to gate electrostatic potential. In a graphene sheet with a channel electrostatic potential V_S and the total charge density Q, the quantum capacitance is defined as $C_Q = dQ/dV_S$. Assuming a uniform channel potential, the quantum capacitance for 2D graphene can be obtained by [31]:

$$C_Q = \frac{2q^2 kT}{\pi(\hbar v_F)^2} \ln\left[2\left(1 + \cosh\frac{qV_{ch}}{KT}\right)\right] \tag{1.21}$$

Xu et al. [61] have shown that Eq. (1.21) has excellent agreement with experimental results at large channel potential, while the quantum capacitance deviates from theory and has a finite value due to residual carrier near Dirac point [62]. In general, the quantum capacitance of a clean channel at finite temperature is a function of its density of states (DOS), $D(E)$, and can be determined using Eq. (1.22) [63].

$$C_Q = q^2 \int_{-\infty}^{+\infty} D(E)\left(-\frac{\partial f(E - E_f)}{\partial E}\right) dE \tag{1.22}$$

2.8 Scattering mechanism

2.8.1 Long- and short-range scattering

The phonon scattering is negligible at low temperatures, and the mobility determines mostly by the dominant scattering mechanism of long-range scattering (Coulomb scattering) and short-range scattering [31]. Lattice imperfections, edge roughness, and point defects are intrinsic sources of short-range scattering in graphene sheets. The Coulomb scattering is mostly because of trapped charges in the graphene-substrate interface [64]. These impurities cause long-range variations of the electrostatic potential, which are screened by the carrier transport in graphene sheet, leading to degradation of the mobility, shift of Dirac point, and increase in the minimum conductivity plateau width. A small carrier density, corresponding to small gate voltage, can contribute to the long-range scattering mechanism, while the short-range scattering can be dominating in much higher carrier densities, leading to sublinear and eventually a constant conductivity [65]. By applying a semi-classical theory to randomly distributed charged impurities with density n_i, Adam et al. [32] predicted that charged-impurity scattering is proportional to $\frac{\sqrt{n}}{n_i}$. The conductivity at high carrier density ($n \gg n_i$) is given by [23]:

$$\sigma_i = \frac{Ce^2}{h}\frac{n}{n_i} \tag{1.23}$$

where C is a dimensionless parameter related to the scattering strength.

The short-range scattering is usually associated with vacancies in graphene flakes, which can produce mid-gap states in graphene [66]. The vacancies can be modeled as a deep circular potential well of radius R, and the implications of this strong disorder are given in Eq. (1.24).

$$\sigma_d = \frac{2e^2}{\pi h}\frac{n}{n_d} \ln^2\left(\sqrt{\pi n}R\right) \tag{1.24}$$

where n_d is the defect density.

We can conclude from Eq. (1.24) that the conductivity is roughly linear in n. Furthermore, the scattering time of short-range mechanism can be calculated [33,64] and is proportional to $1/\sqrt{n}$, which leads to inverse proportionality of mobility to the carrier density. Chen et al. [67] deposited controlled potassium dopants on a clean graphene surface for the charged impurity and showed that the long-range Coulomb scattering is responsible for the linear dependence of conductivity on carrier density. Bolotin et al. [13] verified the importance of long-range Coulomb scattering by removing the impurity of the suspended graphene by current-induced heating, which resulted in a significant improvement of carrier mobility.

2.8.2 Phonon scattering

Unlike long- and short-range scattering, phonons are considered as intrinsic scattering source because they limit the mobility at finite temperature even when there is no extrinsic scattering. This occurs as a result of electron–phonon scattering due to lattice vibration and dominates the extrinsic scattering mechanisms at finite temperature, thereby limiting the carrier mobility in graphene. The acoustic and optical phonon scatterings can induce electronic transitions within a single valley (intra-valley) or between different valleys (inter-valley). In intra-valley transitions, the low-energy phonons contribute to elastic process of acoustic phonon scatterings, while high-energy and low momentum phonons contribute to optical phonon scatterings. In inter-valley transitions, both the energy and momentum of acoustic or optical phonons are high, which can be a dominant scattering process at high temperatures [42]. In general, the dominant scattering mechanisms are changed by increasing temperature to acoustic phonons scattering and eventually optical phonons scattering at higher temperatures.

The acoustic phonons can contribute in quasi-elastic scattering since their energies are usually much less than the Fermi energy of electrons in graphene [37]. The resistivity of graphene that is limited by the acoustic phonon can be investigated by defining two transport regimes based on the characteristic temperature of Bloch-Grüneisen [68] T_{BG} as follows:

$$k_B T_{BG} = 2k_F v_{ph} \tag{1.25}$$

where k_B is the Boltzmann constant, k_F is the Fermi wave vector with reference to K points in BZ, and v_{ph} is the sound velocity, which is 20 km/s for acoustic phonons in graphene [69]. The Bloch-Grüneisen temperature can be calculated as $54\sqrt{n}\ K$ with density measured in units of $n = 10^{12}\,\mathrm{cm}^{-2}$ [37,52]. In Bloch-Grüneisen regime ($T < < T_{BG}$), phonons reduces exponentially with decreasing temperature and carrier density dependence of resistivity is $\rho_{BG} \approx T^4$ and $\rho_{BG} \approx n^{-3/2}$, while at temperature higher than T_{BG}, the resistivity of graphene is linear in temperature ($\rho_{BG} \approx T$) and independent of carrier density. In this regime, the low-field mobility of acoustic phonons is given by [52]:

$$\mu_{ac} = \frac{e\rho_m \hbar v_F^2 v_{ph}^2}{\pi k_B} \frac{1}{D_{ac}^2 n T} \tag{1.26}$$

where $\rho_m = 7.66 \times 10^{-11}\,\text{kg/cm}^{-2}$ is the graphene mass density and D_{ac} is a deformation potential, which has been reported between 10 and 30 eV in earlier.

The optical phonons have relatively lower contribution on the low-field mobility since the acoustic phonons and impurity scatterings are dominant scattering mechanisms in graphene. The effects of intrinsic optical phonons are mostly due to inelastic scatterings in high field and/or at high temperature, which is important in determining the velocity saturation in graphene [52]. The low-field mobility limited by intrinsic optical phonons in graphene can be obtained by [70]:

$$\mu_{op} = \frac{e\rho_m v_F^2 \omega_{op}}{2\pi D_{op}^2} \frac{1}{n N_{op}} \tag{1.27}$$

where D_{op} is the mode-specific optical deformation potential of graphene, ω_{op} is the optical phonon frequency, and N_{op} is the phonon occupation. The strongest intrinsic electron–phonon coupling in graphene is because of the zone-edge transverse optical mode, which has the energy around $\hbar\omega_{op} \approx 160\,\text{meV}$ and deformation potential $D_{op} = 25.6\,\text{eV/A}$ [70].

2.8.3 Surface polar phonons

Surface polar phonons (SPP) is another source of inelastic scatterings in graphene, which is the coupling of electrons in graphene to thermally excited SPP on the substrate [34]. The SPP effect in graphene FET is much more important than in a conventional MOSFET because of its extremely low vertical dimension. Consequently, SPP can induce higher electric field on the nearby sheet to manipulate the electron transport in graphene [71]. The strength of the dielectric polarization field, which depends on phonon frequencies ($\omega_{so,v}$) and dielectric constants in the substrate and gate materials, is given by the Fröhlich coupling [72] as follows:

$$F_v^2 = \frac{\hbar\omega_{so,v}}{2\pi} \left[\frac{1}{\varepsilon_\theta + \varepsilon_{env}} - \frac{1}{\varepsilon_0 + \varepsilon_{env}} \right] \tag{1.28}$$

where $\hbar\,\omega_{so,v}$ is the surface phonon energy, ε_θ and ε_0 are the dielectric constants of polar substrate in high and low frequencies, and ε_{env} is the environment screening above the polar dielectric. The low field mobility of graphene limited by SPP phonons can be calculated by [52].

$$\mu_{SPP}^{-1} = \sum_{v,n} \left(\sqrt{\frac{\beta}{\hbar\omega_v}} \frac{\hbar v_F}{e^2} \frac{e v_F}{F_v^2} \frac{\exp(k_0 z_0)}{N_{SPP,v}\sqrt{n}} \right)^{-1} \tag{1.29}$$

where $k_0 \approx \sqrt{(2\omega_{SO,v}/v_F)^2 + \alpha n}$. The parameters $\alpha \approx 10.5$ and $\beta \approx 0.153 \times 10^{-4}\,\text{eV}$ are the fitting parameters, $N_{SPP,v}$ is SPP occupation number, and z_0 is the van der Waal distance between the polar substrate and the graphene sheet. The coupling of carriers in graphene to SPP phonons on SiO_2 substrate is much stronger than the intrinsic acoustic phonons of graphene since the van der Waal distance of the substrate is small $z_0 \approx 3.5°$ A [33].

2.9 Low-field mobility

In diffusive regime, the low-field mobility of substrate-supported graphene can be degraded by the scattering mechanisms such that the effective mobility of graphene is extracted through Matthiessen's rule [52] by adding the inverse of the mobilities of scattering mechanisms as follows:

$$\mu_{\text{eff}}^{-1} = \mu_{\text{ac}}^{-1} + \mu_{\text{op}}^{-1} + \mu_{\text{SPP}}^{-1} \tag{1.30}$$

When graphene is suspended, the extrinsic source of scattering such as charged impurities, substrate polar phonons, and ripples can be eliminated, achieving the mobilities in excess of 200,000 $cm^2 V^{-1} s^{-1}$ for gapless large-area graphene [13], which is much higher than Si (~1000 $cm^2 V^{-1} s^{-1}$). While a binary semiconductor such as InSb can only exhibit high mobility (~77,000 $cm^2 V^{-1} s^{-1}$) at very low carrier concentration, graphene with 2D electron gas can exhibit similar high mobility for high carrier concentration. Phonon scattering becomes the dominant scattering mechanism as carrier density and temperature are increased. We can then model the mobility as follows [34]:

$$\mu(n, T) = \frac{\mu_0}{1 + (n/n_{\text{ref}})^{\alpha}} \times \frac{1}{1 + (T/T_{\text{ref}} - 1)^{\beta}} \tag{1.31}$$

where $\mu_0 = 4650$ cm^2/V s, $n_{\text{ref}} = 1.1 \times 10^{13}$ cm^{-2}, $T_{\text{ref}} = 300$ K, $\alpha = 2.2$, and $\beta = 3$. Fig. 1.6A and B shows the relationship between temperature and carrier density for saturation velocity and low-field mobility, respectively [31]. In conventional semiconductors, hole mobility is significantly lower than the electron mobility, while it can be more than electron mobility

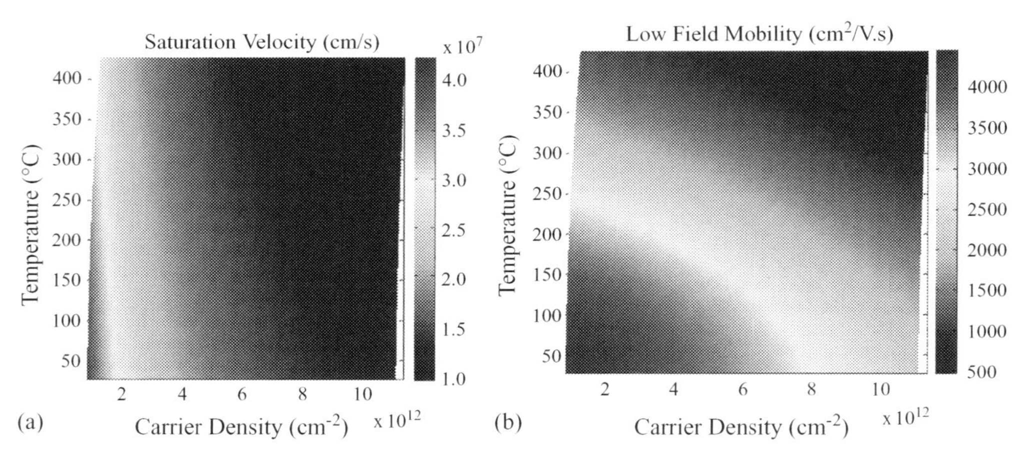

FIG. 1.6 Graphene as a function of carrier density and temperature: (A) saturation velocity and (B) low-field mobility. *Reproduced with permission from A. Srivastava, Graphene and other graphene materials technology and beyond, in: A. Srivastava, S. Mohanty (Eds.), Advanced Technologies for Next Generation Integrated Circuits, IET Press, 2020.*

in graphene. Low-field mobility of short-channel device decreases with the channel length since the transport regime changes from diffusive to quasi-ballistic [51].

2.10 Thermal conductivity

The thermal conductivity is dependent on both electron and phonon according to the kinetic theory of gases. The expression for thermal conductivity is given in Eq. (1.32):

$$k \sim c_{ph} C_V l_{ph} \tag{1.32}$$

where C_V is the specific heat per unit volume, c_{ph} is the specific heat of phonon, and l_{ph} is the phonon mean free path. Since c_{ph} is very large in graphene, one can expect a large thermal conductivity [23]. Balandin et al. [31] and Ghosh et al. [73] reported that $k \cong 3080 - 5150$ W/mK for an experiment conducted on a set graphene flakes at near room temperature and a phonon mean free path of $l_{ph} \cong 775$ nm. Since a high thermal conductivity facilitates the diffusion of heat to the contacts and allows for more compact circuits, these results indicate that graphene is a good candidate for applications to electronic devices [23].

2.11 Optical properties

Graphene has high transparency, high carrier mobility, low reflectance, and near-ballistic transport at room temperature, and this makes it a promising choice for transparent electrodes. The properties of graphene favor it an attractive choice for use in optoelectronic devices [23]. The parameter such as light transmittance T is a good metric to classify optical materials. The Fresnel equation for a thin film, in Eq. (1.33), is used to obtain light transmittance for free-standing graphene.

$$T = (1 + 0.5\pi\alpha)^{-2} \approx 1 - \pi\alpha \approx 0.977 \tag{1.33}$$

where $\alpha = \frac{e^2}{4\pi\varepsilon_0 \hbar c} = \frac{G_0}{\pi\varepsilon_0 c}$. G_0 is the optical conductance. A typical single-layer graphene has a light absorbance of $A = 1 - T \approx 2.3\%$. As shown in Fig. 1.7 [74], the absorption spectrum of graphene from the ultraviolet to infrared is notably constant around 2–3% absorption as compared to some other materials. It further confirms graphene as a good optoelectronic material and has found its way into many optoelectronic and photonic devices such as photodetectors, touch screens, smart windows, and saturable absorbers.

2.12 Joule heating

The generated self-heating of graphene on the substrate at high bias can limit the saturation velocity [75] such that the experimental measurement of the saturation velocity in graphene at room temperature is significantly smaller than the theoretical value due to scattering mechanisms by intrinsic phonon, surface polar phonon, acoustic phonon, and charged impurities. Perebeinos et al. [52] have modeled the generated Joule heating of the field (E) in Boltzmann transport calculation by the following equation:

$$T = T_{amb} + jE/r \tag{1.34}$$

FIG. 1.7 Transmittance as a function of wavelength for graphene, ITO and two other metal oxides, as well as SWNTs [74].

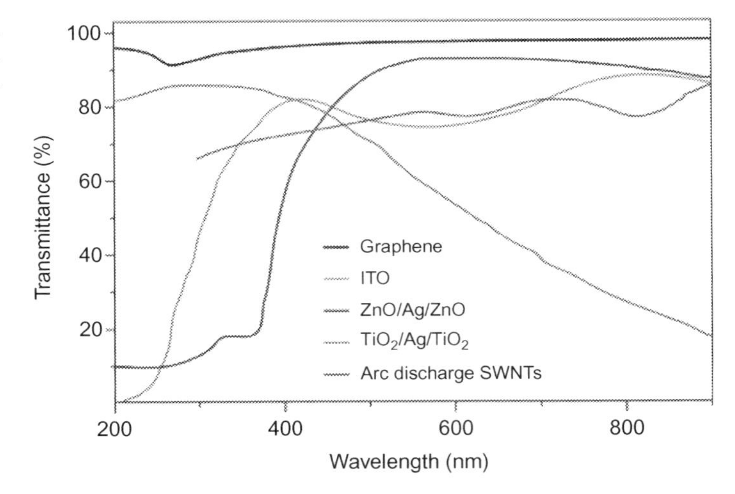

where T_{amb} is the ambient temperature and $j = j(T)$ is the current density, which solves self-consistently to consider the Joule losses. The parameter $r = K/h$ models thermal conductance of dissipated heat by the substrate, where K and h are the thermal conductivity and thickness of the insulating substrate. The current density has been found to drop by approximately a factor of four due to the self-heating on SiO_2 substrate where $r \approx 0.47\,\mathrm{kW/(K\,cm^2)}$ for the insulator height $h = 300$ nm. The heat generated in a graphene channel has to cross the graphene/substrate interface, thereby experiencing the thermal contact resistance, known as Kapitza resistance. The high SPP scattering in SiO_2 substrate minimizes the effect of the thermal contact resistance. In contrast, the thermal contact resistance of SiC, h-BN, and HfO_2 substrates is much higher than the substrate thermal conductance since the SPP scattering in graphene/substrate interface is lower than SiO_2 and the substrate thermal conductance is much higher than SiO_2 substrate. For instance, the graphene/substrate contact resistance of SiC substrate is more than a factor of ten larger than the SiO_2 substrate [76]. Thus, for increasing the saturation velocity, the heat dissipation limited by the graphene/substrate interface is an important factor to manipulate in substrate with high thermal conductance [77, 78], while the thermal conductivity and thickness of the substrate are the dominant factors for the heat dissipation in SiO_2 substrate. Dorgan et al. [34] have modeled the average temperature of the graphene on SiO_2 substrate considering thermal resistances of the graphene/SiO_2 boundary (R_B), SiO_2 substrate (R_{Ox}), and Si wafer (R_{si}) as follows:

$$\Delta T = T - T_0 \approx P(R_B + R_{ox} + R_{Si}) \tag{1.35}$$

where P is the power delivered to graphene sheet, $R_B = 1/(hA)$, $R_{ox} = t_{ox}/K_{ox}A$, and $R_{Si} \approx 1/2K_{Si}A^{1/2}$. $h \approx 10^8\,\mathrm{Wm^{-2}K^{-1}}$ is the thermal conductance of graphene/SiO_2 boundary, $A = LW$ is the graphene channel area, and K_{ox} and K_{Si} are the thermal conductivities of SiO_2 and doped Si wafer. For graphene on SiO_2 substrate with the thickness of 300 nm, the Joule-heating contribution of the graphene/SiO_2 boundary, SiO_2 substrate, and Si wafer has been reported to 4%, 84%, and 12%, respectively.

3 Synthesis of graphene and fabrication technology

Since performance of graphene is largely dependent on the contamination and structural defects from processing or the transfer process, appropriate growth methods should be chosen based on device quality, scale, processing, and architecture [79]. There is also a great desire to synthesizing graphene samples with high carrier mobility and low density of defects. Fig. 1.8 summarizes some of the methods used for graphene synthesis and deposition of graphene.

3.1 Mechanical exfoliation

Mechanical exfoliation is one of the popular methods of graphene synthesis, which uses highly oriented pyrolytic graphite (HOPG) crystal as precursor. This was developed by Novoselov et al. [3] whereby they subjected HOPG dry etching (oxygen plasma etching) to create mesas which are later pressed into a layer of photoresist. The HOPG on baked photoresist was cleaved and the flakes of graphite were peeled from mesas using the adhesive tape method. Subsequently, the graphite flakes were treated with acetone and collected over Si/SiO_2 substrate. In 2009, Geim [6] employed the scotch tape method to generate flakes up to 1 mm in length, which have excellent quality and well suited for fundamental research. However, the process is limited to small sizes and cannot be scaled for industrial production [23]. In general, large-area graphene fabrication using mechanical cleaving is a serious challenge, which limits the feasibility of this process for industrialization. Furthermore, graphene synthesized with mechanical exfoliation cannot be produced with high accuracy and repeatability. In 2014, Zheng et al. [80] addressed the problems associated with by exfoliating graphene from HOPG utilizing a sharp single-crystal diamond wedge. This method has a big limitation because shear stress creates defects that eventually distort the reliable placement of the graphene flakes after exfoliation.

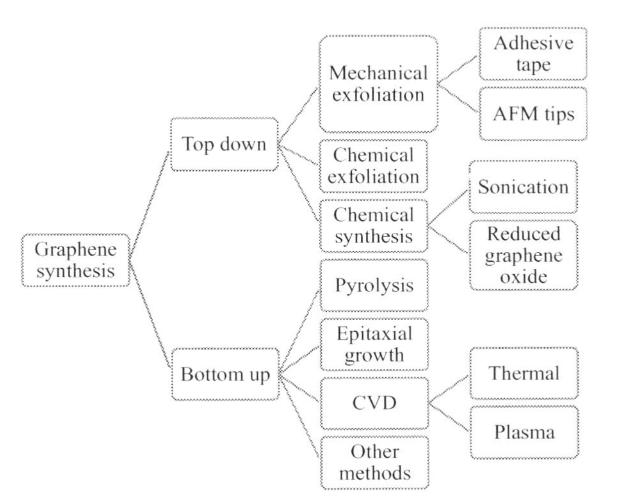

FIG. 1.8 Synthesis methods for graphene.

3.2 Chemical exfoliation

This involves using liquid exfoliation of graphite to produce graphene. Hernandez et al. [81] reported the exfoliation of pure graphite in *N*-methyl-pyrrolidone by a simple sonication process. They achieved this by sonicating powdered graphite in a range of solvents using a low-power sonic bath. The reported exfoliated graphene films showed high-quality synthesis with a yield of ~1%. A research group led by Coleman [82] extended the work but focused on solvents known to exfoliate nanotubes such as *N*-methyl-pyrrolidone (NMP) and dimethylformamide (DMF). The undispersed material was removed by centrifugation, and the measured disperse concentration reached maximum at tensions close to 40 mJ/m^2. For better understanding, they calculated the enthalpy of mixing (ΔH_{Mix}) for graphene flakes dispersed and exfoliated in solvents as follows:

$$\frac{\Delta H_{Mix}}{V} \approx \frac{2}{T_{NS}}\left(\sqrt{E_{S,S}} - \sqrt{E_{S,G}}\right)^2 \o_G \qquad (1.36)$$

where T_{NS} is the nanosheet thickness, $\o_G$ is the dispersed graphene volume fraction, and $E_{S,S}$ and $E_{S,G}$ are the solvent and graphene surface energy, respectively. Although it is clear that single layer and few layers of graphene are obtained from graphite exfoliated in solvents to give monolayer and few layers of graphene, it is not yet clear whether the graphene is pristine.

3.3 Chemical synthesis

Direct graphene synthesis using electrochemical methods was reported by Liu et al. [83]. In a typical electrochemical functionalization of graphite, 10 mL of 1-octyl-3-methyl-imidazolium hexafluorophosphate and 10 L of water served as the electrolytes with two graphite electrodes. A static potential of 15 V was applied for 6 h to the setup in Fig. 1.9 and resulted in the formation of a black precipitate of graphene nanosheets (GNS) at room temperature. The formation of GNS was investigated using Raman spectroscopy, which is one of the key analytical techniques used in characterization. The Raman spectra of natural graphite and graphene nanosheet are presented in Fig. 1.10. The peak at 1353 cm^{-1} indicates vibrations of carbon atoms with dangling bonds in plane terminations of disordered graphite, while the peak at 1593 cm^{-1} corresponds to a mode of graphite with vibrations of sp^2-bonded carbon atoms in 2D honeycomb

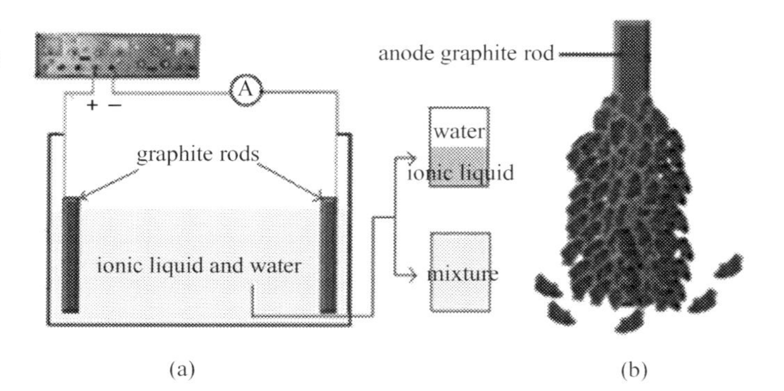

FIG. 1.9 (A) Experimental setup diagram and (B) exfoliation of the graphite anode [83].

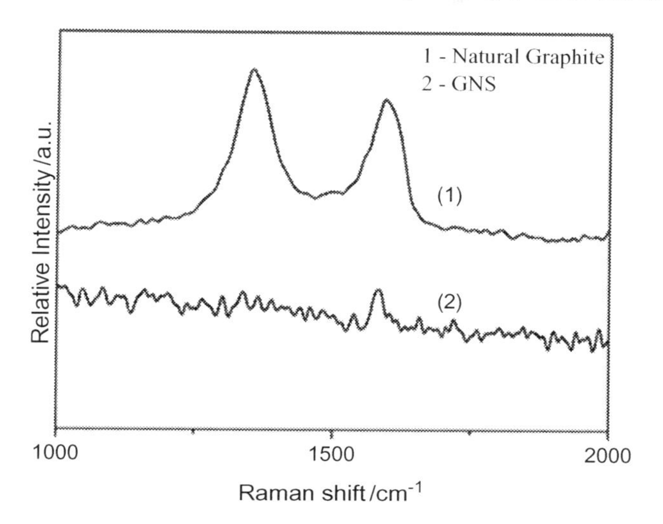

FIG. 1.10 Raman spectra (solid scan, excitation wavelength 514.5 nm) [83].

structure such as in graphite layer. The method is environment-friendly and leads to the production of a colloidal suspension of imidazolium ion-functionalized graphene sheets by direct electrochemical treatment of graphite.

3.4 Pyrolysis

The laboratory reagents ethanol and sodium are reacted in a ratio of 1:1 to give an intermediate product, which is then pyrolyzed to yield a fussed array of graphene sheets. Mohammad et al. [84] have demonstrated that single-layer graphene can be synthesized by low-temperature flash pyrolysis of a solvothermal product of sodium and ethanol, followed by gentle sonication of the nanoporous carbon product. The graphene sheets produced herein were characterized by a number of physical methods that clearly demonstrate the nanostructured form of the carbon product. Representative TEM was used to take measurements, and the crystallinity of the sheets was determined with selected area electron diffraction (SAED). This yielded graphene sheets with dimensions of up to 10 μm.

3.5 Chemical vapor deposition

In 2006, Somani et al. [85] first attempted for CVD-grown graphene on Ni using camphor (terpenoid), a white transparent solid of chemical formula ($C_{10}H_{16}O$) as the precursor material. However, using TEM, they found that the planar few-layer graphene consists of ~35 layers of stacked single graphene sheets with an interlayer distance of 0.34 nm. Using methane (CH_4), Li et al. [86,87] studied the growth of large scale (1 cm^2) single-layer graphene on Ni and Cu substrates, which is so far the widely used method employed for obtaining CVD graphene. Further, they developed a graphene transfer method by solution etching of Cu and then transferring of the floated graphene onto any substrate. Bae et al. [88] in 2010 produced a

30-in. scaled graphene sheet using roll-to-roll production on a Cu substrate and transferred by wet chemical etching of Cu.

A typical CVD process for the deposition of graphene consists of four steps: (a) adsorption and catalytic decomposition of precursor gas, (b) diffusion and dissolution of decomposed carbon species on the surface and metal bulk, (c) dissolved carbon atoms segregation onto metal surface, and (d) surface nucleation and growth of graphene [89]. However, in case of metals having poor carbon affinity such as copper, the decomposition of carbon precursor is followed by direct formation of graphene on copper where dissolution and subsequent segregation of carbon atoms are prohibited. The low solubility of the carbon in copper also makes the growth process predominantly self-limiting to single-layer graphene [86]. The most common carbon precursor for graphene growth is methane (CH_4), which has a strong C—H bond (440 kJ mol^{-1}). For this strong C—H bond in methane, its thermal decomposition occurs at very high temperature (>1200 °C). However, such a high temperature is not easily obtained in typical thermal CVD setup. In order to reduce the decomposition temperature of methane, different transition metal catalysts (e.g., Fe, Co, Ni, Cu) are widely used and the growth of graphene on such metals can be obtained at low temperatures (<900 °C).

During the annealing step, the catalyst surface is covered with molecular hydrogen, which can be referred as dissociative chemisorption of H_2 on the metal surface [89]. Compared to Ni, Cu shows higher hydrogen solubility. This process is followed by the catalytic decomposition of the carbon precursors on the metal surface. At this stage, the competitive process between the dissociative chemisorption of H_2 and physical adsorption and dehydrogenization of CH_4 on catalyst surface occurs. With suitable choice of thermodynamic parameters, the chemical potentials of surface carbon atoms are maintained lower than the carbon in gas phases, which further helps to form stable graphitic rings and grow into large graphitic structures up to graphene formation [89]. Once such nucleation of graphene structure is stable on the metal surface, the growth mechanism is followed by the attachment of carbon species onto graphene edges. The quality, uniformity, and surface coverage on metal substrate depend on suitable choice of high temperature, pressure, and exposure time. As the growth time increases, the individual graphene domains progressively increase in size and coalesce into a continuous layer. Nevertheless, after the growth and formation of a continuous layer, further exposure to carbon precursor does not lead to the deposition of multilayered graphene due to the self-limiting process as described earlier in case of copper substrate. It is important to note that the graphene growth on copper is surface-related and does not occur due to outdiffusion from bulk. Using the isotope labeling, Li et al. [87] demonstrated that the Raman modes of ^{12}C and ^{13}C isotopes differ in energy, which provided a substantial understanding of the gradual increment of the graphene layer growth laterally on copper surface providing critical structural information of graphene growth.

Fig. 1.11A shows floating graphene film on Cu etchant Fe $(NO_3)_3$ after Cu has been fully etched. Prior to that, CVD graphene was grown on Cu foil of 25 μm thickness. Fig. 1.11B shows the floating graphene transferred on SiO_2 substrate. The optical contrast confirms the single-layer graphene compared to SiO_2. Fig. 1.11C shows Raman spectroscopy of graphene transferred on SiO_2 substrate. A 632-nm laser is used for Raman spectroscopy. The graphene on Cu has been deposited using the Benchtop nanoCVD-8G System (Moorfield Nanotechnology Ltd., UK). A small D peak and a dominant 2D peak compared to the G peak confirm the growth of single-layer graphene on Cu foil.

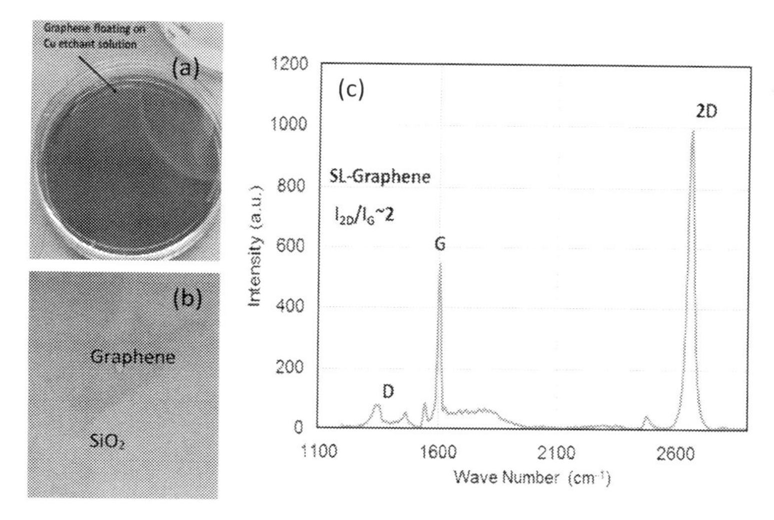

FIG. 1.11 Chemical vapor deposition: (A) CVD-grown graphene floating on Cu etchant after Cu has been fully etched, (B) transferred on SiO_2, and (C) Raman spectroscopy of single-layer graphene after transferred on SiO_2 [90].

Chen et al. [91] proposed that by switching hydrogen pressure between high and low would result in the growth of bilayer graphene. Similar results were produced by Lu et al. [92] whereby simply controlling the hydrogen pressure, bilayer graphene has been grown. Following the work in [91,92], the chamber pressure has been modified for obtaining multilayer graphene using the cold wall-resistive heater nanoCVD-8G system. However, compared to the earlier reported growth time, the process adopted here not only requires less time but also becomes economical. Fig. 1.12A shows the optical image of a copper foil processed under the similar growth as described in the work of Bointon et al. [93] for a chamber pressure of 20 Torr during the growth period. With the carbon precursor $CH_4=10\%$, $H_2=5\%$, and $Ar=85\%$ for 120 s and a chamber pressure of 20 Torr at 1000 °C, both the bilayer and multilayer graphene have been observed along with single layer. The Raman spectroscopy performed at different areas, as observed in Fig. 1.12A, confirms the observation of bilayer and multilayer graphene on copper foil, which are shown in Fig. 1.12B and Fig. 1.12C, respectively. Note that with a growth time of only 120 s, the total processing time for such graphene sheet on copper foil was only 20 min, which is shorter than the earlier reported growth time in the work of Bointon et al. [93]. From the optically contrast image of Fig. 1.12A, difference in numbers of layers of graphene can be easily understood as well. Compared to the lighter area, the darker area represents more number of graphene layers. The Raman peaks studied in the comparatively less dark area and shown in Fig. 1.12B reveal that graphene is bilayer (BL-graphene) with an extensive level of defects or hydrogenated edges. A strong D peak compared to both *G* and 2D peak is a characteristic feature of a graphene film with defects or halogen-terminated edges. An I_{2D}/I_G ratio near 1 also reveals that the area is bilayer [94].

Furthermore, Raman analysis of the darker region confirms that the graphene is multilayer as shown in Fig. 1.12C. With an I_{2D}/I_G ratio of nearly 0.25, the graphene in the region is more than 10 layers and similar to graphitic carbon [94]. The D peak for this region is low, which informs comparatively less defects or hydrogen-terminated edges compared to the Raman spectra in Fig. 1.12B.

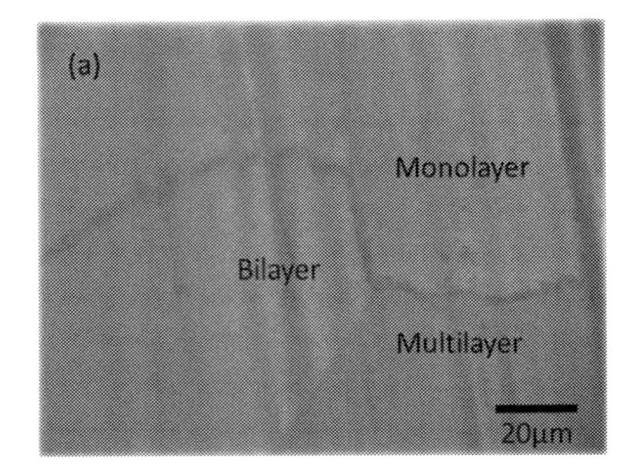

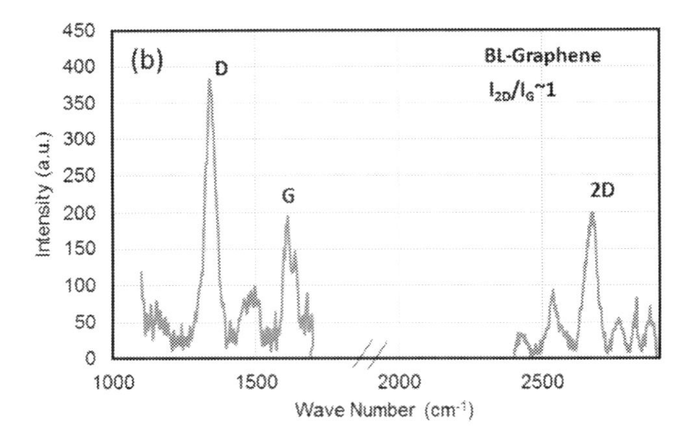

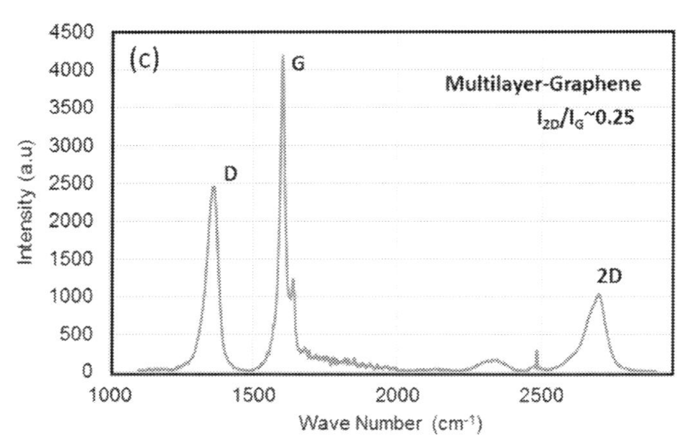

FIG. 1.12 Raman spectroscopy of graphene grown on copper foil at 20 Torr pressure: (A) optical image showing three different regions, (B) Raman peaks for bilayer graphene area, and (C) Raman peaks for multilayer graphene area [93].

3.6 Graphene oxide (GO)

The historical background of graphene goes back to Brodie [95] in 1859 who discovered the lamellar structure of thermally reduced graphene oxide (GO), a multilayer carbon oxide material often used as an analogy to graphene. Kohlschütte and Haenni [96] in 1919 studied the properties of graphene oxide papers, a composite material with graphene skeleton. Three decades later, Ruess and Vogt [97] in 1948 reported the first transmission emission microscopy of few layers of graphite dry residue, which is structurally a multilayer graphene. This remained the best observation of graphene for several decades. The theoretical groundwork of graphene also goes back to Wallace [19] who in 1947 first described the zone structure, number of free electrons, and conductivity of a single hexagonal layer of graphite. Zhu et al. [24] showed the formation of graphene oxide nanoplatelets from sonicated graphene oxide. Afterward, a reducing agent was used to remove oxygen. The alternative method developed has to do with GO being produced as a precursor for graphene synthesis. It was then sonicated in water and spin-coated to form a single- or double-layer graphene oxide.

3.7 Other methods

Berger et al. [98,99] produced few-layer graphene by thermal decomposition of SiC. It involves epitaxial growth on SiC substrate by desorption of Si. Another method was employed by Zhan et al. [100] to create the layer-by-layer growth of graphene with molecular beam deposition. They made use of ethylene gas as the starting material, which was later broken down at high temperature of about 1200 °C before deposited on a Ni substrate. There are few other means to produce graphene such as electron beam irradiation of PMMA nanofibers [101], thermal fusion of PAHs [102], conversion of nanodiamond [103], and arc discharge of graphite [104].

4 Bandgap engineering, G FET/GNR FET, and GNR interconnects

One of the major interests of graphene developments is for nanoelectronics applications. This is because of its high intrinsic mobility and switching speed. Even though the integration of graphene into transistor results in excellent three terminal devices, the ratio of I_{ON} to I_{OFF} is considerably low as compared to MOSFET. Although this is a major concern in the field of electronics, the graphene-based devices are still preferred to high electron mobility transistors (HEMTs) for the replacement of RF-based devices. Graphene is a monolayer of carbon atoms in a two-dimensional hexagonal lattice with semi-metallic nature, in which electrons behave as massless Dirac fermions resulting in high carrier velocities and current density. However, the application of large-area graphene is limited for integrated circuits and a narrow stripes of graphene, known as GNR, is a promising alternative as a replacement of transistor channel [105,106] and interconnect [107] for the next-generation VLSI circuits. Since its discovery, researchers have been dreaming to make *graphene* a semiconductor material—a key requirement for making semiconductor chips. Feng Wang and his co-workers at the UC Berkeley and Lawrence Berkeley National Laboratory (US Department of Energy, Lawrence Berkeley National Laboratory News Center: News Release, July 10, 2009) have succeeded in creating a tunable energy bandgap, though small, but a major breakthrough in realization of *semiconductor graphene* for making transistors, switches, lasers, and several types of solid-state devices. Fig. 1.13 explains the formation of tunable energy bandgap in *graphene*. Though graphene created excitement in the field of electronics and numerous applications, the problems started showing up. Lack of bandgap became a serious barrier for digital electronics and opening of bandgap became very problematic than initially thought. However, narrow strips of graphene demonstrated the necessary bandgap needed for semiconductor electronics, which became known as GNR. Experimentally feasibility of *GNR* has also been demonstrated for making transistors and interconnects. Some of the recently reported devices are room-temperature ballistic transport field-effect transistors (FETs), single-electron transistors, spin transistors, and solar batteries. Researchers at MIT have already demonstrated a *graphene chip*, which could reach 1000 GHz (MIT Technical Talk, April 1, 2009).

According to Prof. Novoselov, one of the inventors of graphene "Being able to control the resistivity, optical transmittance and a material's work function would all be important for photonic devices like solar cells and liquid crystal displays, for example, and altering mechanical properties and surface potential is at the heart of designing composite materials. Chemical modification of graphene—with graphene as its first example—uncovers a whole new dimension of research. The capabilities are practically endless."

4.1 Bandgap engineering in graphene

Graphene is a zero-bandgap semiconductor or a semi-metal. This results in transistors made of graphene difficult to turn off. In order to obtain appropriate switching behavior using graphene-based transistors, a significant bandgap is required, which leads to the study of bandgap engineering of graphene. In this section, some methods for obtaining a bandgap in graphene are discussed. Fig. 1.14 shows a summary of different ways of obtaining bandgap in graphene.

Castro et al. [25] reported that if an electric field is applied to a bilayer graphene vertically, then this opens a bandgap, making graphene as a field-tunable semiconductor. Both the theoretical and experimental considerations have shown that a field of few 10^4 kV/cm could open a

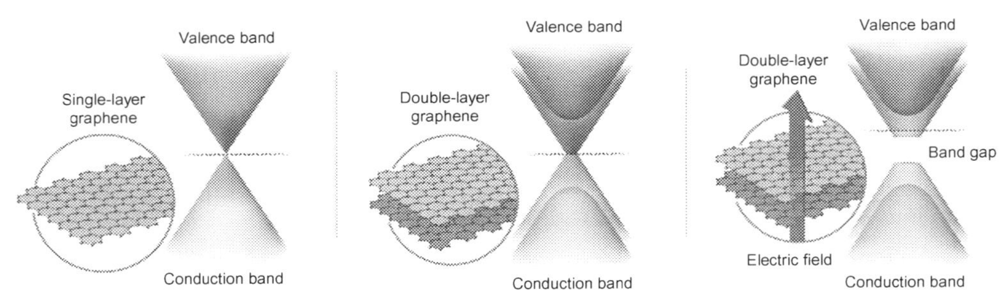

FIG. 1.13 Left and middle—no bandgap, right—generation of bandgap after the application of perpendicular electrical field to the layers of graphene [*US DoE News Release*, July 10, 2009].

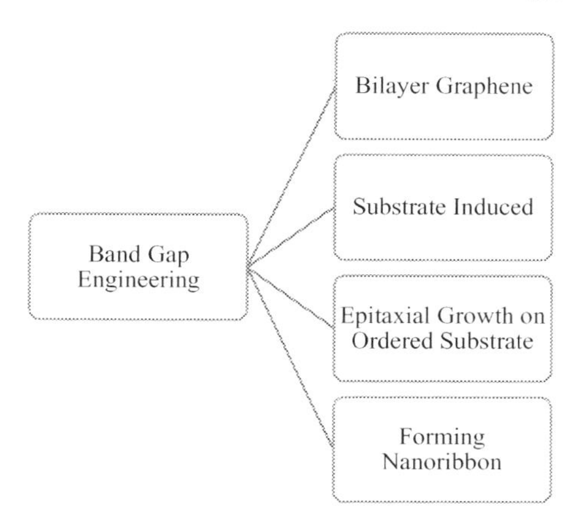

FIG. 1.14 Energy bandgap engineering methods used for graphene.

bandgap of 250 meV. Recently, two unconventional methods have been reported: (1) graphene growth on MgO [108] and (2) by irradiation of graphene with an ion beam [109]. Being atomically thick, interaction of graphene with underneath substrate plays a critical role on graphene electronic properties. Giovannetti et al. [110] reported in 2007 the ab initio density functional theory (DFT)-based electronic structure calculation of graphene on hex-boron nitride substrate, resulting in a bandgap of 53 meV at the Dirac point. Recently, Nevius et al. [111] reported the growth of semiconducting graphene on highly ordered SiC substrate along the <0001> direction of the SiC hexagonal crystal pack (HCP). They measured the so far recorded highest bandgap of single-layer graphene of 0.55 eV using angle-resolved photoemission spectroscopy (ARPES), which is a direct method to measure energy bandgaps in materials. The most commonly known method for opening a bandgap in graphene is to confine the infinite graphene sheet into a narrow ribbon where length is much greater than width. Due to the quantum confinement of the electrons into a nanoribbon, a measurable finite bandgap can be opened [112,113]. The required bandgap of several hundred meV can be introduced by quantum confinement of carriers in one-dimensional graphene, called GNR.

In GNR, the bandgap is directly proportional to the inverse of the width [112]. It has been predicted that GNR with width scaled down to 2 nm should provide a gap in excess of 1 eV [112]. It is important to note that the origin of bandgap is still under debate. Apart from considering the lateral confinement as the origin of bandgap, it has been suggested that other notable effect such as Coulomb blockade is responsible for the formation of such bandgap [113]. Han et al. [114] experimentally demonstrated lithographically patterned GNR with width-dependent bandgap. One of the most effective methods for obtaining GNR is to unzip a single-walled carbon nanotube with bottom-up chemical approach [115]. Compared to lithographically patterned GNR, this method provides smooth defect-free GNR [116].

Energy bandgap in GNR is also dependent on the edge types along which the transport occurs. Fig. 1.15 shows a top view of a GNR with two types of edges, i.e., armchair and zigzag edges. Localized edge states at the Fermi level are observed in zigzag edge nanoribbon, whereas such edge states are absent in armchair edge nanoribbons. These localized states are important as these infer to localized wave functions at the GNR edges and contribute

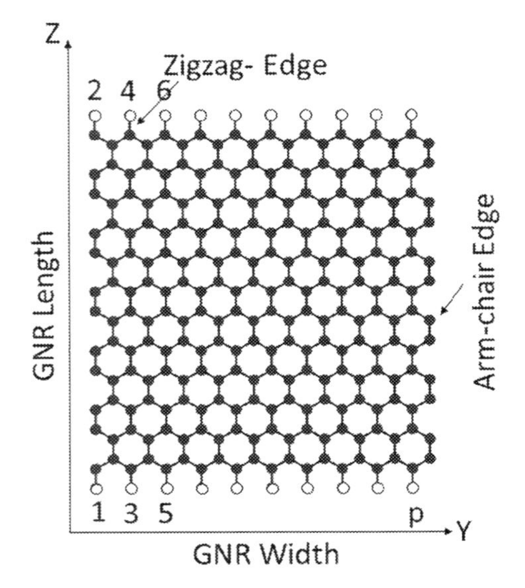

FIG. 1.15 Graphene nanoribbon (GNR) where p is an integer denoting the pth atom along the width [118].

to antibonding properties of GNR and electronic structure [18]. For the GNR shown in Fig. 1.15, the nanoribbon width varies along the Y direction and length along Z direction. The variable p is an integer. The numbering of atoms (1, 2, 3… p) along the GNR width is also shown in Fig. 1.15. The notation of chirality used for GNR is expressed as (p,0), where p is the number of carbon atoms on each ring of unrolled nanotube. Generally, "p" is defined in terms of any of the configurations from $3N$, $3N+1$, or $3N+2$ along the GNR width. It should be noted that p is the total number of atoms considering both sides of the nanoribbons, whereas N is an integer. Therefore, in a (4,0) armchair GNR, $p=4$ with $3N+1$ configuration considering $N=1$. However, in a (5,0) armchair GNR, $p=5$ with a $3N+2$ configuration considering $N=1$. For (6,0) armchair GNR, $p=6$ with $3N$ configuration considering $N=2$ [117].

Energy bandgap of GNR, both armchair and zigzag, differs depending on the method of calculation. Electronic structure of GNR is modeled traditionally by the simple tight binding (TB) approximation based on π-bonded p_z-orbital electrons or usually studied by Dirac equation of massless particle considering effective speed of light ($\sim 10^6$ m/s). Such assumptions lead to conclude armchair GNR to be either metallic or semiconducting. Results obtained by TB approximation considering the nearest-neighbor hopping integral of 2.7 eV show that armchair GNR is metallic for $p=3N+2$ and semiconducting for both $p=3N$ and $p=3N+1$ configurations [119]. Basically, the hierarchy of energy bandgap is maintained as $\Delta_{3N+1} > \Delta_{3N} > \Delta_{3N+2}$(=0 eV), with Δ being the energy gap and N is an integer. Fig. 1.16 shows the width-dependent bandgap, calculated using nearest-neighbor tight-binding Hamiltonian considering p_z orbital encoded in "CNT bands," available in the open-source simulation framework Nanohub [105]. In Fig. 1.16A and C, both (4,0) and (6,0) are semiconducting. Zero bandgap is observed for (5,0) GNR in Fig. 1.16B, which is a $3N+2$ configuration for $N=1$.

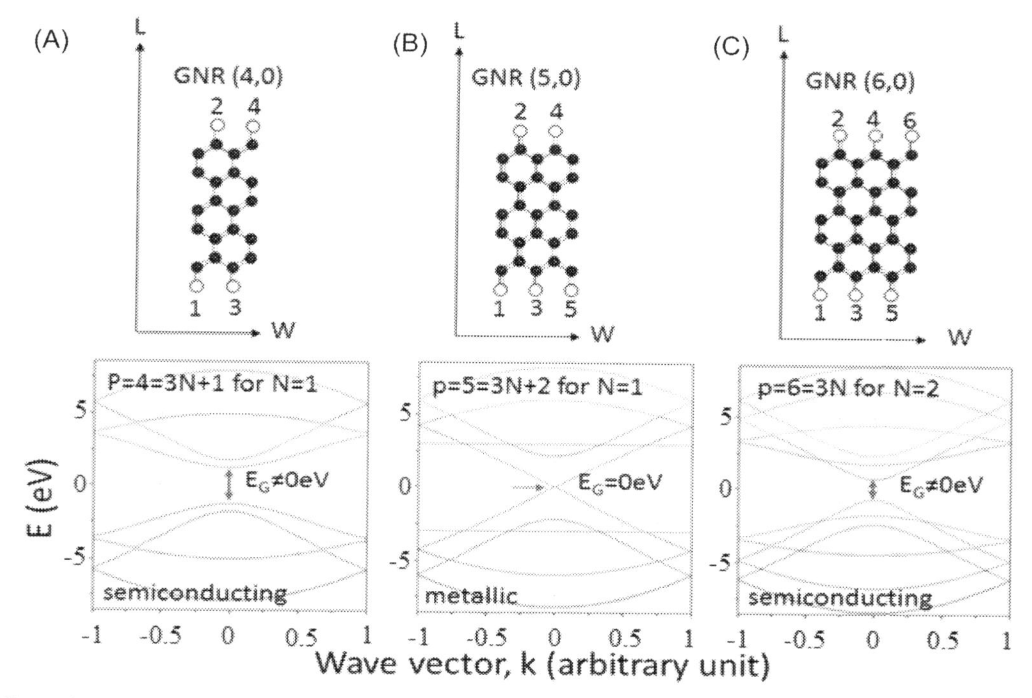

FIG. 1.16 Width-dependent bandgap of graphene nanoribbon with an increase in number of atoms. (A) Energy band diagram for (4,0) GNR, which is a $3N+1$ configuration for $N=1$ and semiconducting; (B) energy band diagram for (5,0) GNR, which is a $3N+2$ configuration for $N=1$ and metallic; and (C) energy band diagram for (6,0) GNR, which is a $3N$ configuration for $N=2$ and semiconducting. L denotes the length and W denotes the width of GNR. The numbers shown for chirality of GNR are depicted along the width of GNR. *Reproduced with permission from A. Srivastava, Graphene and other graphene materials technology and beyond, in: A. Srivastava, S. Mohanty (Eds.), Advanced Technologies for Next Generation Integrated Circuits, IET Press, 2020.*

4.2 Graphene FET

In order to have a functional field-effect transistor with graphene, there is a need to open up the bandgap between its valence and conduction bands. This is highly desirable for current saturation and appropriate voltage and current gains. Several geometries of GFET have been proposed based on the expected performance of the device and applications [120,121]. Fig. 1.17 presents one of the most common geometries because of the ease at which the gate coupling can be increased and optimizes the graphene dielectric interface to reduce scattering and make the conduction and valence states continuous. The GFET in Fig. 1.17 uses a bilayer graphene channel with a large back-gate voltage to induce an electric field that opens a bandgap. The back-gated FET structure is sufficient for proof of concept but not useful for practical applications due to large parasitic capacitances and poor gate control [122]. In 2007, the first top-gated graphene FET was demonstrated and fabricated with exfoliated graphene, grown on either copper or nickel and epitaxial graphene [123]. The top-gated structure represents the preferred option for realistic applications. However, GFET is not suitable

FIG. 1.17 Diagram of a B-bilayer graphene FET with back contact to create a pinch-off region and voltage gain [121].

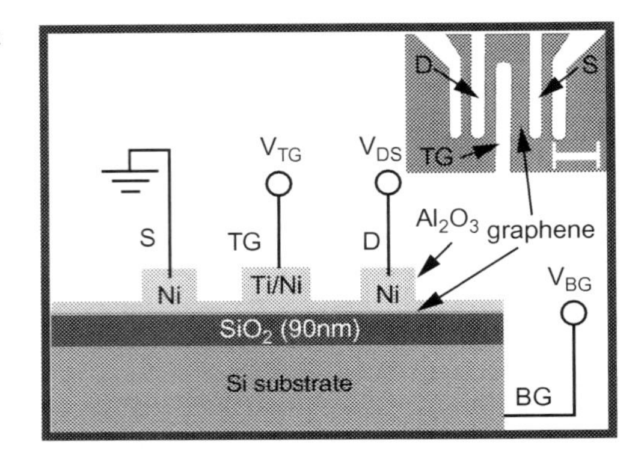

for logic applications because of poor I_{ON}/I_{OFF} ratio. For using graphene in transistor-level operation, it is necessary to have a finite bandgap of the material and GNR helps in this regard significantly.

4.3 GNR-based transistors

Zhang et al. [124] reported a basic structure of a p-i-n n-type armchair GNR (a-GNR) tunnel FET of 20 nm channel length and 4.9 nm channel width. Fahad et al. [125] presented an extensive study of single-gate a-GNR TFET shown in Fig. 1.18. Fig. 1.18A shows the vertical cross section of p-type a-GNR TFET with 1 nm SiO_2 top gate dielectric. The channel length is 20 nm with 5 nm of source and drain extension, making the total length of GNR 30 nm. An n-type a-GNR TFET and associated energy band diagram of p-i-n GNR TFET (n-type GNR TFET where both V_{GS} and V_{DS} are "+" ve) are presented in Fig. 1.18B and D, respectively. Fig. 1.18C shows the energy band diagram of n-i-p GNR TFET (p-type GNR TFET where both V_{GS} and V_{DS} are "−" ve). It should be noted that in both Figs. 1.18C and D, solid line is for OFF state, whereas dashed line is for ON state. OFF state is defined as $|V_{DS}|=0.1$ V and $|V_{GS}|=0$ V, and ON state is defined as $|V_{DS}|=0.1$ V and $|V_{GS}|=0.1$ V. Semiconducting a-GNR (20, 0) has a bandgap of 0.289 eV for its corresponding 4.9 nm width.

We already discussed that narrow GNR has the adequate bandgap, which can decrease the off-state current and consequently increase the I_{ON}/I_{OFF} ratio [126]. In earlier work [127], the GNR FET structure in Fig. 1.19 was used in our simulation. The GNRs are sandwiched between two thin insulator layers in a double metal gate topology in order to maximize the electrostatic control of the gate electrostatic on the channel. The dielectric layer is assumed aluminum nitride (AlN) with the relative dielectric permittivity $k=9$ as it reduces the phonon scattering in epitaxial graphene [128] and its thin film production is cost-efficient with good reproducibility and uniformity [129,130]. The extensions of source and drain regions are heavily doped with n-type dopants in order to form ohmic junctions between graphene and source (or drain) contacts minimizing contact resistances [31]. The length of the metallic gates is assumed equal to the length of intrinsic GNR channel, and the pitch size between ribbons is kept equal to the width of ribbons. The oxide thickness t_{ox}, the physical gate length L_G, and the power supply voltage V_{DD} are the variable parameters, which have been assigned based on the roadmap presented in ITRS report [53].

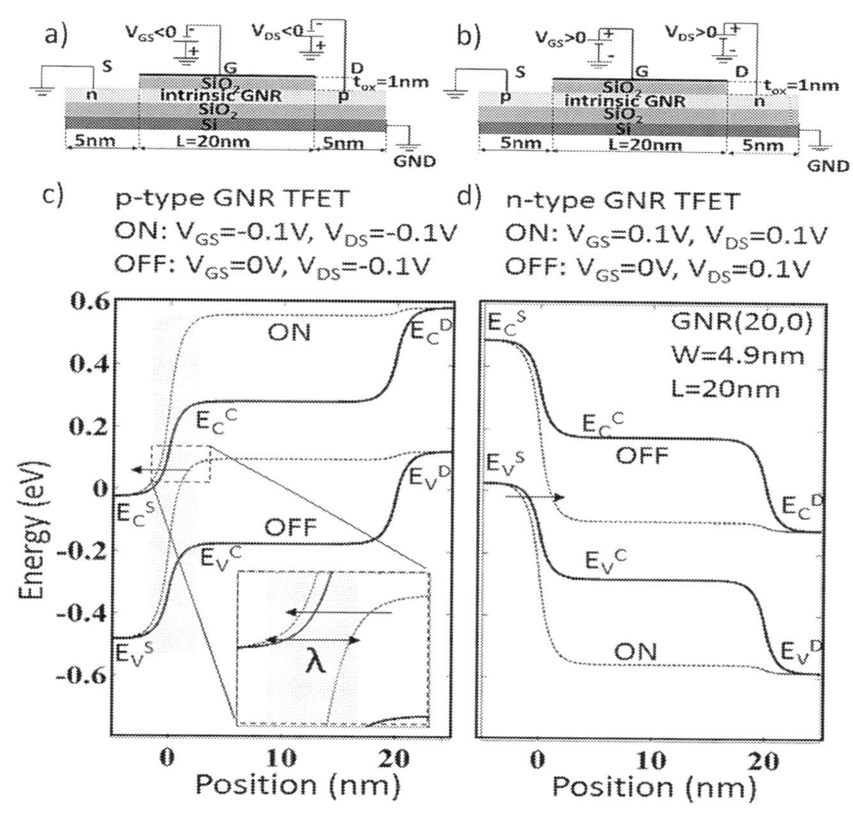

FIG. 1.18 (A) p-type a-GNR TFET, (B) n-type a-GNR TFET, (C) energy band diagram of p-type GNR TFET, and (D) energy band diagram of n-type GNR TFET [125].

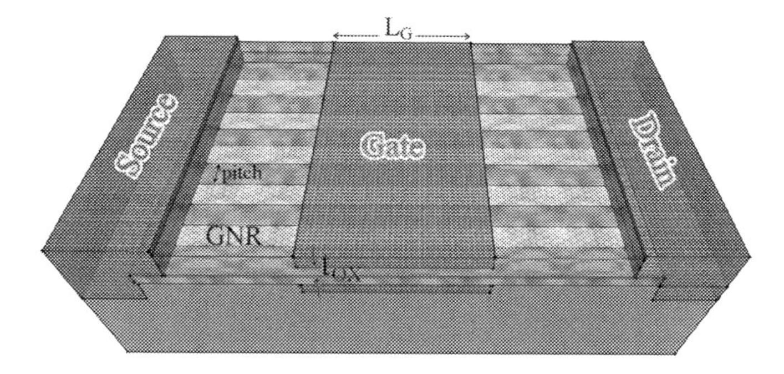

FIG. 1.19 3D schematic view of GNRFET structure with five parallel graphene nanoribbons in connection with two wide metallic contacts [127].

The I_{DS} versus V_{DS} for different V_{GS} of GNR FET is shown in Fig. 1.20A, where the channel is GNR(6,0) with the gate length of $L_G = 5$ nm. The strong saturation region indicates good MOSFET-type device behavior of GNR FET, where the saturation slope is significantly determined by GNR width rather than the gate length [131]. The transfer characteristics I_D–V_G for

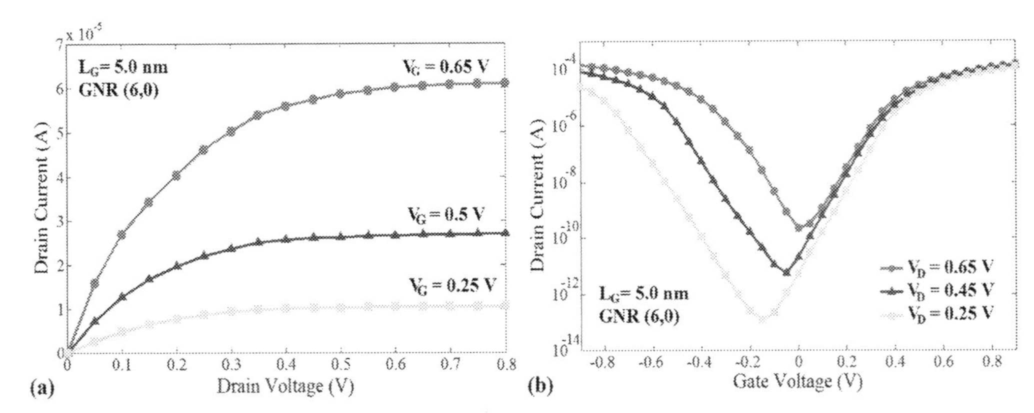

FIG. 1.20 (A) Drain current versus drain voltage characteristics of GNRFET for varying gate voltages and (B) drain current versus gate voltage characteristics of GNRFET for varying drain voltages [127].

different drain voltages is shown in Fig. 1.20B. For a given drain voltage, the minimum current is obtained at charge neutrality point corresponding to a gate electrostatic potential at which the contribution of electron current becomes equal to hole current. The accumulation of holes in the channel is increased by increasing drain voltage due to an increase in band-to-band tunneling from the source contact to the channel, resulting in larger minimum current and shift of charge neutrality point to positive gate voltages.

4.4 Graphene nanoribbon and carbon nanotubes interconnects

At present, electronic information technology has become an important drive force that promotes social and economic progress. Integrated circuit (IC) as a core and foundation of the electronic information technology has a great influence on the daily life of human being. The semiconductor technology and IC industry have become an important symbol to embody a country's comprehensive scientific and technological capability [132]. In order to improve circuit's performance and increase the number of transistors on a chip, microelectronic devices have been continuously reduced in dimension according to Moore's law and scaling rule [133]. According to the 2013 International Technology Roadmap for Semiconductors (ITRS 2013), the feature size of semiconductor devices will reduce to 10 nm in 2025 [53] in very large-scale integrated circuits (VLSIs). For the first-generation interconnect material aluminum (Al) [134], an increase in electric resistance and capacitance due to increasing wire length and decreasing wire interval as dimension scales down had led to large signal delays [135] and poor tolerance to electromigration (EM) [136]. Because of its lower resistivity and higher melting point (1083 °C vs. 660 °C of Al) and longer EM lifetime [137], copper (Cu) has replaced Al as an interconnect material in the 180-nm technology node [138] and beyond. But as interconnects scale down to the 45 nm and beyond technology generations, Cu interconnect is also facing similar problems with those of Al interconnects encountered, including an increase in resistivity due to size effect [139], an increase in power consumption [140], delay [141], and EM distress [142]. GNR is a potential candidate for interconnect applications because it overcomes all the problems associated with Cu and Al. GNRs interconnect usually generates bandgap that is smaller than a Cu interconnect with the similar nanometer diameter, such that the first subband of GNRs with narrow bandgap can be sufficiently populated at room temperature. Therefore, small bandgap of GNRs makes negligible degradation in conductance properties.

Although it is experimentally demonstrated that monolayer GNR exhibits much better scalability [107], larger EM reliability [143], and current-carrying capacity ($\sim$100 A/cm^2 [108]) compared with Cu interconnect, its performance is limited by its atomically thin 2D geometry [143]. Stacking graphene layers in the 3D configuration can improve the GNR performance for interconnect applications by lowering interconnect resistance [144,145]. Kondo et al. [146,147] showed that intercalation of iron chloride molecules between sheets of graphene decreases the resistivity of MLGNR interconnects by an order of magnitude and can be as low as that of Cu interconnects. By fabricating MLGNR interconnects with a width of 20 nm and stacking 11 graphene layers, they demonstrated that MLGNR interconnects have high current reliability and the variation of its resistivity by narrowing down the width of MLGNR interconnect is negligible. The performance of stacked non-interacting multilayer GNRs with smooth edge and Fermi energies above 0.2 eV outperforms Cu interconnects by scaling down the cross-sectional dimensions and increasing the interconnect length [148]. Multilayer graphene nanoribbon (MLGNR) requires to be electrically decoupled from each other by breaking the van der Waals force between layers to preserve exceptional conducting behavior of each individual graphene monolayer as shown in Fig. 1.21 [149].

Although we also have CNT interconnects, GNR interconnects are preferred because they have a higher degree of interaction with the substrate compared to CNT interconnects [150–153]. This can lead to smaller heat-limited current degradation due to higher heat dissipation in GNR interconnects [154], while it can also work against its performance by increasing scattering due to the disorders in substrate or insulating dielectrics [155,156]. Thus, the performance of GNR interconnects is more substrate dependent than that of Cu or CNT interconnects. CNTs can be regarded as rolling up one or several graphene sheets into one-dimensional seamless cylindrical structure, called single-walled CNT (SWCNT) and multiple-walled CNT (MWCNT), respectively. The SWCNT and MWCNT were discovered in 1993 [157,158] and 1991 [15], respectively. The diameter of SWCNTs may vary from 0.4 to 4 nm with a typical diameter of 1.4 nm and outer diameters of MWCNTs may vary from several nm to 100 nm, and the spacing between the adjacent walls is about 0.34 nm [159,160].

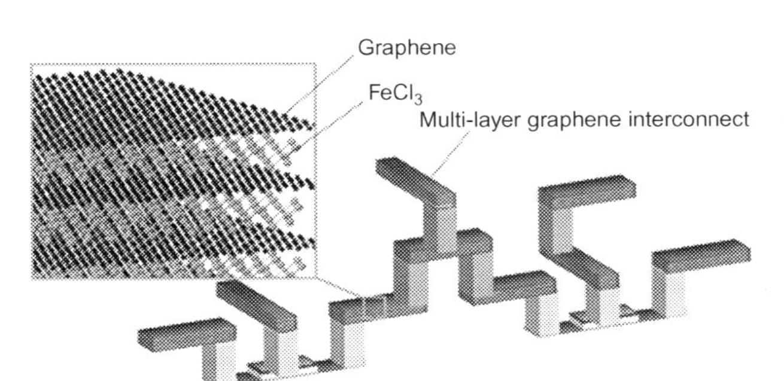

FIG. 1.21 Schematic diagram of LSI using MLGNR interconnects. *Reprinted with permission from Advanced Industrial Science and Technology (AIST), Development of Technology for Producing Micro-Scale Interconnect from Multi-Layer Graphene, http:/www.aist.go.jp/, 2013.; AIST, http://www.aist.go.jp.*

4.5 The challenges for graphene nanoribbon interconnects

There are several limitations and challenges to implement GNRs in current technology. In the first place, wafer-scale high-quality graphene is required to be synthesized on arbitrary substrates, which is suitable for patterning in the form of GNR interconnects. In general, the method of graphene preparation and the type of its substrate determine the weight of scattering sources for electrons in large-area graphene [20]. Acoustic and optical phonons in graphene, substrate polar phonons, and Coulomb scatterings due to charged impurities are the most prominent sources of scattering in graphene at a finite temperature [161].

5 G/GNR-based inductor design for voltage-controlled oscillator

A voltage-controlled oscillator (VCO) is the most important building block in the phase-locked loop (PLL) [162]. It generates desired frequency range depending on the input tuning voltage. Quite often, it determines the overall performance of PLL in terms of phase noise, tuning range, and power consumption. In addition, it dominates almost all spectral purity performance of a frequency synthesizer. The phase noise of VCO causes reciprocal mixing, which leads to interference and hinders the performance of the entire PLL system. It is, however, desirable to design a voltage-controlled oscillator with reduced phase noise, wide tuning range, high stability, and low power consumption. LC oscillator is a preferred option because it offers lower-phase noise. As shown in Fig. 1.22, an LC VCO consists of three

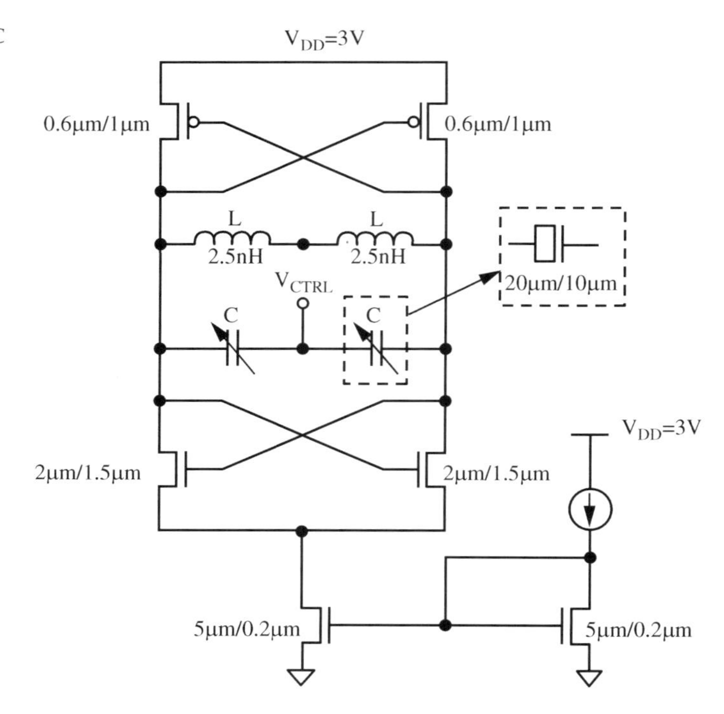

FIG. 1.22 Circuit diagram of a LC VCO [164].

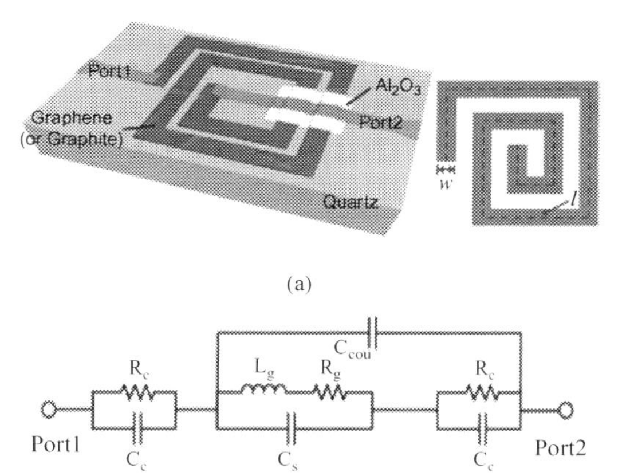

FIG. 1.23 The structure of graphene (or graphite) inductor: (A) schematic view of a 2-turn graphene (or graphite) inductor on quartz substrate and (B) an equivalent circuit model of the inductor [167].

components: LC tank, tail bias transistor, and cross-coupled differential pair. LC tank is made by an inductor and a capacitor connected in cascade or in parallel.

Fig. 1.22 shows the circuit diagram of a complementary cross-coupled differential LC oscillator with tail current. The symmetrical design of the VCO gives good phase noise performance and large voltage swing [163]. To achieve a better performance VCO, we can design graphene inductor (carbon-based inductor), which may serve as an important passive component.

Sarkar et al. [164,165] discussed theoretical works and application of graphene for on-chip inductor. Graphene is planar in nature and has large relaxation time, which aids its patterning into square helical structures with no metal contacts required. In addition, graphene has high energy efficiency and high-quality factors. This makes it suitable for designing high Q-factor inductor. In 2014, Li et al. [166] did the experimental study of practical graphene on-chip inductors. Furthermore, they designed, fabricated, characterized, and proposed a circuit model for graphene inductor based on its extracted parameters. The work was extended by Wang et al. [167] and fabricated CVD monolayer graphene and graphite inductors with high Q-factors, as shown in Fig. 1.23.

In Fig. 1.23, R_g, L_g, and C_g denote the resistance, inductance, and inter-turn capacitance of graphene (or graphite) coil, respectively. R_C and C_C represent the contact resistance and capacitance, respectively. C_{Coup} is the coupling capacitance. In 2021, Adesina et al. [168] designed a low-phase noise VCO in TSMC 0.18 μm n-well CMOS process technology using high Q-factor graphene-based inductor.

6 Conclusion

Scaled-down CMOS technology is expected to encounter several challenging issues in near future; graphene has emerged as one of the promising alternative materials for post-CMOS technology due to large carrier mobilities, high carrier velocity, high thermal conductivity, and planar structure. Owing to its zero-bandgap property, graphene is not yet suitable for digital applications. However, finite bandgap can be obtained in the form of GNR, which is formed by reducing the width of 2D graphene sheets to few nm. Invariably, we have quantum confinement of carriers in quasi-one-dimensional graphene and quantum transport properties strongly depend on edge effect. GFET, GNR TFET, and GNR interconnects can be a viable option for low-power high-performance integrated circuit design. Also, graphene inductors are suitable for LC VCO design because of their high-quality factors.

Acknowledgments

The part of the work is supported by the United States Air Force Research Laboratory under Agreement Number FA9453-18-1-0103. The U.S. Government is authorized to reproduce and distribute reprints for Government purposes notwithstanding any copyright notation thereon. The part of the work is also supported under Norfolk State University, Virginia NSF CREST-CREAM HRD-157771.

References

[1] Israel Science, Technology Directory, List of Elements of the Periodic Table—Sorted by Abundance in Earth's Crust, Israel Science and Technology Homepage (http://www.science.co.il/PTelements.asp?s=Earth). (1999).
[2] A. Srivastava, Graphene and other graphene materials technology and beyond, in: A. Srivastava, S. Mohanty (Eds.), Chapter 1 in Advanced Technologies for Next Generation Integrated Circuits, IET Press, 2020.
[3] K.S. Novoselov, A.K. Geim, S.V. Morozov, D. Jiang, Y. Zhang, S.V. Dubonos, et al., Electric field effect in atomically thin carbon films, Science 306 (5696) (2004) 666–669.
[4] K.S. Novoselov, D. Jiang, F. Schedin, T.J. Booth, V.V. Khotkevich, et al., Two-dimensional atomic crystals, Proc. Natl. Acad. Sci. U. S. A. 102 (30) (2005) 10451–10453.
[5] S. Bharech, R. Kumar, A review on the properties and applications of graphene, J. Mater. Sci. Mech. Eng. 2 (10) (2015) 70–73.
[6] A.K. Geim, Graphene: status and prospects, Science 324 (5934) (2009) 1530–1534.
[7] K.S. Novoselov, A.K. Geim, S.V. Morozov, D. Jiang, M.I. Katsnelson, I.V. Grigorieva, et al., Two-dimensional gas of massless Dirac fermions in graphene, Nature 438 (7065) (2005) 197–200.
[8] M.I. Katsnelson, K.S. Novoselov, A.K. Geim, Chiral tunnelling and the Klein paradox in graphene, Nat. Phys. 2 (9) (2006) 620–625.
[9] K.S. Novoselov, Z. Jiang, Y. Zhang, S.V. Morozov, H.L. Stormer, U. Zeitler, et al., Room-temperature quantum Hall effect in graphene, Science 315 (5817) (2007) 1379.
[10] K.I. Bolotin, K.J. Sikes, J. Hone, H.L. Stormer, P. Kim, Temperature-dependent transport in suspended graphene, Phys. Rev. Lett. 101 (9) (2008), 096802.
[11] R. Prasher, Graphene spreads the heat, Science 328 (5975) (2010) 185–186.
[12] J.C. Meyer, A.K. Geim, M.I. Katsnelson, K.S. Novoselov, T.J. Booth, S. Roth, The structure of suspended graphene sheets, Nature 446 (7131) (2007) 60–63.
[13] K.I. Bolotin, K.J. Sikes, Z. Jiang, M. Klima, G. Fudenberg, J. Hone, et al., Ultrahigh electron mobility in suspended graphene, Solid State Commun. 146 (9-10) (2008) 351–355.
[14] H.W. Kroto, J.R. Heath, S.C. O'Brien, R.F. Curl, R.E. Smalley, C60: buckminsterfullerene, Nature 318 (14) (1985) 162–163.
[15] S. Iijima, Helical microtubules of graphitic carbon, Nature 354 (6348) (1991) 56–58.
[16] G.W. Semenoff, Condensed-matter simulation of a three-dimensional anomaly, Phys. Rev. Lett. 53 (26) (1984) 2449–2452.
[17] S. Mouras, A. Hamm, D. Djurado, J.C. Cousseins, Synthesis of first stage graphite intercalation compounds with fluorides, J. Flourine Chem. 24 (5) (1987) 572–582.
[18] K. Nakada, M. Fujita, G. Dresselhaus, M.S. Dresselhaus, Edge state in graphene ribbons: nanometer size effect and edge shape dependence, Phys. Rev. B 54 (24) (1996) 17954–17961.
[19] P.R. Wallace, The band theory of graphite, Phys. Rev. 71 (9) (1947) 622–634.
[20] A.C. Neto, F. Guinea, N. Peres, K.S. Novoselov, A.K. Geim, The electronic properties of graphene, Rev. Mod. Phys. 81 (1) (2009) 109.

[21] C. Stampfer, S. Fringes, J. Güttinger, F. Molitor, C. Volk, B. Terrés, et al., Transport in graphene nanostructures, Front. Phys. 6 (3) (2011) 271–293.
[22] D. Chung, Review graphite, J. Mater. Sci. 37 (2002) 1475–1489.
[23] D.R. Cooper, B. D'Anjou, N. Ghattamaneni, B. Harack, M. Hilke, A. Horth, et al., Experimental review of graphene, Int. Scholar. Res. Notes—ISRN Condens. Matter Phys. 501686 (2012) 1–56.
[24] Y. Zhu, S. Murali, W. Cai, X. Li, J.W. Suk, J.R. Potts, et al., Graphene and graphene oxide: synthesis, properties, and applications, Adv. Mater. 22 (35) (2010) 3906–3924.
[25] E.V. Castro, K.S. Novoselov, S.V. Morozov, N.M.R. Peres, J.M.B.L. dos Santos, J. Nilsson, et al., Biased bilayer graphene: semiconductor with a gap tunable by the electric field effect, Phys. Rev. Lett. 99 (2007), 216802.
[26] A. Suzuki, M. Tanabe, S. Fujita, Electronic band structure of graphene based on the rectangular 4-atom unit cell, J. Mod. Phys. 8 (2017) 607–621.
[27] Z. Li, E.A. Henriksen, Z. Jiang, Z. Hao, M.C. Martin, P. Kim, et al., Dirac charge dynamics in graphene by infrared spectroscopy, Nat. Phys. 4 (7) (2008) 532–535.
[28] P.E. Trevisanutto, C. Giorgetti, L. Reining, M. Ladisa, V. Olevano, Ab initio G W many-body effects in graphene, Phys. Rev. Lett. 101 (22) (2008), 226405.
[29] J. González, F. Guinea, M. Vozmediano, Unconventional quasiparticle lifetime in graphite, Phys. Rev. Lett. 77 (17) (1996) 3589.
[30] C. Hwang, D.A. Siegel, S.K. Mo, W. Regan, A. Ismach, Y. Zhang, et al., Fermi velocity engineering in graphene by substrate modification, Sci. Rep. 2 (2012) 590.
[31] A. Srivastava, Y.M. Banadaki, Graphene transistors—present and beyond, in: S. Mohanty, A. Srivastava (Eds.), Chapter 4 in Nano-CMOS and Post-CMOS Electronics: Devices and Modeling, IET Press, 2016.
[32] S. Adam, E. Hwang, V. Galitski, S.D. Sarma, A self-consistent theory for graphene transport, Proc. Natl. Acad. Sci. 104 (47) (2007) 18392–18397.
[33] W. Zhu, V. Perebeinos, M. Freitag, P. Avouris, Carrier scattering, mobilities, and electrostatic potential in monolayer, bilayer, and trilayer graphene, Phys. Rev. B 80 (23) (2009), 235402.
[34] V.E. Dorgan, M.H. Bae, E. Pop, Mobility and saturation velocity in graphene on SiO_2, Appl. Phys. Lett. 97 (8) (2010), 082112.
[35] J.H. Chen, W. Cullen, C. Jang, M. Fuhrer, E. Williams, Defect scattering in graphene, Phys. Rev. Lett. 102 (23) (2009), 236805.
[36] Y. Zhang, V.W. Brar, C. Girit, A. Zettl, M.F. Crommie, Origin of spatial charge inhomogeneity in graphene, Nat. Phys. 5 (10) (2009) 722–726.
[37] E. Hwang, S.D. Sarma, Acoustic phonon scattering limited carrier mobility in two-dimensional extrinsic graphene, Phys. Rev. B 77 (11) (2008), 115449.
[38] E.V. Castro, H. Ochoa, M. Katsnelson, R. Gorbachev, D. Elias, K. Novoselov, et al., Limits on charge carrier mobility in suspended graphene due to flexural phonons, Phys. Rev. Lett. 105 (26) (2010), 266601.
[39] A.S. Mayorov, R.V. Gorbachev, S.V. Morozov, L. Britnell, R. Jalil, L.A. Ponomarenko, et al., Micrometer-scale ballistic transport in encapsulated graphene at room temperature, Nano Lett. 11 (6) (2011) 2396–2399.
[40] F. Giannazzo, V. Raineri, E. Rimini, Transport Properties of Graphene with Nanoscale Lateral Resolution, Scanning Probe Microscopy in Nanoscience and Nanotechnology, vol. 2, Springer, 2011, pp. 247–285.
[41] N.M. Peres, The transport properties of graphene, J. Phys. Condens. Matter 21 (32) (2009), 323201.
[42] S.D. Sarma, S. Adam, E. Hwang, E. Rossi, Electronic transport in two-dimensional graphene, Rev. Mod. Phys. 83 (2) (2011) 407.
[43] S.V. Morozov, K.S. Novoselov, M.I. Katsnelson, F. Schedin, D.C. Elias, J.A. Jaszczak, A.K. Geim, Giant intrinsic carrier mobilities in graphene and its bilayer, Phys. Rev. Lett. 100 (2008), 016602.
[44] A.K. Geim, K.S. Novoselov, The rise of graphene, Nat. Mater. 6 (3) (2007) 183–191.
[45] S. Datta, Electronic Transport in Mesoscopic Systems, Cambridge University Press, Cambridge, 1997.
[46] J. Martin, N. Akerman, G. Ulbricht, T. Lohmann, J. Smet, K. Von Klitzing, et al., Observation of electron–hole puddles in graphene using a scanning single-electron transistor, Nat. Phys. 4 (2) (2007) 144–148.
[47] J. Tworzydło, B. Trauzettel, M. Titov, A. Rycerz, C.W. Beenakker, Sub-Poissonian shot noise in graphene, Phys. Rev. Lett. 96 (24) (2006), 246802.
[48] Y.W. Tan, Y. Zhang, K. Bolotin, Y. Zhao, S. Adam, E. Hwang, et al., Measurement of scattering rate and minimum conductivity in graphene, Phys. Rev. Lett. 99 (24) (2007), 246803.
[49] M. Bresciani, A. Paussa, P. Palestri, D. Esseni, L. Selmi, Low-field mobility and high-field drift velocity in graphene nanoribbons and graphene bilayers, IEEE Int. Electron Dev. Meet. (IEDM) (2010) 724–727.
[50] R. Shishir, D. Ferry, Velocity saturation in intrinsic graphene, J. Phys. Condens. Matter 21 (34) (2009), 344201.

[51] I. Meric, C.R. Dean, A.F. Young, N. Baklitskaya, N.J. Tremblay, C. Nuckolls, et al., Channel length scaling in graphene field-effect transistors studied with pulsed current-voltage measurements, Nano Lett. 11 (3) (2011) 1093–1097.
[52] V. Perebeinos, P. Avouris, Inelastic scattering and current saturation in graphene, Phys. Rev. B 81 (19) (2010), 195442.
[53] L. Wilson, International Technology Roadmap for Semiconductors (ITRS), 2013. http://www.itrs.net/.
[54] K. Nagashio, T. Nishimura, K. Kita, A. Toriumi, Metal/graphene contact as a performance killer of ultra-high mobility graphene analysis of intrinsic mobility and contact resistance, IEEE Int. Electron Dev. Meet. (IEDM) (2009) 1–4.
[55] B. Huard, N. Stander, J. Sulpizio, G.D. Goldhaber, Evidence of the role of contacts on the observed electron-hole asymmetry in graphene, Phys. Rev. B 78 (12) (2008), 121402.
[56] E.J. Lee, K. Balasubramanian, R.T. Weitz, M. Burghard, K. Kern, Contact and edge effects in graphene devices, Nat. Nanotechnol. 3 (8) (2008) 486–490.
[57] K. Nagashio, T. Nishimura, K. Kita, A. Toriumi, Contact resistivity and current flow path at metal/graphene contact, Appl. Phys. Lett. 97 (14) (2010), 143514.
[58] P. Blake, R. Yang, S. Morozov, F. Schedin, L. Ponomarenko, A. Zhukov, et al., Influence of metal contacts and charge inhomogeneity on transport properties of graphene near the neutrality point, Solid State Commun. 149 (27) (2009) 1068–1071.
[59] W.S. Leong, H. Gong, J.T. Thong, Low-contact-resistance graphene devices with nickel-etched-graphene contacts, ACS Nano 8 (1) (2013) 994–1001.
[60] R. Grassi, T. Low, A. Gnudi, G. Baccarani, Contact-induced negative differential resistance in short-channel graphene FETs, IEEE Trans. Electron Dev. 60 (1) (2013) 140–146.
[61] H. Xu, Z. Zhang, L.M. Peng, Measurements and microscopic model of quantum capacitance in graphene, Appl. Phys. Lett. 98 (13) (2011), 133122.
[62] K. Nagashio, T. Nishimura, A. Toriumi, Estimation of residual carrier density near the Dirac point in graphene through quantum capacitance measurement, Appl. Phys. Lett. 102 (17) (2013), 173507.
[63] S. Datta, Quantum Transport: Atom to Transistor, Cambridge University Press, Cambridge, 2005.
[64] D.B. Farmer, V. Perebeinos, Y.M. Lin, C. Dimitrakopoulos, P. Avouris, Charge trapping and scattering in epitaxial graphene, Phys. Rev. B 84 (20) (2011), 205417.
[65] F. Chen, J. Xia, D.K. Ferry, N. Tao, Dielectric screening enhanced performance in graphene FET, Nano Lett. 9 (7) (2009) 2571–2574.
[66] T. Stauber, N. Peres, F. Guinea, Electronic transport in graphene: a semiclassical approach including midgap states, Phys. Rev. B 76 (20) (2007), 205423.
[67] J.H. Chen, C. Jang, M. Ishigami, S. Xiao, W. Cullen, E. Williams, et al., Diffusive charge transport in graphene on SiO_2, Solid State Commun. 149 (27) (2009) 1080–1086.
[68] H. Stormer, L. Pfeiffer, K. Baldwin, K. West, Observation of a Bloch-Grüneisen regime in two-dimensional electron transport, Phys. Rev. B 41 (2) (1990) 1278.
[69] T. Fang, A. Konar, H. Xing, D. Jena, High-field transport in two-dimensional graphene, Phys. Rev. B 84 (12) (2011), 125450.
[70] J. Chauhan, J. Guo, High-field transport and velocity saturation in graphene, Appl. Phys. Lett. 95 (2) (2009), 023120.
[71] I.T. Lin, J.M. Liu, Surface polar optical phonon scattering of carriers in graphene on various substrates, Appl. Phys. Lett. 103 (8) (2013), 081606.
[72] S. Wang, G. Mahan, Electron scattering from surface excitations, Phys. Rev. B 6 (12) (1972) 4517.
[73] S. Ghosh, I. Calizo, D. Teweldebrhan, E.P. Pokatilov, D.L. Nika, A.A. Balandin, W. Bao, F. Miao, C.N. Lau, Extremely high thermal conductivity of graphene: prospects for thermal management applications in nanoelectronic circuits, Appl. Phys. Lett. 92 (15) (2008), 151911.
[74] F. Bonaccorso, Z. Sun, T. Hasan, A.C. Ferrari, Graphene photonics and optoelectronics, Nat. Photonics 4 (9) (2010) 611–612.
[75] Y. Banadaki, K. Mohsin, A. Srivastava, A graphene field effect transistor for high temperature sensing applications, in: Proc. SPIE 9060, Nanosensors, Biosensors, and Info-Tech Sensors and Systems, San Diego, California, USA, 2014, p. 90600F-F-7.
[76] F. Schwierz, Graphene transistors: Status, prospects, and problems, Proc. IEEE 101 (7) (2013) 1567–1584.
[77] W. Cai, A.L. Moore, Y. Zhu, X. Li, S. Chen, L. Shi, et al., Thermal transport in suspended and supported monolayer graphene grown by chemical vapor deposition, Nano Lett. 10 (5) (2010) 1645–1651.

[78] R. Mao, B.D. Kong, K.W. Kim, T. Jayasekera, A. Calzolari, M.B. Nardelli, Phonon engineering in nanostructures: controlling interfacial thermal resistance in multilayer-graphene/dielectric heterojunctions, Appl. Phys. Lett. 101 (11) (2012) 113111.
[79] C. Max, Lemme, current status of graphene transistors, Solid State Phenom. 156-158 (2010) 499–509.
[80] Y. Zheng, P. Zhiwei, C. Gilberto, L. Jian, X. Changsheng, Z. Haiqing, Y. Yang, R. Gedeng, R.O. Abdul-Rahman, L.G.S. Errol, H.H. Robert, J.Y. Miguel, M.T. James, Rebar graphene, ACS Nano 8 (5) (2014) 5061–5068.
[81] Y. Hernandez, V. Nicolosi, M. Lotya, F.M. Blighe, Z. Sun, S. De, et al., High-yield production of graphene by liquid-phase exfoliation of graphite, Nat. Nanotechnol. 3 (2008) 563–568.
[82] N.C. Jonathan, Liquid exfoliation of defect-free graphene, ACS 46 (1) (2013) 14–22.
[83] N. Liu, F. Luo, H. Wu, Y. Liu, C. Zhang, J. Chen, One-step ionic-liquid-assisted electrochemical synthesis of ionic-liquid-functionalized graphene sheets directly from graphite, Adv. Funct. Mater. 18 (2008) 1518–1525.
[84] C. Mohammad, T. Pall, J.A. Stride, Gram-scale production of graphene based onsolvothermal synthesis and sonication, Nat. Nanotechnol. 4 (1) (2009) 30–33.
[85] P.R. Somani, S.P. Somani, M. Umeno, Planer nano-graphenes from camphor by CVD, Chem. Phys. Lett. 430 (1-3) (2006) 56–59.
[86] X. Li, W. Cai, J. An, S. Kim, J. Nah, D. Yang, et al., Large-area synthesis of high-quality and uniform graphene films on copper foils, Science 324 (5932) (2009) 1312–1314.
[87] X. Li, W. Cai, L. Colombo, R.S. Ruoff, Evolution of graphene growth on Ni and Cu by carbon isotope labeling, Nano Lett. 9 (12) (2009) 4268–4272.
[88] S. Bae, H. Kim, Y. Lee, X. Xu, J.-S. Park, Y. Zheng, et al., Roll-to-roll production of 30-inch graphene films for transparent electrodes, Nat. Nanotechnol. 5 (2010) 574–578.
[89] K. Yan, L. Fu, H. Peng, Z. Liu, Designed CVD growth of graphene via process engineering, Acc. Chem. Res. 46 (10) (2013) 2263–2274.
[90] X.L. Chen, Photo-Effect on Current Transport in Back Gate Graphene Field Effect Transistor, M.S. (EE), August 2017, BS 2012, Shadong University, China, 2017.
[91] W. Chen, P. Cui, W. Zhu, E. Kaxiras, Y. Gao, Z.Z. Zhang, Atomistic mechanisms for bilayer growth of graphene on metal substrates, Phys. Rev. B 91 (4) (2015), 045408.
[92] Q. Liu, Y. Gong, J.S. Wilt, R. Sakidja, J. Wu, Synchronous growth of AB-stacked bilayer graphene on Cu by simply controlling hydrogen pressure in CVD process, Carbon 93 (11) (2015) 199–206.
[93] T.H. Bointon, M.D. Barnes, S. Russo, M.F. Craciun, High quality monolayer graphene synthesized by resistive heating cold wall chemical vapor deposition, Adv. Mater. 27 (28) (2015) 4200–4206.
[94] A.C. Ferrari, J.C. Meyer, V. Scardaci, C. Casiraghi, M. Lazzeri, F. Mauri, et al., Raman spectrum of graphene and graphene layers, Phys. Rev. Lett. 97 (18) (2006), 187401.
[95] B.C. Brodie, On the atomic weight of graphite, Phil. Trans. Roy. Soc. Lond. 149 (1859) 249–259.
[96] V. Kohlschütte, P. Haenni, Zur kenntnis des graphitischen kohlenstoffs und der graphitsäure, Z. Anorg. Allg. Chem. 105 (1919) 121–144.
[97] G. Ruess, F. Vogt, Höchstlamellarer kohlenstoff aus graphitoxyhydroxyd, Monatshefte für Chemie und verwandte Teile anderer Wissenschaften 78 (1948) 222–242.
[98] C. Berger, Z. Song, T. Li, X. Li, A.Y. Ogbazghi, R. Feng, Z. Dai, A.N. Marchenkov, E.H. Conrad, P.N. First, W.A. de Heer, Ultrathin epitaxial graphite: 2D electron gas properties and a route toward graphene-based nanoelectronics, J. Phys. Chem. B 108 (52) (2004) 19912–19916.
[99] C. Berger, et al., Electronic confinement and coherence in patterned epitaxial graphene, Science 312 (5777) (2006) 1191–1196.
[100] N. Zhan, M. Olmedo, G. Wang, J. Liu, Layer-by-layer synthesis of large-area graphene films by thermal cracker enhanced gas source molecular beam epitaxy, Carbon 49 (6) (2011) 2046–2052.
[101] H.G. Duan, E.Q. Xie, L. Han, Z. Xu, Turning PMMA nanofibers into graphene nanoribbons by in situ electron beam irradiation, Adv. Mater. 20 (17) (2008) 3284–3288.
[102] X. Wang, L. Zhi, N. Tsao, Z. Tomovic, J. Li, K. Muellen, Graphene-based electrode materials for rechargeable lithium batteries, Angew. Chem. Int. Ed. 47 (16) (2008) 2990–2992.
[103] K.S. Subrahmanyam, S.R.C. Vivekchand, A. Govindaraj, C.N.R. Rao, A study of graphenes prepared by different methods: characterization, properties and solubilization, J. Mater. Chem. 18 (13) (2008) 1517–1523.
[104] K.S. Subrahmanyam, L.S. Panchakarla, A. Govindaraj, C.N.R. Rao, Simple method of preparing graphene flakes by an arc-discharge method, J. Phys. Chem. C 113 (11) (2009) 4257–4259.

[105] Y.M. Banadaki, A. Srivastava, Scaling effects on static metrics and switching attributes of graphene nanoribbon FET for emerging technology, IEEE Trans. Emerg. Topics Comput. 3 (4) (2015) 458–469.
[106] B. Obradovic, R. Kotlyar, F. Heinz, P. Matagne, T. Rakshit, M.D. Giles, M.A. Stettler, D.E. Nikonov, Analysis of graphene nanoribbons as a channel material for field-effect transistors, Appl. Phys. Lett. 88 (14) (2006), 142102.
[107] S. Rakheja, V. Kumar, A. Naeemi, Evaluation of the potential performance of graphene nanoribbons as on-chip interconnects, Proc. IEEE 101 (7) (2013) 1740–1765.
[108] J.A. Kelber, M. Zhou, S. Gaddam, F.L. Pasquale, L.M. Kong, P.A. Dowben, Direct graphene growth on oxides: Interfacial interactions and band gap formation, ECS Trans. 45 (2012) 49–61.
[109] S. Nakaharai, T. Iijima, S. Ogawa, S. Suzuki, K. Tsukagoshi, S. Sato, N. Yokoyama, Electrostatically-reversible polarity of dual-gated graphene transistors with He ion irradiated channel: toward reconfigurable CMOS applications, in: Tech. Digest IEEE Int. Elect. Dev. Meeting, San Francisco, California, United States of America, 2012, pp. 4.2.1–4.2.4.
[110] G. Giovannetti, P.A. Khomyakov, G. Brocks, P.J. Kelly, J. van den Brink, Substrate-induced band gap in graphene on hexagonal boron nitride: ab initio density functional calculations, Phys. Rev. B 76 (7) (2007), 073103.
[111] M.S. Nevius, M. Conrad, F. Wang, A. Celis, M.N. Nair, A. Taleb-Ibrahimi, et al., Semiconducting graphene from highly ordered substrate interactions, Phys. Rev. Lett. 115 (13) (2015), 136802.
[112] L. Yang, C.H. Park, Y.W. Son, M.L. Cohen, S.G. Louie, Quasiparticle energies and band gaps in graphene nanoribbons, Phys. Rev. Lett. 99 (18) (2007), 186801.
[113] F. Sols, F. Guinea, A.H.C. Neto, Coulomb blockade in graphene nanoribbons, Phys. Rev. Lett. 99 (16) (2007), 166803.
[114] M.Y. Han, B. Özyilmaz, Y. Zhang, P. Kim, Energy band-gap engineering of graphene nanoribbons, Phys. Rev. Lett. 98 (20) (2007), 206805.
[115] L. Xie, H. Wang, C. Jin, X. Wang, L. Jiao, K. Suenaga, H. Dai, Graphene nanoribbons from unzipped carbon nanotubes: atomic structures, Raman spectroscopy, and electrical properties, J. Am. Chem. Soc. 133 (27) (2011) 10394–10397.
[116] D.V. Kosynkin, A.L. Higginbotham, A. Sinitskii, J.R. Lomeda, A. Dimiev, B.K. Price, J.M. Tour, Longitudinal unzipping of carbon nanotubes to form graphene nanoribbons, Nature 458 (2009) 872–876.
[117] M.S. Fahad, Modeling of Two Dimensional Graphene and Non-Graphene Material Based Tunnel Field Effect Transistors for Integrated Circuit Design, Louisiana State University, Baton Rouge, 2017 (Ph.D. (Electrical Engineering) Dissertation).
[118] B.K. Kaushik, M.K. Majumder, Carbon Nanotube Based VLSI Interconnects Analysis and Design, Springer, New Delhi, 2015.
[119] Y.W. Son, M.L. Cohen, S.G. Louie, Energy gaps in graphene nanoribbons, Phys. Rev. Lett. 97 (21) (2006), 216803.
[120] F. Schwierz, Graphene transistors, Nat. Nanotechnol. 5 (7) (2010) 487–496.
[121] N.S. Bart, F. Gianluca, S. Daniel, N. Daniel, K. Heinrich, Current saturation and voltage gain in bilayer graphene field effect transistors, Nano Lett. 12 (3) (2012) 1324–1328.
[122] R. Dharmendar, et al., Graphene field-effect transistors, J. Phys. D. Appl. Phys. 44 (31) (2011), 313001.
[123] M.C. Lemme, T.J. Echtermeyer, M. Baus, H. Kurz, A graphene field-effect device, IEEE Elect. Dev. Lett. 28 (4) (2007) 282–284.
[124] Q. Zhang, T. Fang, H. Xing, A. Seabaugh, D. Jena, Graphene nanoribbon tunnel transistors, IEEE Elect. Dev. Lett. 29 (12) (2008) 1344–1346.
[125] M.S. Fahad, A. Srivastava, A.K. Sharma, C. Mayberry, Analytical current transport modeling of graphene nanoribbon tunnel field-effect transistors for digital circuit design, IEEE Trans. Nanotechnol. 15 (1) (2016) 39–50.
[126] X. Wang, Y. Ouyang, X. Li, H. Wang, J. Guo, H. Dai, Room-temperature all-semiconducting sub-10-nm graphene nanoribbon field-effect transistors, Phys. Rev. Lett. 100 (20) (2008) 20680.
[127] Y.M. Banadaki, Physical Modeling of Graphene Nanoribbon Field Effect Transistor Using Non-Equilibrium Green Function Approach for Integrated Circuit Design, Louisiana State University, Baton Rouge, 2016 (Ph. D. (Electrical Engineering) Dissertation).
[128] A. Konar, T. Fang, D. Jena, Effect of high-K gate dielectrics on charge transport in graphene-based field effect transistors, Phys. Rev. B 82 (11) (2010), 115452.
[129] H. Owlia, P. Keshavarzi, Investigation of the novel attributes of a double-gate graphene nanoribbon FET with AlN high-κ dielectrics, Superlattice. Microst. 75 (2014) 613–620.

[130] J.G. Oh, S.K. Hong, C.K. Kim, J.H. Bong, J. Shin, S.Y. Choi, et al., High performance graphene field effect transistors on an aluminum nitride substrate with high surface phonon energy, Appl. Phys. Lett. 104 (19) (2014), 193112.
[131] I. Imperiale, S. Bonsignore, A. Gnudi, E. Gnani, S. Reggiani, G. Baccarani, Computational study of graphene nanoribbon FETs for RF applications, in: IEEE International Electron Devices Meeting (IEDM), San Francisco, California, USA, 2010, pp. 32.3.1–32.3.4.
[132] G.E. Moore, Cramming more components onto integrated circuits, Electron. Mag. 38 (8) (1965) 114–117.
[133] A. Srivastava, X.H. Liu, Y.M. Banadaki, Overview of carbon nanotube interconnects, in: A. Todri-Sanial, J. Dijon, A. Maffucci (Eds.), Chapter 2 in Carbon Nanotubes for Interconnects: Process, Design, and Applications, Springer International Publishing, 2017.
[134] L.L. Vadasz, A.S. Grove, T.A. Rowe, G.E. Moore, Silicon-gate technology, IEEE Spectr. 6 (10) (1969) 28–35.
[135] J.D. Meindl, Beyond Moore's law: the interconnect era, Comput. Sci. Eng. 5 (1) (2003) 20–24.
[136] R. Solanki, B. Pathangey, Atomic layer deposition of copper seed layers, Electrochem. Solid-State Lett. 3 (10) (2000) 479–480.
[137] C.K. Hu, J.M.E. Harper, Copper interconnections and reliability, Mater. Chem. Phys. 52 (1) (1998) 5–16.
[138] M. Yamada, H. Yagi, S. Sugatani, M. Miyajima, D. Matsunaga, T. Hosoda, H. Kudo, N. Misawa, T. Nakamura, Cu interconnect technologies in Fujitsu and problems in installing Cu equipment in an existing semiconductor manufacturing line, in: Proceedings of the IEEE 1999 International Interconnect Technology Conference, IEEE, 1999, p. 115.
[139] G. Steinlesberger, M. Engelhardt, G. Schindlera, W. Steinhogl, A.V. Glasow, K. Mosig, E. Bertagnolli, Electrical assessment of copper damascene interconnects down to sub-50 nm feature sizes, Microelectron. Eng. 64 (1-4) (2002) 409–416.
[140] International Technology Roadmap for Semiconductors (ITRS), 2007. http://www.itrs.net/Links/2007ITRS/Home2007.html.
[141] K.H. Koo, P. Kapur, K.C. Saraswat, Compact performance models and comparisons for gigascale on-chip global interconnect technologies, IEEE Trans. Electron Dev. 56 (9) (2009) 1787–1798.
[142] L. Arnaud, F. Cacho, L. Doyen, F. Terrier, D. Galpin, C. Monget, Analysis of electromigration induced early failures in Cu interconnects for 45 nm node, Microelectron. Eng. 87 (3) (2010) 355–360.
[143] T. Yu, E. Kim, N. Jain, Y. Xu, R. Geer, B. Yu, Carbon-based interconnect: performance, scaling and reliability of 3D stacked multilayer graphene system, in: IEEE International Electron Devices Meeting (IEDM), Washington, DC, 2011, pp. 7.5.1–7.5.4.
[144] Q. Yuan, Z. Xu, B.I. Yakobson, F. Ding, Efficient defect healing in catalytic carbon nanotube growth, Phys. Rev. Lett. 108 (24) (2012), 245505.
[145] C. Soldano, A. Mahmood, E. Dujardin, Production, properties and potential of graphene, Carbon 48 (8) (2010) 2127–2150.
[146] D. Kondo, H. Nakanoa, B. Zhou, I. Kubota, K. Hayashia, K. Yagi, Fabrication and evaluation of 20-nm-wide intercalated multi-layer graphene interconnects and 3D interconnects composed of graphene and vertically aligned CNTs, Int. Semiconduct. Dev. Res. Symp. (ISDRS) (2013) 11–13.
[147] D. Kondo, H. Nakano, B. Zhou, I. Kubota, K. Hayashi, K. Yagi, M. Takahashi, M. Sato, S. Sato, N. Yokoyama, Intercalated multi-layer graphene grown by CVD for LSI interconnects, IEEE Int. Interconnect Technol. Conf. (IITC) (2013) 1–3.
[148] L. Jiao, X. Wang, G. Diankov, H. Wang, H. Dai, Facile synthesis of high-quality graphene nanoribbons, Nat. Nanotechnol. 5 (2010) 321–325.
[149] Advanced Industrial Science and Technology (AIST), Development of Technology for Producing Micro-Scale Interconnect from Multi-Layer Graphene, 2013. http://www.aist.go.jp/.
[150] E. Pop, Energy dissipation and transport in nanoscale devices, Nano Res. 3 (2010) 147–169.
[151] A. Liao, R. Alizadegan, Z.Y. Ong, S. Dutta, F. Xiong, K.J. Hsia, E. Pop, Thermal dissipation and variability in electrical breakdown of carbon nanotube devices, Phys. Rev. B 82 (20) (2010), 205406.
[152] K.M. Mohsin, A. Srivastava, A.K. Sharma, C. Mayberry, A thermal model for carbon nanotube interconnects, Nano 3 (2) (2013) 229–241.
[153] K.M. Mohsin, Y.M. Banadaki, A. Srivastava, Metallic single-walled, carbon nanotube temperature sensor with self heating, in: Proc. of SPIE 9060, Nanosensors, Biosensors, and Info-Tech Sensors and Systems, 2014, p. 906003-1-7.

[154] A.D. Liao, J.Z. Wu, X. Wang, K. Tahy, D. Jena, H. Dai, E. Pop, Thermally limited current carrying ability of graphene nanoribbons, Phys. Rev. Lett. 106 (25) (2011), 256801.
[155] P. Hale, S. Hornett, J. Moger, D. Horsell, E. Hendry, Hot phonon decay in supported and suspended exfoliated graphene, Phys. Rev. B 83 (12) (2011), 121404.
[156] M.Y. Han, J.C. Brant, P. Kim, Electron transport in disordered graphene nanoribbons, Phys. Rev. Lett. 104 (5) (2010), 056801.
[157] S. Iijima, T. Ichihashi, Single-shell carbon nanotubes of 1-nm diameter, Nature 363 (1993) 603–605.
[158] D.S. Bethune, C.H. Kiang, R. Beyers, Cobalt-catalyzed growth of carbon nanotubes with single-atomic layer walls, Nature 363 (1993) 605–607.
[159] R. Saito, G. Dresslhaus, M.S. Dresselhaus, Physical Properties of Carbon Nanotubes, Imperial College Press, London, UK, 1998.
[160] M.S. Dresselhaus, G. Dresselhaus, P. Avouris, Carbon Nanotube: Synthesis, Properties, Structure, and Applications, Springer Verlag, 2001.
[161] A. Srivastava, J.M. Manulanda, Y. Xu, A.K. Sharma, Carbon-Based Electronics: Transistors and Interconnects at the Nanoscale, Pan Stanford Publishing, Singapore, 2015.
[162] E. Xiao, J.S. Yuan, H. Yang, Hot-carrier and soft-breakdown effects on VCO performance, IEEE Trans. Microw. Theory Tech. 50 (11) (2002) 2453–2458.
[163] Y. Liu, Phase Noise in CMOS Phase-Locked Loop Circuits, Ph.D. (Electrical Engineering) Dissertation, Louisiana State University, 2012.
[164] D. Sarkar, et al., High-frequency behavior of graphene-based interconnects—Part I: impedance modeling, IEEE Trans. Electron Dev. 58 (3) (2011) 843–852.
[165] D. Sarkar, et al., High-frequency behavior of graphene-based interconnects—Part II: impedance analysis and implications for inductor design, IEEE Trans. Electron Dev. 58 (3) (2011) 853–859.
[166] X. Li, J. Kang, X. Xie, W. Liu, D. Sarkar, J. Mao, K. Banerjee, in: Graphene Inductors for High-Frequency Applications Design, Fabrication, Characterization, and Study of Skin Effect, IEEE Int. Electron Device Meeting, San Francisco, California, United States of America, 2014, pp. 5.4.1–5.4.4.
[167] Z. Wang, Q. Zhang, P. Peng, Z. Tian, L. Ren, X. Zhang, R. Huang, J. We, Y. Fu, Q-factors of CVD monolayer graphene and graphite inductors, J. Phys. D. Appl. Phys. 50 (34) (2017) 345103.
[168] N.O. Adesina, M.A. Ullah Khan, A. Srivastava, High Q-factor graphene-based inductor CMOS LC voltage controlled oscillator for PLL applications, in: 2021 IEEE Canadian Conference on Electrical and Computer Engineering (CCECE), 2021, pp. 1–7.

CHAPTER

2

Kinks in buckled graphene uncompressed and compressed in the longitudinal direction

Ruslan D. Yamaletdinov[a,b] *and Yuriy V. Pershin*[c]

[a]Nikolaev Institute of Inorganic Chemistry SB RAS, Novosibirsk, Russia [b]Boreskov Institute of Catalysis SB RAS, Novosibirsk, Russia [c]Department of Physics and Astronomy, University of South Carolina, Columbia, SC, United States

1 Introduction

Graphene kinks and antikinks are topological states of buckled graphene membranes introduced by us in Ref. [1]. Up to the moment, only the studies of kink–antikink scattering [1] and radiation–kink interaction [2] have been reported. However, these two publications already suggest a rich physics of topological excitations in buckled graphene and similar structures. The remarkable characteristics of graphene kinks and antikinks—the nanoscale size, stability, high propagation speed, and low energy dissipation—make them well positioned for applications in nanoscale motion.

Although graphene kinks are reminiscent the ones in the classical scalar ϕ^4 theory [3], they are considerably more complex and require different approaches for their understanding. The complexity stems form the fact that the graphene membranes are two-dimensional objects described by multiple degrees of freedom. Importantly, even when the membrane motion is confined to one dimension, the transverse degrees of freedom cannot be neglected. The transverse degrees of freedom "dress" the longitudinal kinks in different ways leading to several types of graphene kinks compared to the single kink type in the classical scalar ϕ^4 theory.

To be more specific, consider a buckled graphene membrane (nanoribbon) over a trench assuming that the trench length L (in the x-direction) is much longer than its width d (in the y-direction). The buckled graphene has two stable configurations minimizing its energy: the uniform buckled up and buckled down states. Graphene kinks and antikinks are

https://doi.org/10.1016/B978-0-12-819514-7.00007-5

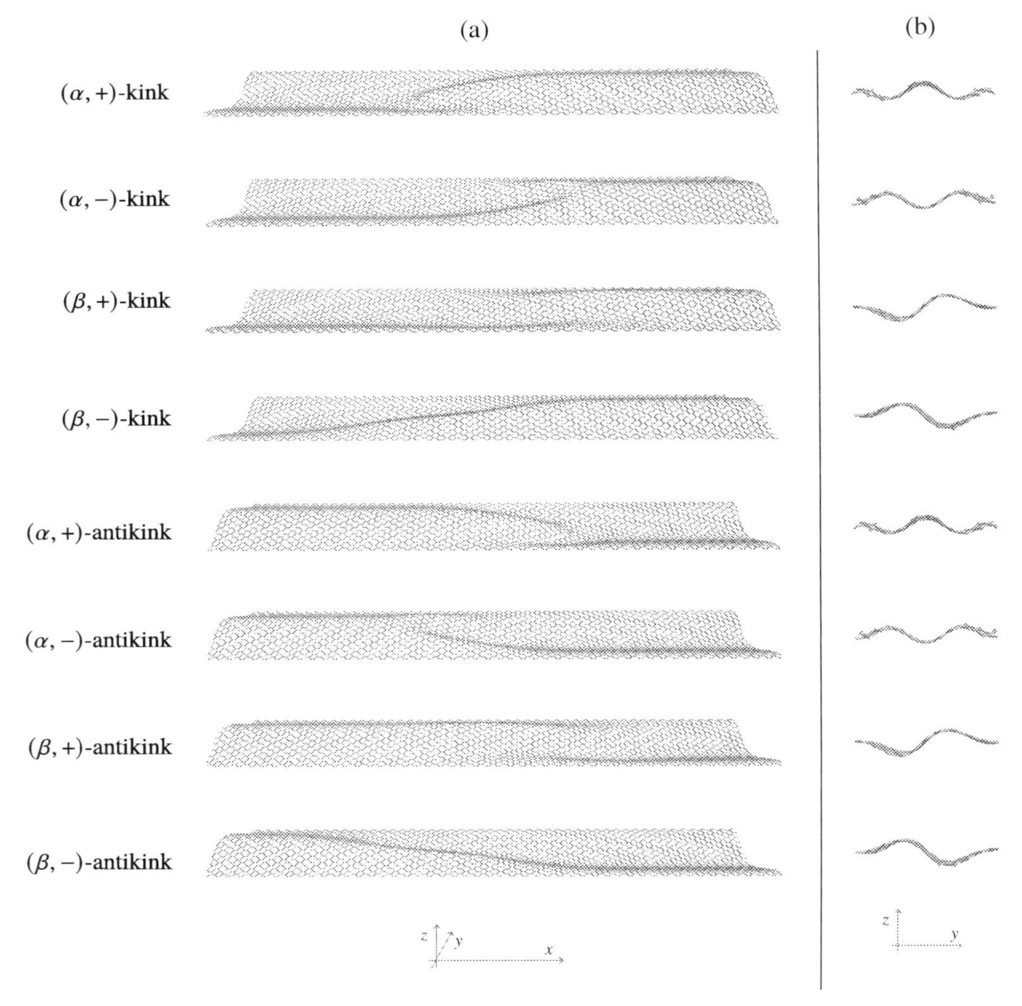

FIG. 2.1 Kinks and antikinks of buckled graphene (A), and their transverse cross sections (B).

excited states of membrane connecting these uniform states, see Fig. 2.1. Fundamentally, kinks and antikinks differ by topological charge [3, 4]

$$Q_{\text{top}} = \frac{1}{2z_0}(z(x \gg x_0, 0) - z(x \ll x_0, 0)) \tag{2.1}$$

such that $Q_{\text{top}} = 1$ for kinks and $Q_{\text{top}} = -1$ for antikinks. In Eq. (2.1), x_0 is the position of kink or antikink, $z(x, y)$ is membrane deflection from the (x, y)-plane, and $z_0 = |z(x \gg x_0, 0)| = |z(x \ll x_0, 0)|$ is the deflection in the uniform buckled up or down state. The topological charge is defined by the boundary conditions (buckling directions very far left and very far right from the kink or antikink in Fig. 2.1), and does not depend on the transverse cross section of membrane at any point. The states connecting the uniform buckled down membrane at $x \ll x_0$ and buckled up membrane at $x \gg x_0$ are kinks. The opposite states are antikinks.

Using molecular dynamics simulations, we have identified four types of kinks and four types of antikinks, see Fig. 2.1. Let us assign $\alpha(\beta)$-type to the kinks and antikinks with the symmetric (asymmetric) transverse cross section (in $y - z$ plane), and use + (−) for kinks and antikinks with unturned (turned) transverse cross section (more details are given below). We note that all α-type kinks and antikinks can be obtained from a single one via symmetry transformations. For instance, the transformation $z \to -z$ transforms (α, +)-kink into (α, −)-antikink and vice versa, the transformation $x \to -x$ transforms (α, +)-kink into (α, +)-antikink, etc. The same is true for β-type kinks and antikinks. We emphasize that the longitudinal cross section (in $x - z$ plane at $y = 0$) defines whether the state is a kink or antikink, while the transverse cross section (in $y - z$ plane) defines its type $\alpha(\beta)$, + (−). Additional details are provided in Section 3.1.

Our previous work on graphene kinks [1, 2] was based on molecular-dynamics simulations, which is a versatile tool to investigate mechanical properties of nanostructures. The success and easiness to use of molecular dynamics are related to the atomistic approach to describe the system of interest, availability of potentials that realistically describe interactions between atoms, and modest requirement to computational resources (compared to the density functional theory calculations). During the last decade, the systems comprising millions on atoms have been routinely modeled on modest computer clusters. Some billion-atom molecular dynamics simulations have been reported recently [5, 6]. We note that the elasticity theory (more specifically, the classical nonlinear theories for plates [7] with strain limitations) offers an alternative framework to the description of graphene elasticity [8, 9]. An interesting future problem is to describe analytically the shape of graphene kinks, and find analytical expression for kink energy.

In this chapter we review our previous results on graphene kinks [1, 2] and extend them in two directions. First, we explore how the energy of graphene kinks relates to the membrane width, degree of buckling, and kink velocity. Second, we investigate the effect of longitudinal stress on the dynamics of graphene kinks.

The chapter is organized as follows. Section 2 provides some preliminary information including (*i*) the classical scalar ϕ^4 model, which sets the framework for the discussion, and (*ii*) details of molecular dynamics simulations. Next, in Section 3 we consider graphene kinks in longitudinally uncompressed graphene. Here, we introduce the nomenclature of graphene kinks, report our new results on kink energetics, as well as briefly review some of our past findings [1, 2]. Section 4 reports few selected results on graphene kinks in longitudinally compressed membranes. Finally, the conclusions and outlook are presented in Section 5.

2 Preliminaries

2.1 ϕ^4 model

To set the stage for a discussion of graphene kinks, let us briefly review kinks in the ϕ^4 classical scalar field theory [3, 4, 10]. For the last several decades this model has been widely used in several diverse branches of physics including statistical mechanics, condensed matter, topological quantum field theory [11], and cosmology [12]. In the area of condensed matter physics, ϕ^4-kinks have been used to describe domain walls in ferroeletrics [13–15] and ferromagnets, proton transport in hydrogen-bonded chains [16–18], and charge-density waves [19, 20]. We note that in recent years, the focus of attention has shifted to higher-ϕ models, such as ϕ^6 [21, 22], ϕ^8 [23, 24], ϕ^{10}, and ϕ^{12} [25]. Their stables states, however, are quite different compared to the ones that we observe in graphene.

The ϕ^4 model is based on the Lagrangian

$$\mathcal{L}=\frac{1}{2}\int \mathrm{d}\tilde{x}\left[\left(\frac{\partial\phi}{\partial\tilde{t}}\right)^2-\left(\frac{\partial\phi}{\partial\tilde{x}}\right)^2-\frac{1}{2}(1-\phi^2)^2\right], \tag{2.2}$$

where $\phi(\tilde{x},\tilde{t})$ is a real scalar field, $\tilde{x}$ and $\tilde{t}$ are dimensionless spatial coordinate and time, respectively. We use tildes to distinguish dimensionless quantities from the dimensional ones; for an example of ϕ^4 model written in dimensional units, see Ref. [26]. Lagrangian (2.2) leads to the Euler-Lagrange equation of motion of the form

$$\frac{\partial^2\phi}{\partial\tilde{t}^2}-\frac{\partial^2\phi}{\partial\tilde{x}^2}=\phi-\phi^3. \tag{2.3}$$

Eq. (2.3) has two stable constant solutions $\phi=\pm 1$, and one unstable $\phi=0$. The kink and antikink solutions connect the two stable ground state values, and, therefore, they are topologically stable. Mathematically, these are given by

$$\phi_{K(A)}=\pm\tanh\left(\frac{\tilde{x}-\tilde{V}\tilde{t}-\tilde{x}_0}{\sqrt{2(1-\tilde{V}^2)}}\right), \tag{2.4}$$

where $\pm$ signs correspond to the kink and antikink, respectively, $\tilde{V}$ is the dimensionless velocity ($|\tilde{V}|<1$), and $\tilde{x}_0$ is the dimensionless position of the kink/antikink center at the initial moment of time $\tilde{t}=0$. According to Eq. (2.4), ϕ^4 kinks and antikinks move without dissipation at a constant velocity. Moreover, their kinetic energy is expressed by a relativistic formula [26]. In the past, the above model was applied to different physical situations, where the system can be effectively described by a 1D model with a double well potential. The physical meaning of $\phi(\tilde{x},\tilde{t})$ is application-specific. For instance, in Ref. [15], ϕ was used to describe the magnetic order parameter, while in the present Chapter it is used to represent the membrane deflection.

A very famous result in ϕ^4 theory is the resonant structure of the kink–antikink scattering [27]. Numerical simulations have shown that intervals of initial velocity for which the kink and antikink capture one another alternate with regions for which kink and antikink separate infinitely [27], see Fig. 2.2. The main feature of the latter regions is two- and higher-bounce resonances in which kink and antikink form a quasi-stable state for some short period of time. Within this process, the kinetic energy is first transferred into internal oscillation mode(s), and realized at a later time (such as on the second bounce). The early work on two-bounce resonance is attributed to Campbell and coauthors [27]. Higher-order resonances were identified by Anninos et al. [28].

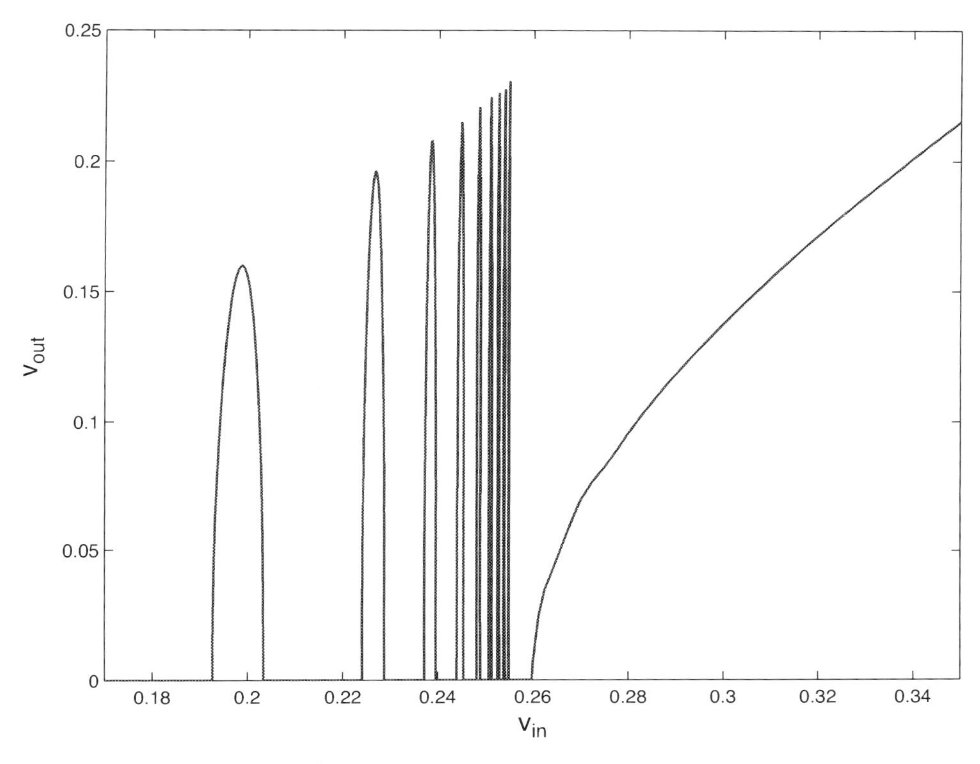

FIG. 2.2 Kink–antikink collision in ϕ^4 model: Final velocity as a function of the initial kink velocity [29]. *Reprinted with permission from R.H. Goodman, R. Haberman, Kink-Antikink collisions in the* ϕ^4 *equation: the n-bounce resonance and the separatrix map, SIAM J. Appl. Dyn. Syst. 4 (2005) 1195–1228.*

Below we demonstrate that the solutions (2.4) of Eq. (2.3) are relevant to graphene kinks and antikinks. Eq. (2.4) can be considered as zero-order approximation to the deflection of the central line ($y = 0$) of membrane atoms from (x, y)-plane. Specifically, the graphene kinks are qualitatively related to the ϕ^4 kinks through the following correspondence: the x-coordinate (along the membrane) corresponds to $\tilde{x}$, the central line deflection $z(x, y = 0, t)$ plays the role of $\phi(\tilde{x},\tilde{t})$, and the graphene kink velocity V corresponds to $\tilde{V}$. In Section 3.2 we show that the relativistic formula fits well the kinetic energy of graphene kinks. Moreover, preliminary data indicates the possibility of bounce resonances in graphene kink–antikink scattering (under certain conditions).

2.2 Molecular dynamics simulations

Our results on graphene kinks were obtained using NAMD2 [30]—a highly scalable massively parallel classical MD code[a] —with optimized CHARMM-based force field [31] for

[a] NAMD was developed by the Theoretical and Computational Biophysics Group in the Beckman Institute for Advanced Science and Technology at the University of Illinois at Urbana-Champaign.

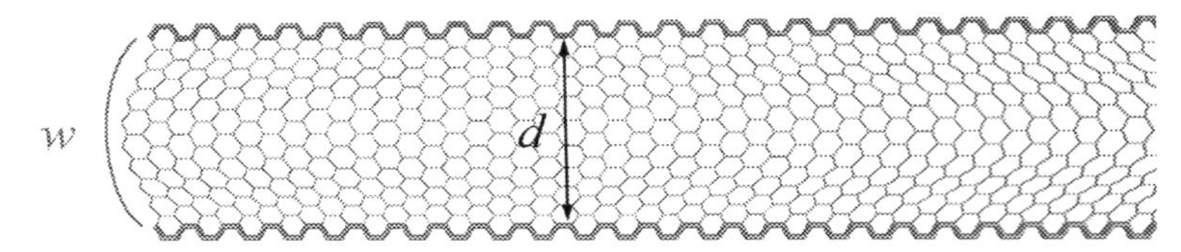

FIG. 2.3 Geometry of buckled graphene nanoribbon. The *red* (dark gray in print version) armchair edges represent the location of frozen atoms.

graphene atoms (see description below and Ref. [32]). Some of our results were verified with a more comprehensive Tersoff potential [33] using LAMMPS code [34]. This method has confirmed the validity of our simulations with NAMD2.

We simulated the dynamics of graphene nanoribbons (membranes) of a length L and width w. We used clamped boundary conditions for the longer armchair edges and free boundary conditions for the shorter edges, see Fig. 2.3. To implement the clamped boundary conditions, the two first lines of carbon atoms of longer edges were fixed. Membranes were buckled by changing the distance between the fixed sides from w to $d < w$. In what follows for the sake of clarity we use "ring" units of width and length. In our geometry one ring has the size (width × length) of $\sqrt{3}a_0 \times 3a_0 = 2.46$ Å × 4.26 Å.

To simulate the graphene dynamics we used the optimized CHARMM-based force field [31], which includes 2-body spring bond, 3-body angular bond (including the Urey-Bradley term), 4-body torsion angle, and Lennard-Jones potential energy terms [35]. All force-field constants have been optimized to match known properties of graphene. In particular, the AB stacking distance and energy of graphite [36] have been used for choose the Lennard-Jones coefficients. The remaining parameters were optimized to reproduce the in-plane stiffness (E_{2D} = 342 N/m), bending rigidity (D = 1.6 eV) and equilibrium bond length (a = 1.421 Å) of graphene. All simulations were performed with 1 fs time step. The van der Waals interactions were gradually cut off starting at 10 Å from the atom until reaching zero interaction 12 Å away.

Several methods were used to explore the properties of graphene kinks:

- Method 1. To determine configurations and energies (Sections 3.1 and 3.2), the Langevin dynamics of the initially flat compressed nanoribbon was simulated for 20 ps at $T = 293$ K using a Langevin damping parameter of 0.2 ps^{-1} in the equations of motion. This simulation stage was followed by 10000 steps of energy minimization. The temperature of 293 K was used in our initial simulations and resulted in very good results in terms of the configuration representativeness.
- Method 2. To generate moving kinks (Sections 3.2 and 3.3), we applied a downward (in $-z$ direction) force to the atoms located at the distance up to about 3 rings from the shorter edge or edges. As the initial configuration, we used an optimized membrane in the uniform buckled up state. The dynamics was simulated for 20 ps at $T = 0$ K without any temperature or energy control.
- Method 3. To simulate the radiation–kink interaction (Section 3.4), we applied a sinusoidal force to the atoms located at the distance up to about 3 rings from the shorter edge. As the initial configuration, we used an optimized membrane with a stationary kink located in the middle part of membrane. We performed a series of computations with x-, y- or, z- directed

force with amplitudes and vibration period in ranges 0–160 pN/atom and 0–160 fs, respectively. The dynamics was simulated for 30 ps at $T = 0$ K without any temperature or energy control.

The interested reader can contact the authors directly for examples of input files of their molecular dynamics simulations.

3 Kinks in longitudinally uncompressed graphene

3.1 Types of graphene kinks

Our studies have revealed that there exists four types of graphene kinks and four types of antikinks, which are summarized in Fig. 2.1. In the *symmetric* kinks (referred as α-kinks), the cross section in the transverse to the trench direction is symmetric (see Fig. 2.4A). Very approximately, close to the kink center x_0, the membrane deflection in symmetric kinks and antikinks as a function of y can be represented by

$$z_\alpha(x \approx x_0, y) \overset{\propto}{\sim} \pm \cos\left(\frac{3\pi y}{d}\right). \tag{2.5}$$

The "+" and "−" α-kinks correspond to $\pm$ in the above equation. According to this terminology, Fig. 2.4A represents an (α, +)-kink.

Similarly, the *nonsymmetric* kinks (referred as β-kinks) are characterized by a nonsymmetric cross section. An example of nonsymmetric kink is shown in Fig. 2.4B. Again, very approximately, in the vicinity of the center the cross section of β-kinks and antikinks is given by

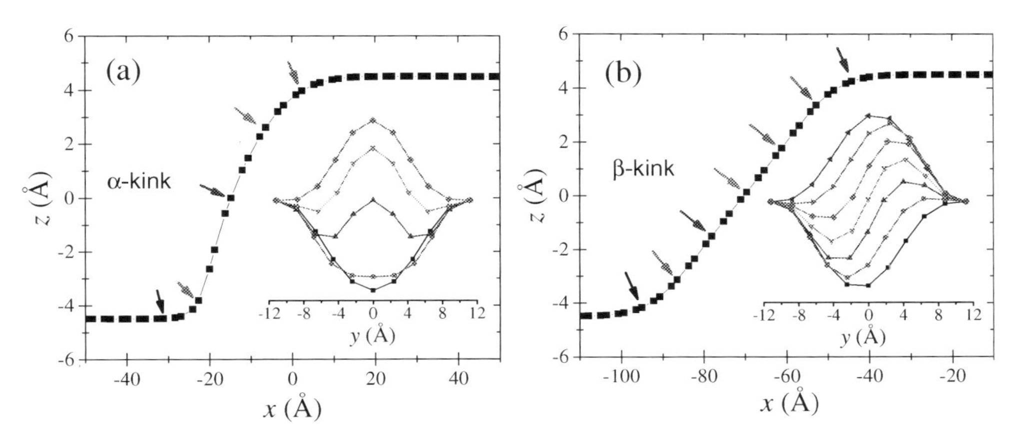

FIG. 2.4 (A) Symmetric and (B) nonsymmetric kinks. The *arrows* show the positions of cross sections (*insets*). These configurations were obtained using Method 1 in Section 2.2 for $d/w = 0.9$, $w = 9$ rings, and $L = 100$ rings.

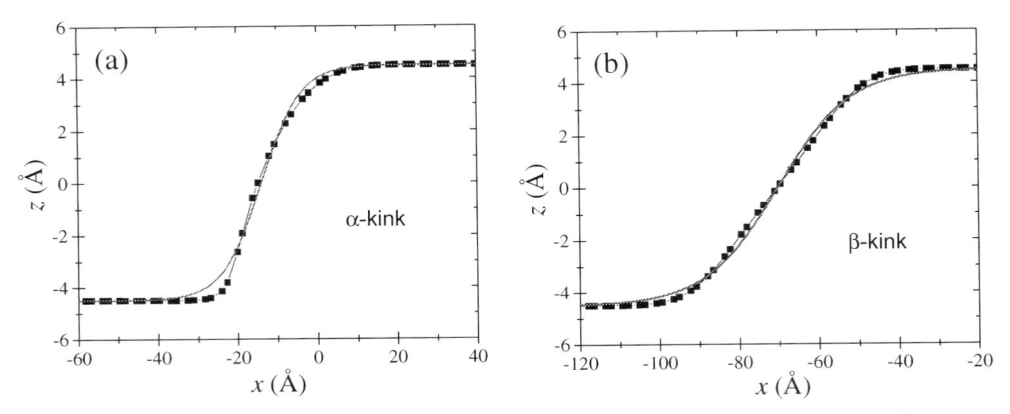

FIG. 2.5 Comparison of the graphene and ϕ^4 kinks (*red solid line*; gray in print version). To plot ϕ^4 kinks we used Eq. (2.4) with a scaled amplitude, $V=0$, and x scaled by 7 Å in (A) and 13 Å in (B). The graphene kinks are the same as in Fig. 2.4.

$$z(x\approx x_0,y) \underset{\sim}{\propto} \pm \sin\left(\frac{2\pi y}{d}\right). \tag{2.6}$$

The "+" and "−" β-kinks correspond to $\pm$ in the above equation. The kink in Fig. 2.4B is thus a (β, +)-kink.

In Fig. 2.5, the graphene and ϕ^4 kinks (Eq. 2.4) are superimposed for comparison. This plot indicates that the longitudinal cross sections of α- and β-kinks can not be ideally described by the ideal ϕ^4 model. We also emphasize that the β-kinks are approximately two times wider in x-direction compared to α-kinks. Moreover, α-kinks are slightly nonsymmetric in the longitudinal direction.

3.2 Kink energy

Stationary kinks

The energies of stationary kinks were calculated using the dynamics/energy minimization approach for several values of w and d/w (Method 1 in Section 2.2). The use of finite temperature dynamics in Method 1 enables sampling the stable membrane conformations and their energies [37]. An example of our results is presented in Fig. 2.6. Here, the final conformation energies are plotted for 100 independent runs. Due to the stochastic nature of molecular dynamics simulations at finite temperature, the results are different in different runs.

Importantly, the final state energies in Fig. 2.6 are discrete. The lowest possible energy $N=0$ is the ground state energy in which the membrane is uniformly buckled up or down. The excited state energies correspond to the membrane with N kinks and/or antikinks. While each kink/antikink contributes approximately the same amount of energy to the total energy, there is a difference in the energies of α- and β-kinks. Due to this difference, the lines with the same N are splitted. When kinks/antikinks are sufficiently far from the edges, and spaced apart enough to neglect their interaction, the total energy can be written as

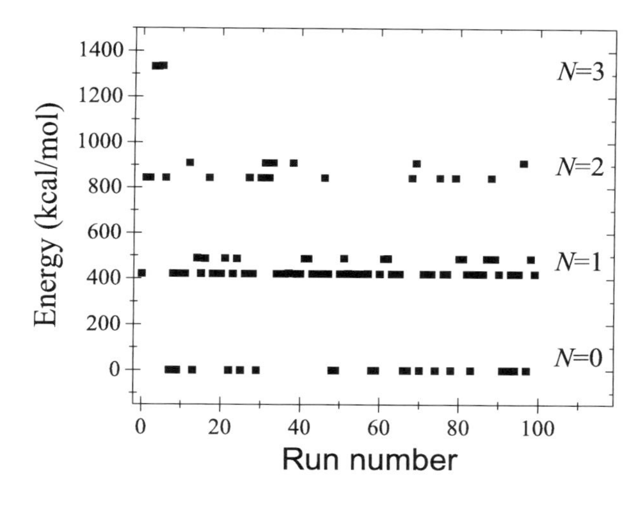

FIG. 2.6 Final energy, counted from the energy of the uniform buckled state, calculated using Method 1 in 100 independent runs. Here, N is the number of kinks and antikinks in the final conformation for each run. This plot was obtained for a $w = 13$ rings membrane.

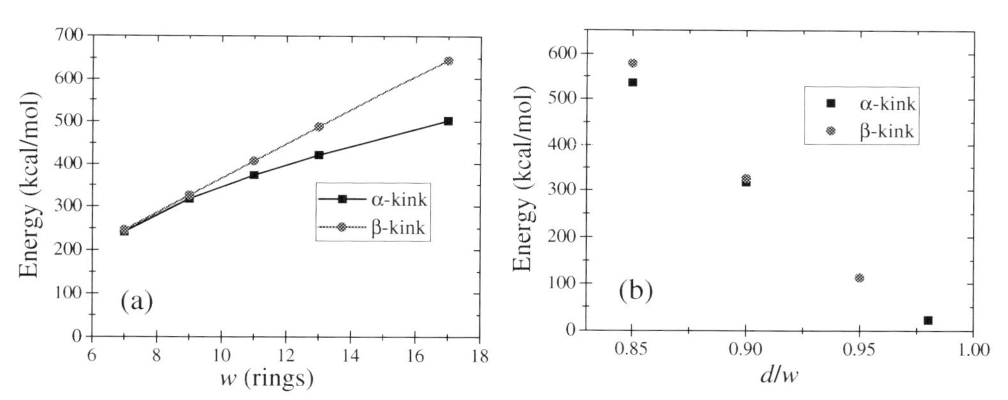

FIG. 2.7 Energies of symmetric and nonsymmetric kinks plotted at fixed (A) $d/w = 0.9$ and (B) $w = 9$ rings. In (B), we observed only nonsymmetric kinks at $d/w = 0.95$, and only symmetric ones at $d/w = 0.98$ (in both cases, in 100 runs).

$$E_{tot} = E_0 + (N_{K,\alpha} + N_{A,\alpha})E_\alpha + (N_{K,\beta} + N_{A,\beta})E_\beta, \tag{2.7}$$

where E_0 is the ground state energy, $E_{\alpha(\beta)}$ is the energy of $\alpha(\beta)$-kink at rest, $N_{K(A),\alpha}$ is the number of α-kinks (antikinks), and $N_{K(A),\ \beta}$ is the number of β-kinks (antikinks), and $N = N_{K,\alpha} + N_{A,\alpha} + N_{K,\beta} + N_{A,\beta}$. Clearly, the level splitting increases with N as $N + 1$. We note that in some calculations the minimization occasionally resulted in a kink or antikink trapped at the boundary with energy different than $E_{\alpha(\beta)}$. Such cases were rejected by manual inspection.

It is interesting to investigate how the kink energy changes with w and d/w. For this purpose, we performed Method 1 simulations for selected values of w and d/w (similar to the ones reported in Fig. 2.6) and extracted kink energies from these simulations. Fig. 2.7A shows that at a fixed ratio of $w/d = 0.9$ the energy of α-kink is always smaller than the energy of β-kink. This plot demonstrates that the energy of β-kink increases almost linearly with the channel width, while the increase of α-kink energy is nonlinear and not so fast.

Fig. 2.7B demonstrates that at a fixed w, the kink energies decrease with increase of d/w. This is expected behavior. Unfortunately, the fine understanding of Fig. 2.7 features is not possible based solely on molecular dynamics simulations. The combination of analytical and numerical approaches would be an ideal route to accomplish this goal.

Moving kinks

Another nontrivial task is to understand the properties of moving kinks. The classical scalar ϕ^4 theory predicts that the kink kinetic energy is expressed by the relativistic formula [26]

$$E_k = mC^2\left(\frac{1}{\sqrt{1-V^2/C^2}} - 1\right), \tag{2.8}$$

where m is the effective kink mass, C and V are the characteristic and kink velocities, respectively. In this subsection we demonstrate numerically that the relativistic expression (2.8) provides a much better fit to the kink kinetic energy compared to the classic expression. This strongly indicates that the kinetic energy of the kink is described by the relativistic expression [26]. One of the major consequences is that the velocity of kink cannot exceed its "speed of light" C, which has no relation to the real speed of light, but enters in the same way in the kinetic energy of the kink as the real speed of light in the kinetic energy of particles moving at relativistic velocities.

The moving kinks were generated using Method 2 from Section 2.2 (for their generation a downward force was applied to a group of atoms near the left short edge of buckled up membrane), and their kinetic energies and velocities were extracted from molecular dynamics simulations. We have observed that (α, $-$)-kinks are created when the pulling force exceeds a threshold value. Moreover, the initial speed of graphene kinks depends on the pulling force (saturating at about 5 km/s in $w = 9$ rings, $d/w = 0.9$ membrane [1]) and stays practically unchanged when the kink moves along the membrane. Unfortunately, Method 2 generates a significant amount of noise that contributes to the total kinetic energy of membrane. This makes challenging the precise measurement of the kink kinetic energy. To reduce the noise, after 5 ps of the initial dynamics, the atomic coordinates and velocities in the regions beyond the moving kink (starting at $\pm$ 20 Å from the kink center) were set to the values in the optimized buckled up or down membrane, and the pulling force was removed. Fig. 2.8 shows the numerically found kinetic energy of kink as a function of its velocity fitted with the classical ($E_k = mV^2/2$) and relativistic Eq. (2.8) expressions.

It is interesting that both the kink mass m and characteristic speed C strongly correlate with each other: they both increase as the buckling increases. The numerical values of these parameters are reasonable. While the characteristic speed C is of the order of the speed of sound in flat graphene (9.2–18.4 km/s [38]), the effective kink mass m is comparable to the mass of several atoms (one carbon atom mass equals 12.0107 a.m.u.).

3.3 Kink–antikink scattering

In Ref. [1] we studied the kink–antikink scattering in a $w = 9$ rings graphene membrane at $d/w = 0.9$. Using Method 2 (see Section 2.2 and Ref. [1]), kinks and antikinks with the speeds in the interval from 3–5 km/s were created at the opposite sides of membrane and collided at

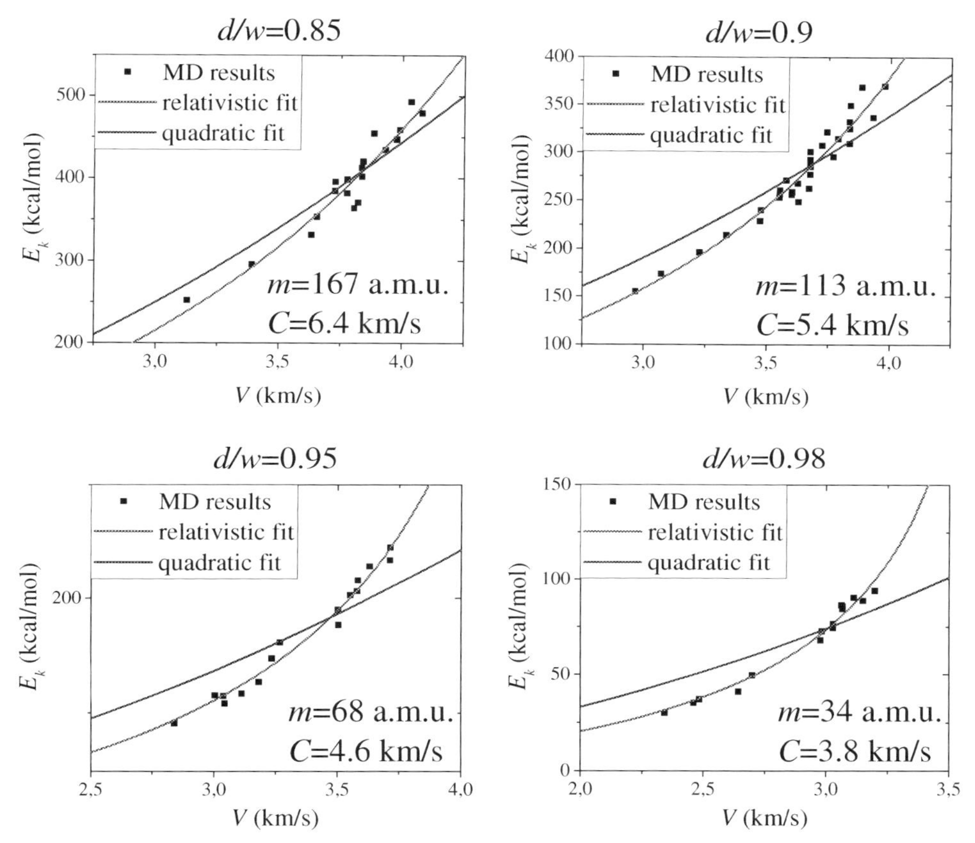

FIG. 2.8 Graphene kink kinetic energy in $w = 9$ rings membrane at several values of d/w. The relativistic kinetic energy expression (*red line*; gray in print version) much better fits the MD data. The values of m and C are given for the relativistic fit (Eq. 2.8).

the center. The results were compared with those for the classical ϕ^4 model in which, depending on the initial kink velocity, the kink–antikink collision leads either to the reflection, or annihilation with formation of a long-radiating bound state, or reflection through two- or several-bounce resonance collisions (see Fig. 2.2 and Refs. [10, 28, 39]).

To study the kink–antikink scattering, we have performed a series of calculations in which the force used to generate moving kinks and antikinks was varied. It was observed that the decaying bound state is formed in the collisions of slower moving kinks and antikinks. An example of such situation is presented in Fig. 2.9A where the collisions occurs at the speeds of 2.9 km/s. We also observed that the collision of fast moving kink and antikink leads to their immediate reflection (see Fig. 2.9B).

In Ref. [1] we did not observe, however, a series of resonances below the critical velocity separating the annihilation and reflection regimes (for ϕ^4 model resonances, see Fig. 2.2). At the same time, the behavior similar to the two-bounce reflection was spotted close to the

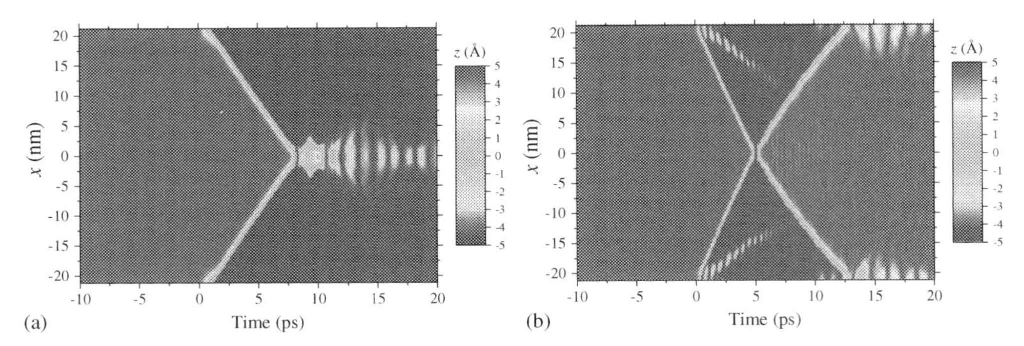

FIG. 2.9 Collision of (A) slower and (B) faster moving kink and antikink. *Reprinted with permission from R.D. Yamaletdinov, V.A. Slipko, Y.V. Pershin, Kinks and antikinks of buckled graphene: a testing ground for the φ^4 field model, Phys. Rev. B 96 (2017) 094306.*

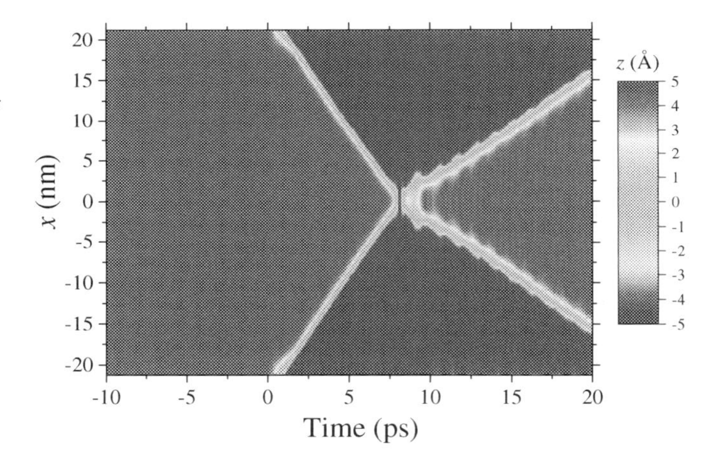

FIG. 2.10 A resonance phenomenon in the kink–antikink scattering. *Reprinted with permission from R.D. Yamaletdinov, V.A. Slipko, Y.V. Pershin, Kinks and antikinks of buckled graphene: a testing ground for the φ^4 field model, Phys. Rev. B 96 (2017) 094306.*

critical velocity, see Fig. 2.10. However, in some preliminary simulations of membranes with smaller d/w we spotted a reflection window below the critical velocity. Further work is required to evaluate and refine this result as the effect may be related to noise (which is a significant side-effect in this type of simulations). Generally, the deviation of the scattering in graphene compared to the one in ϕ^4 model can be explained by a larger number of energy relaxation channels/degrees of freedom in graphene due to its two-dimensional nature.

3.4 Radiation–kink interaction

Recently the radiation–kink interaction was investigated by us in Ref. [2]. The interest in this topic has been motivated by an unexpected theoretical prediction of a *negative radiation pressure effect* (NRP) in ϕ^4 field model [40]. In the standard linearized scattering theory the radiation–kink interaction is described by a second-order term $\sim A^2|R^2|$, where A is the radiation amplitude, and R is the reflection coefficient. A distinctive feature of ϕ^4 model is the absence of the radiation–kink interaction up to forth order in A, so that the force F experienced by the kink $F \sim \pm A^4$. The NRP effect corresponds to the minus sign (the force is directed toward the radiation source), and can be explained as follows. In ϕ^4 field model, the radiation–kink interaction is nonlinear and may lead to frequency doubling. As the doubled-frequency (transmitted) waves carry more momentum than the incident ones, the kink must accelerate toward the radiation source to compensate the momentum surplus.

In graphene, the NRP effect is quite complex as the role of radiation is played by phonons, which can be of several types [41]. Moreover, the boundary scattering as well as phonon–kink interaction may lead to the conversion between different types of phonons—an effect, which is beyond the scope of this investigation. A qualitative description of the phonon–kink scattering can be obtained in terms of a multichannel scattering model [2], in which the force F is written as

$$F \sim \mathcal{P}_i + \sum_j \frac{k_j}{k_i}(|R_{ij}|^2 - |T_{ij}|^2)\mathcal{P}_i, \tag{2.9}$$

where k_i is the wave number, and it is assumed that the incoming radiation is contained in the i-th channel. In Eq. (2.9), the first term is responsible for the absorption of the i-th component of incoming momentum flux $\mathcal{P}_i$, while the second term describes the emission of the absorbed radiation (from channel i) into the reflected and transmitted modes in j-th channels (with the scattering probabilities $|R_{ij}|^2$ and $|T_{ij}|^2$, respectively). A favorable condition for NRP is when the scattering probabilities $|R_{ij}|^2$ are small, and one of $|T_{ij}|^2$ into a certain high-momentum channel is large.

The negative radiation pressure effect in buckled graphene was studied using Method 3 in Section 2.2. To generate phonons, a sinusoidal force was applied to a group of atoms near the left short edge of membrane (the radiation source). The kink was initially placed at a distance of 200 Å from the radiation source, and its position as a function of time was recorded. Due to the large space of parameters, we performed a series of MD simulations for a single buckled membrane ($w = 9$ rings, $d/w = 0.9$) varying the amplitude, frequency and direction of sinusoidal force. For additional information, see Section 2.2 or Ref. [2]. The direction of coordinate axes can be found in Fig. 2.4.

The positive and negative radiation pressure effects (PRP and NRP) are exemplified in Fig. 2.11. The difference between these effects is in the direction of kink displacement: in the case of PRP the radiation pushes the kink away from the radiation source (located in the vicinity of $x \sim 0$ in Fig. 2.11), while in the case of NRP the radiation pulls the kink toward the radiation source. The ϕ^4 model predicts a very narrow frequency window for NRP [40]. To understand how the radiation pressure depends on the driving force parameters, an extensive scan in the force amplitude-force period space for forces in the x-, y-, and z-directions was performed. The results are shown in Fig. 2.12. Here, the blue regions correspond to the NRP effect (except of $F_x \sim 150$ pN/atom $T \sim 350$ fs region in the left plot), while the red regions—to the PRP effect. We emphasize that the type of effect has a complex dependence on the driving force parameters, and is different for F_x, F_y, and F_z excitations because of the different types of phonons produced by forces in different directions.

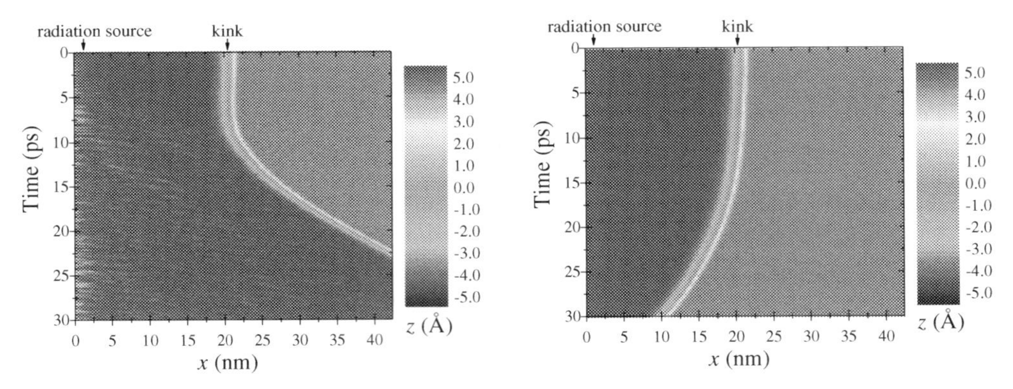

FIG. 2.11 Positive (left) and negative (right) radiation pressure effect. In both cases the radiation is caused by an external sinusoidal force in y-direction (80 pN/atom) with the period of $T = 220$ fs (left) and $T = 190$ fs (right). *Reprinted with permission from R.D. Yamaletdinov, T. Romańczukiewicz, Y.V. Pershin, Manipulating graphene kinks through positive and negative radiation pressure effects, Carbon 141 (2019) 253–257.*

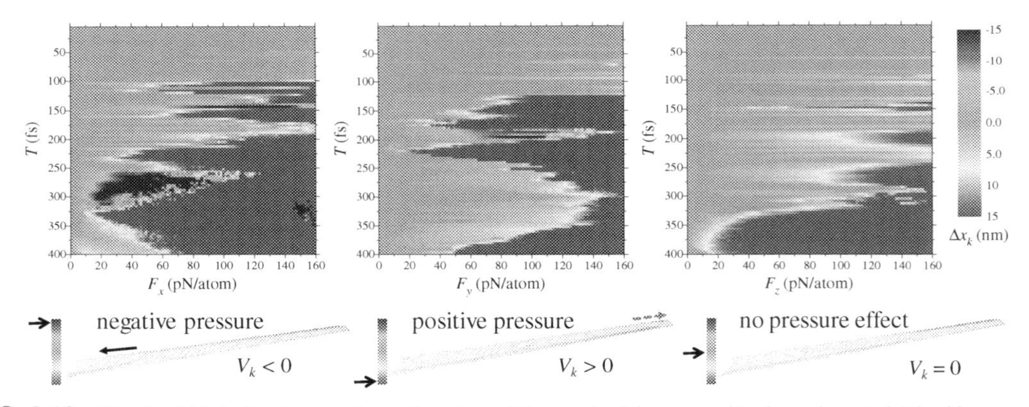

FIG. 2.12 The final kink displacement as a function of the applied force amplitude and period. The *blue* regions correspond to the negative pressure effect, the *red* (gray in print version) ones—to the positive radiation pressure effect. *Reprinted with permission from R.D. Yamaletdinov, T. Romańczukiewicz, Y.V. Pershin, Manipulating graphene kinks through positive and negative radiation pressure effects, Carbon 141 (2019) 253–257.*

The negative radiation pressure effect that has been observed in our MD results can be explained as follows. Assume that the incoming radiation has the wavelength λ_{in}. The nonlinear interaction with kink scatters the incoming phonons into the modes with the wavelengths $\lambda_{tr,1} = \lambda_{in}$, $\lambda_{tr,2} = \lambda_{in}/2$, etc. The most efficient higher-harmonic generation can be expected when the incoming wave is in resonance with the kink length L_K, namely, $L_K = n\lambda_{in}$ where n is an integer. In such situation, the role of the emission at $\lambda_{tr,\,2}$ is increased, and the kink is pushed in the negative x-direction.

To test the above speculation, it is convenient to consider the scaled difference of atomic positions. For the case of NRP effect, this quantity is plotted in Fig. 2.13 for $F_z = 120$ pN/atom and $T = 144$ fs. Fig. 2.13 clearly demonstrates that the above discussed conditions for NRP

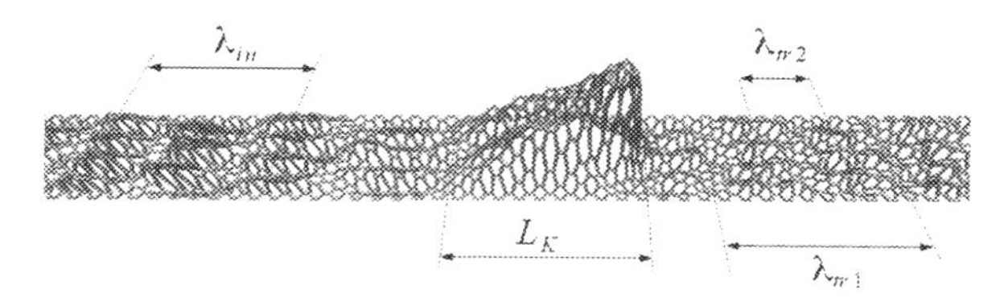

FIG. 2.13 The scaled differences of atomic positions in the NRP regime induced by an $F_z = 120$ pN/atom, $T \doteq 144$ fs force. *Reprinted with permission R.D. Yamaletdinov, T. Romańczukiewicz, Y.V. Pershin, Manipulating graphene kinks through positive and negative radiation pressure effects, Carbon 141 (2019) 253–257.*

effect ($L \approx \lambda_{in} = 42.6$ Å, $\lambda_{tr,\,2} = \lambda_{in}/2$) are met in this specific MD simulation. Similar results were obtained for the driving force in y-direction ($F_y = 74$ pN/atom and $T = 194$ fs). It is important to note that in reality the radiation–kink scattering involves several channels, and single polarized monochromatic incoming waves may be partially transformed to waves with another polarization or/and with higher harmonics. In Ref. [2] we also performed an analysis of vibrational modes of the kink and refer the interested reader to this publication for additional information.

4 Kinks in longitudinally compressed graphene

It is of interest to understand the effect of a longitudinal compression superimposed on the compression in the transverse direction. This is not an abstract question because the longitudinal compression can be introduced during the fabrication stage in an experimental setup. For instance, the buckled graphene can be created by leveraging an intriguing property of graphene known as negative thermal expansion (opposite to most materials, graphene contracts on heating and expands on cooling) [42]. A thermally oxidized silicon wafer with an array of lithographically defined U-shaped grooves can be used as a substrate (the grooves can be formed by chemical or plasma etching). Graphene transferred to the wafer surface at a high temperature will cool, expand in *two* directions, and buckle above the grooves.

Molecular dynamics simulations of longitudinally compressed membranes were performed similarly to the simulations described above with the only difference that the initial longitudinal separation between the atoms were scaled by a factor of l/L, where l is the length of compressed membrane. Some initial results on the properties of longitudinal compressed membranes were obtained using Methods 1 and 2 from Section 2.2. Fig. 2.14 shows examples of optimized geometries found with the help of Method 1 for several selected values of l/L and fixed $d/w = 0.9$.

When the longitudinal compression is relatively small (below a critical value), the membrane is not visibly deformed as we show in Fig. 2.14A. However, above the critical value (which is in the interval {0.95, 0.98} of l/L for our membrane), an instability develops strongly resembling the telephone cord buckling [43, 44] in appearance (Figs. 2.14B and C). In the past, the telephone buckling instability has been observed in the delamination of biaxially compressed thin films [43, 44] and have been extensively studied both experimentally and theoretically. Fig. 2.15 exemplifies the telephone cord delamination using an in-house grown structure. The striking similarity of Fig. 2.15 with Figs. 2.14B and C indicates that the deformation mechanism in the graphene membrane is likely of the same origin. When the

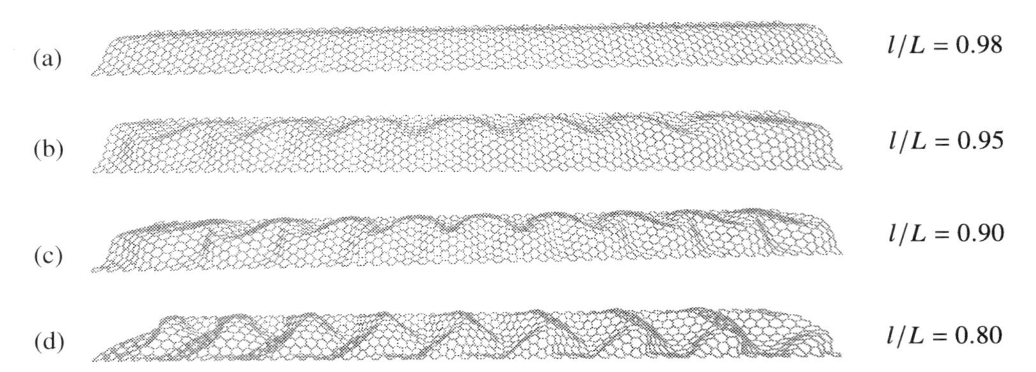

FIG. 2.14 Low energy conformations of a buckled membrane at several degrees of longitudinal compression ($d/w = 0.9$, $w = 9$ rings).

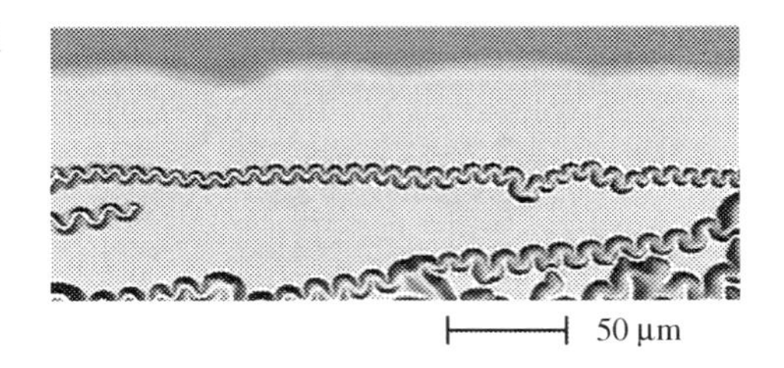

FIG. 2.15 Telephone cord buckling of a biaxially compressed tungsten film grown on silicon oxide.

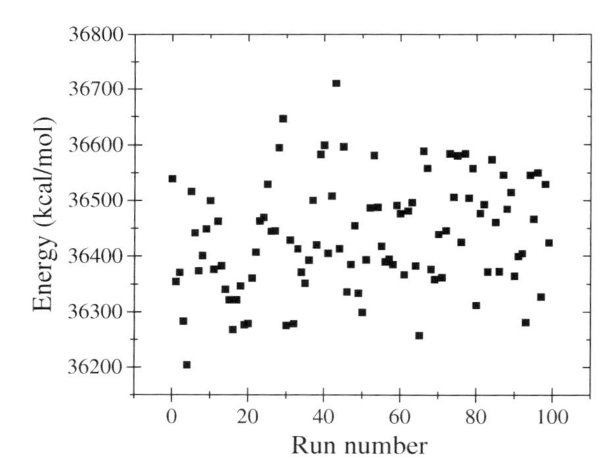

FIG. 2.16 Final configuration energy found in 100 runs of Method 1 simulation ($w = 9$ rings, $d/w = 0.9$, $l/L = 0.8$).

longitudinal compression is very strong, the structure experience the transition to a different shape as shown in Fig. 2.14D.

The longitudinal instability leads to significant variations in the stable conformation energies calculated using Method 1. Fig. 2.16, for instance, demonstrates the absence of distinct energy states in $l/L = 0.8$ membrane in distinction to the longitudinally uncompressed membranes, see Fig. 2.6. While possible explanations include the position-dependence of kink energy in longitudinally deformed membranes, and trapping in local minima within molecular dynamics optimization process, such answers at this time remain purely speculative. We hope to clarify these in the future.

Method 2 was used to generate and study moving kinks in the longitudinally compressed membranes. It was observed that at small levels of compression (below the critical value) kinks move at visibly constant speeds, see Fig. 2.17A. However, kinks experience significant friction when the compression strength is above the threshold. An example of this situation is shown in Fig. 2.17B, where the kink velocity decreases almost to zero due to the friction effect. Moreover, a close examination of this plot reveals that the kink pushes the telephone cord deformation as a whole in front of itself.

To better understand the friction effect caused by the longitudinal buckling, we have performed a series of kink dynamics experiments using five times longer membrane to avoid the interference of the reflected wave (from the right boundary) with the motion of kink within the simulation time. The kink position as a function of the pulling force was found using Method 2. One of the main observations is a considerable stochastic component of unclear origin in the kink dynamics. This observation is clearly manifested in Fig. 2.18 that exhibits the final kink displacement as a function of pulling force. It should be emphasized that in each individual simulation the kink displacement is represented by a smooth curve. However, a small change in the pulling force sometimes leads to significant changes in the dynamical behavior. Moreover, we have attempted to fit the kink dynamics by a classical relativistic model with a velocity-dependent friction force of the form

$$F_d = -a_0 \text{sgn}(V) - a_1 V - a_2 V^2, \tag{2.10}$$

where a_i are fitting parameters, and sgn(·) is the sign function. While for each individual trajectory we have obtained a very nice fitting, the fitting model parameters were inconsistent between different runs implying that the deterministic models are too narrow to account for the details of kink behavior in longitudinally buckled systems.

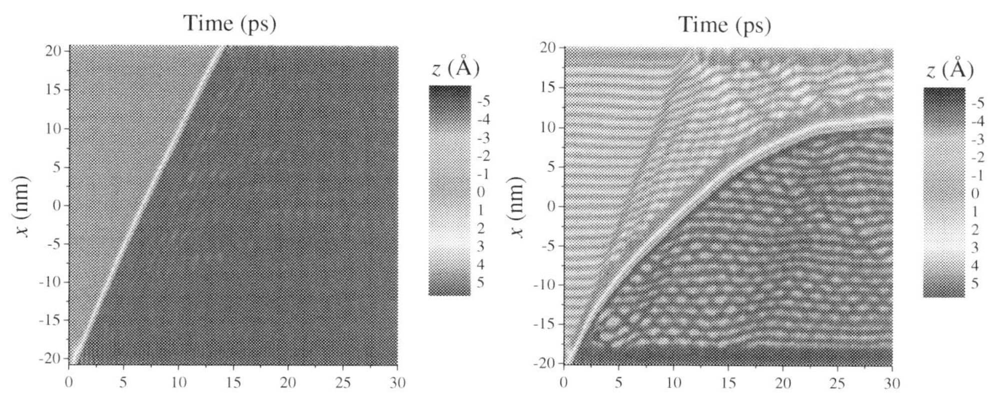

FIG. 2.17 Moving kinks in the longitudinally compressed membranes: (A) $l/L = 0.98$, $d/w = 0.9$, and (B) $l/L = 0.95$, $d/w = 0.9$. The kinks are created by 50 pN/atom force.

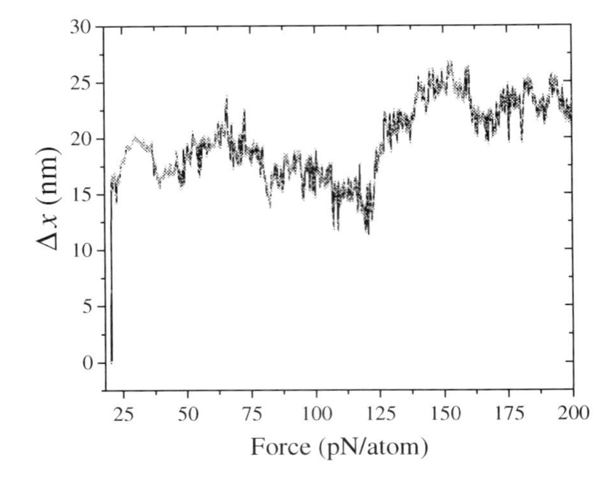

FIG. 2.18 Kink displacement within the simulation time interval (30 ps) as a function of the pulling force.

5 Conclusion and outlook

In this Chapter we have reviewed the properties of graphene kinks, and have extended the previous results in two directions. First, we have explored the energetics of graphene kinks as a function of membrane width and degree of buckling. Second, we have studied the effect of longitudinal compression on kink dynamics. Overall, our recent studies have uncovered a rich physics of topological excitations in buckled graphene membranes, including certain similarity with ϕ^4 kinks, as well as some differences related to two-dimensional nature of graphene. We speculate that the graphene kinks may find applications in nanoscale motion because of their unique characteristics such as topological stability, high propagation speed, etc.

There are several future research directions to our work. First, it would be interesting to pursue the matter of the scattering resonances further, perhaps by more sophisticated calculations in which kinks and antikinks are generated with less noise. In particular, an important question is how the scattering depends on the types of kinks and antikinks involved and their symmetries. Second, an analytical 2D model of (at least) stationary kinks is highly desirable. Such model could help understand the dependence of kink parameters (the energy and shape) on the geometrical and material properties of membrane.

Last but not least, the effects that we simulated on the nanoscale should be also observed on the microscale (possibly in a modified form, e.g., with friction playing more important role) with quasi–two-dimensional (effective zero-thickness) materials. Possible candidates include multilayer graphene, BN, MoS_2, copper oxides [45], etc. Therefore, the predictions that we have made could be verified experimentally immediately.

Acknowledgments

The authors are thankful to T. Romańczukiewicz and V. A. Slipko for their contribution to some of original publications reviewed here. The authors wish to thank J. Kim and T. Datta for their help with obtaining Fig. 2.15. R. D. Yamaletdinov gratefully acknowledges funding from RFBR (grant number 19-32-60012).

References

[1] R.D. Yamaletdinov, V.A. Slipko, Y.V. Pershin, Kinks and antikinks of buckled graphene: a testing ground for the φ^4 field model, Phys. Rev. B 96 (2017) 094306.
[2] R.D. Yamaletdinov, T. Romańczukiewicz, Y.V. Pershin, Manipulating graphene kinks through positive and negative radiation pressure effects, Carbon 141 (2019) 253–257.
[3] E.J. Weinberg, Classical Solutions in Quantum Field Theory: Solitons and Instantons in High Energy Physics, Cambridge University Press, 2012.
[4] T. Vachaspati, Kinks and Domain Walls: An Introduction to Classical and Quantum Solitons, Cambridge University Press, 2006.
[5] Y. Shibuta, S. Sakane, E. Miyoshi, S. Okita, T. Takaki, M. Ohno, Heterogeneity in homogeneous nucleation from billion-atom molecular dynamics simulation of solidification of pure metal, Nat. Commun. 8 (2017) 10.
[6] J. Jung, W. Nishima, M. Daniels, G. Bascom, C. Kobayashi, A. Adedoyin, M. Wall, A. Lappala, D. Phillips, W. Fischer, C.-S. Tung, T. Schlick, Y. Sugita, K.Y. Sanbonmatsu, Scaling molecular dynamics beyond 100,000 processor cores for large-scale biophysical simulations, J. Comput. Chem. 40 (2019) 1919–1930.
[7] S.P. Timoshenko, S. Woinowsky-Krieger, Theory of Plates and Shells, McGraw-Hill, 1959.
[8] K. Samadikhah, J. Atalaya, C. Huldt, A. Isacsson, J. Kinaret, General elasticity theory for graphene membranes based on molecular dynamics, Mater. Res. Soc. Symp. Proc. 1057 (2007). 1057-II10-20.
[9] J.-W. Jiang, J.-S. Wang, B. Li, Young's modulus of graphene: a molecular dynamics study, Phys. Rev. B 80 (2009) 113405.
[10] P.G. Kevrekidis, J. Cuevas-Maraver, A Dynamical Perspective on the ϕ4 Model: Past, Present and Future, vol. 26, Springer, 2019.
[11] N. Manton, P. Sutcliffe, Topological Solitons, Cambridge University Press, 2004.
[12] A. Vilenkin, E.P.S. Shellard, Cosmic Strings and Other Topological Defects, Cambridge University Press, 2000.
[13] S. Aubry, R. Pick, Dynamical behaviour of a coupled double-well system, Ferroelectrics 8 (1974) 471–473.
[14] J.A. Krumhansl, J.R. Schrieffer, Dynamics and statistical mechanics of a one-dimensional model Hamiltonian for structural phase transitions, Phys. Rev. B 11 (1975) 3535.
[15] A. Khare, A. Saxena, Domain wall and periodic solutions of coupled ϕ^4 models in an external field, J. Math. Phys. 47 (2006) 092902.
[16] Y. Kashimori, T. Kikuchi, K. Nishimoto, The solitonic mechanism for proton transport in a hydrogen bonded chain, J. Chem. Phys. 77 (1982) 1904–1907.
[17] E.W. Laedke, K.H. Spatschek, M. Wilkens, A.V. Zolotariuk, Two-component solitons and their stability in hydrogen-bonded chains, Phys. Rev. A 32 (1985) 1161–1179.
[18] M. Peyrard, S. Pnevmatikos, N. Flytzanis, Dynamics of two-component solitary waves in hydrogen-bonded chains, Phys. Rev. A 36 (1987) 903–914.
[19] M.J. Rice, A.R. Bishop, J.A. Krumhansl, S.E. Trullinger, Weakly Pinned Fröhlich charge-density-wave condensates: a new, nonlinear, current-carrying elementary excitation, Phys. Rev. Lett. 36 (1976) 432–435.
[20] M.J. Rice, J. Timonen, Insulator-to-metal transition in doped polyacetylene, Phys. Lett. A 73 (1979) 368–370.
[21] V.A. Gani, A.E. Kudryavtsev, M.A. Lizunova, Kink interactions in the (1+ 1)-dimensional φ^6 model, Phys. Rev. D 89 (2014) 125009.
[22] A.M. Marjaneh, V.A. Gani, D. Saadatmand, S.V. Dmitriev, K. Javidan, Multi-kink collisions in the ϕ^6 model, J. High Energy Phys. 2017 (2017) 28.
[23] V.A. Gani, V. Lensky, M.A. Lizunova, Kink excitation spectra in the (1+ 1)-dimensional φ^8 model, J. High Energy Phys. 2015 (2015) 147.
[24] E. Belendryasova, V.A. Gani, Scattering of the φ^8 kinks with power-law asymptotics, Commun. Nonlinear. Sci. Numer. Simul. 67 (2019) 414–426.
[25] A. Khare, I.C. Christov, A. Saxena, Successive phase transitions and kink solutions in ϕ^8, ϕ^{10}, and ϕ^{12} field theories, Phys. Rev. E 90 (2014) 023208.
[26] J.F. Currie, J.A. Krumhansl, A.R. Bishop, S.E. Trullinger, Statistical mechanics of one-dimensional solitary-wave-bearing scalar fields: exact results and ideal-gas phenomenology, Phys. Rev. B 22 (1980) 477–496, https://doi.org/10.1103/PhysRevB.22.477.
[27] D.K. Campbell, J.F. Schonfeld, C.A. Wingate, Resonance structure in kink-antikink interactions in ϕ^4 theory, Phys. D Nonlinear Phenomena 9 (1983) 1–32.
[28] P. Anninos, S. Oliveira, R.A. Matzner, Fractal structure in the scalar $\lambda(\varphi^2-1)^2$ theory, Phys. Rev. D 44 (1991) 1147–1160.

[29] R.H. Goodman, R. Haberman, Kink-Antikink collisions in the ϕ^4 equation: the n-bounce resonance and the separatrix map, SIAM J. Appl. Dyn. Syst. 4 (2005) 1195–1228.
[30] J.C. Phillips, R. Braun, W. Wand, J. Gumbart, E. Tajkhorshid, E. Villa, C. Chipot, R.D. Skeel, L. Kale, K. Schulten, Scalable molecular dynamics with NAMD, J. Comput. Chem. 26 (16) (2005) 1781–1802.
[31] R.B. Best, X. Zhu, J. Shim, P.E.M. Lopes, J. Mittal, M. Feig, A.D. MacKerell, Optimization of the additive CHARMM all-atom protein force field targeting improved sampling of the backbone φ, ψ and side-chain $\chi(1)$ and $\chi(2)$ Dihedral Angles, J. Chem. Theory Comput. 8 (9) (2012) 3257–3273.
[32] R.D. Yamaletdinov, O.V. Ivakhnenko, O.V. Sedelnikova, S.N. Shevchenko, Y.V. Pershin, Snap-through transition of buckled graphene membranes for memcapacitor applications, Sci. Rep. 8 (2018) 3566.
[33] J. Tersoff, New empirical approach for the structure and energy of covalent systems, Phys. Rev. B 37 (12) (1988) 6991–7000.
[34] S. Plimpton, Fast parallel algorithms for short-range molecular dynamics, J. Comput. Phys. 117 (1995) 1–19.
[35] K. Vanommeslaeghe, E. Hatcher, C. Acharya, S. Kundu, S. Zhong, J. Shim, E. Darian, O. Guvench, P. Lopes, I. Vorobyov, A.D. Mackerell, CHARMM general force field: A force field for drug-like molecules compatible with the CHARMM all-atom additive biological force fields, J. Comput. Chem. 31 (2010) 671.
[36] X. Chen, F. Tian, C. Persson, W. Duan, N.-X. Chen, Interlayer interactions in graphites, Sci. Rep. 3 (2013) 3046.
[37] R.D. Yamaletdinov, Y.V. Pershin, Finding stable graphene conformations from pull and release experiments with molecular dynamics, Sci. Rep. 7 (2017) 42356.
[38] V. Adamyan, V. Zavalniuk, Phonons in graphene with point defects, J. Phys. Condens. Matter 23 (2011) 015402.
[39] D.K. Campbell, J.F. Schonfeld, C.A. Wingate, Resonance structure in kink-antikink interactions in φ^4 theory, Phys. D Nonlinear Phenomena 9 (1983) 1–32.
[40] P. Forgács, A. Lukács, T. Romańczukiewicz, Negative radiation pressure exerted on kinks, Phys. Rev. D 77 (2008) 125012.
[41] D.L. Nika, A.A. Balandin, Two-dimensional phonon transport in graphene, J. Phys. Condens. Matter 24 (2012) 233203.
[42] D. Yoon, Y.-W. Son, H. Cheong, Negative thermal expansion coefficient of graphene measured by Raman spectroscopy, Nano Lett. 11 (2011) 3227–3231.
[43] M.W. Moon, H.M. Jensen, J.W. Hutchinson, K.H. Oh, A.G. Evans, The characterization of telephone cord buckling of compressed thin films on substrates, J. Mech. Phys. Solids 50 (2002) 2355–2377.
[44] Y. Ni, S. Yu, H. Jiang, L. He, The shape of telephone cord blisters, Nat. Commun. 8 (2017) 1–6.
[45] K. Yin, Y.-Y. Zhang, Y. Zhou, L. Sun, M.F. Chisholm, S.T. Pantelides, W. Zhou, Unsupported single-atom-thick copper oxide monolayers, 2D Materials 4 (2016) 011001.

CHAPTER

3

From classical to quantum dynamics of atomic and ionic species interacting with graphene and its analogue[☆]

Sophya Garashchuk[a], *Jingsong Huang*[b], *Bobby G. Sumpter*[b], *and Jacek Jakowski*[b,c,*]

[a]Department of Chemistry & Biochemistry, University of South Carolina, Columbia, SC, United States [b]Center for Nanophase Materials Sciences, Oak Ridge National Laboratory, Oak Ridge, TN, United States [c]Computational Sciences and Engineering Division, Oak Ridge National Laboratory, Oak Ridge, TN, United States

*Corresponding author: E-mail: jakowskij@ornl.gov

1 Introduction

Understanding interactions and chemical activities of graphene and its two-dimensional layered analogues with small molecules or ions offers a broad range of opportunities for designing novel applications such as nanoscale (opto)-electronics, few-particle sensors, catalysis, membranes for efficient gas and liquid sieving, phase separation, energy conversion and storage, and quantum information devices [1]. The semiconducting properties of graphene nanoribbons and hexagonal boron nitride (hBN) nanoribbons are known to depend on their width and edge character [2–5]. Further tailoring of their opto-electronic properties can be achieved by employing physicochemical processes and interactions that alter the number

☆Notice: This chapter has been authored by UT-Battelle, LLC under Contract No. DE-AC05-00OR22725 with the U.S. Department of Energy. The United States Government retains and the publisher, by accepting the article for publication, acknowledges that the United States Government retains a nonexclusive, paid-up, irrevocable, world-wide license to publish or reproduce the published form of this chapter, or allow others to do so, for United States Government purposes. The Department of Energy will provide public access to these results of federally sponsored research in accordance with the DOE Public Access Plan (http://energy.gov/downloads/doe-public-access-plan).

https://doi.org/10.1016/B978-0-12-819514-7.00001-4

of π-conjugated electrons. Even a single vacancy deformation or interaction with a small number of molecules that chemically engage graphene's π-conjugated orbitals can have measurable effects [6–9]. Toward engineering the properties of these materials for desired applications, techniques based on focused ion and electron beams have been demonstrated as effective tools for cleaning, cutting, etching, patterning, and controllable defect formation. Possible defects include Stone–Wales type transformation, lattice dislocations, formation of vacancies and nanopores, substitution of carbon with other atoms, as well as chemical functionalization that transforms sp^2 hybridized carbon atoms into sp^3 centers [1]. To gain fundamental understandings of beam–matter interactions, it is indispensable to conduct multiscale and multiphysics modeling of relevant physicochemical processes on realistic time scales and molecular system sizes.

In this chapter we survey the approaches for first-principles dynamics modeling of the interactions between graphene and its analogues with beams of atomic and ionic species. Methods for ab initio dynamics focusing on nuclear motion (classical and quantum) of graphene and beam species on the ground electronic state are discussed. We begin by sketching a theory of time-dependent field separation as a starting point for a multiscale and multiphysics decomposition of large molecular systems into fragments that are treatable with different theoretical approaches. A quantum trajectory method is then discussed as an approach for treating nuclear quantum effects (NQEs) for selected light nuclei. The density functional tight-binding (DFTB) theory, which is an inexpensive, approximate density functional theory (DFT) used in conjunction with a quantum trajectory, is also outlined. We do not include the effects of beam induced electronic excitation and dynamics on the excited states surfaces, which are addressed separately in chapter "From ground to excited electronic state dynamics of electron and ion irradiated graphene nanomaterials" by Lingerfelt et al. of this book. Following the review of theory we present a few examples of simulations that illustrate a range of physicochemical phenomena caused by interactions between beam and graphene and related computational aspects. We discuss reactivity, scattering, and transmission of atomic and ionic species including Ar cluster ion, H/D, and H^+/D^+ on graphene, focusing on the atomic-scale motion and energy dissipation pathways involved in forming and breaking covalent bonding, the NQEs for light species, and the isotopic substitution effects.

2 Theoretical methods

Today's fabrication and processing of advanced materials are increasingly complex with technological applications heading toward the quantum scale and involving simultaneous manipulation of atoms, electrons, and light (photons). The multiscale and multiphysics character of physics and chemistry behind processes related to fabrication and functioning of such materials often requires computational techniques and methods capable of spanning across several scales of time and space. Multiscale modeling of physicochemical processes on realistic time scales and molecular system sizes often involves partitioning the system into active and inactive (spectator) parts, which can be treated at different levels of theory. Such partitioning can be achieved via time-dependent field separation. Typically for ab initio dynamics modeling of irradiation processes, the inexpensive electronic structure methods are desired such as DFTB theory.

2.1 Multiphysics modeling

There is a wide range of excellent computational chemistry tools that focus on a specific science aspect but neglect other less important parts of the big picture (e.g., the multiphysics). A comprehensive review of dynamics methods/approaches is difficult to achieve in a finite space, so we limit this discussion to representative methods and selected references. A typical approach focuses on one specific aspect (for example the evolution of electrons) while using a simplest possible (and computationally cheap) theoretical framework for the rest of the species. Fig. 3.1 illustrates a multiscale partitioning of H adsorption on graphene for multiphysics treatment on different scales.

For the nuclei, conventional molecular dynamics approaches provide an appropriate description of nuclear dynamics as classical particles in many situations [10, 11]. However, for understanding of chemical processes at low energy or temperature, it is essential to include the quantum-mechanical (**QM**) effects, such as the zero-point energy (**ZPE**), tunneling and nonadiabatic transitions, even in large molecular systems or in condensed phases. For example, QM tunneling is known to dominate proton transfer at low temperature in biological environments. In the area of bond-selective chemistry, the intermode energy flow and the ZPE should be considered for understanding effects of the vibrational excitation on bond breaking. Incorporation of the QM effects into classical trajectory simulation is desirable when describing chemical reactions in large molecular systems. Fully quantum approaches to the time-dependent Schrödinger equation (TDSE), based on a direct product Finite Basis or Discrete Variable representations, are difficult to extend beyond 10–12 degrees of freedom due to the exponential scaling of computational cost [12, 13]. To achieve improved scalability, one can express the nuclear wavefunction in terms of optimized basis functions. Such an approach is used in the Multi-Configurational Time-Dependent Hartree (MCTDH) methods, multiple spawning, coupled and local coherent states approximation and split operator Fourier transform based approaches [14–21]. Nevertheless, even with improved scaling the computational cost is still a limitation for large scale system applications. Therefore, mixed

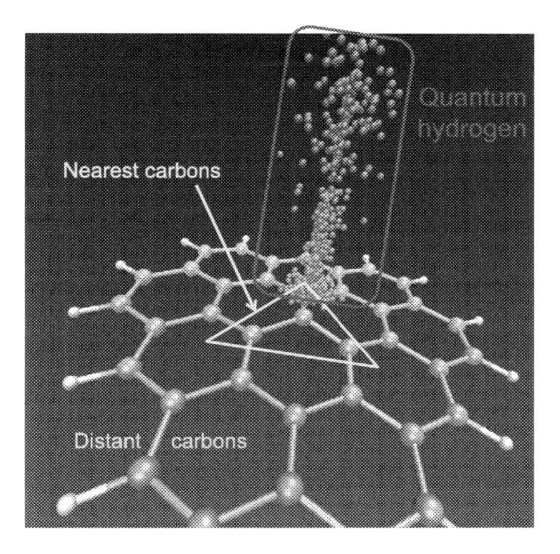

FIG. 3.1 An example of hierarchical multiscale partitioning for multiphysics treatment of nuclear motion for the adsorption of H irradiated on graphene. Fully quantum hydrogen is represented as an ensemble of quantum trajectories (*purple spheres*, gray in print version). Nearest carbons to the adsorption site marked by the *yellow triangle* (*light gray* in print version) can be treated using classical molecular dynamics. Distant, edge carbons are treated as frozen or via a force field.

approaches, in which a selected small number of degrees of freedom are treated quantum mechanically while the remaining nuclei are treated classically, are computationally very attractive [22–27]. Among other approaches are Semiclassical Initial Value Representation, Quasiclassical Trajectory Dynamics, Centroid Molecular Dynamics, Ring Polymer Molecular Dynamics and Path Integral Molecular Dynamics [28–32].

For the electrons, one can either solve the stationary time-independent electronic structure problem, which leads to Born–Oppenheimer molecular dynamics. Alternatively, one directly integrates the time-dependent Schrödinger or Kohn–Sham DFT equations. The two most widely used classes of first-principles molecular dynamics (MD) with time-dependent quantum-mechanical treatment of the electrons are Ehrenfest dynamics and surface hopping dynamics, each having its own limitations. Ehrenfest dynamics offers correct short-term behavior of coherent evolution of the superposition of electronic states in the strong coupling region but the long-time trajectory follows an incorrect averaged state trajectory. The surface hopping approaches are based on stochastic branching of trajectories to recover a correct long-time behavior. In a fully quantum description for all nuclei and electrons the correct approach should retain features of both approaches—coherent mean-field mixing of electronic states in the strong coupling region and the ability of a wavefunction to split into spatially separated branches precluding possibility for interference. These features could be included into mean-field approaches by adding a coherence dephasing term that switches the electronic surfaces during the dynamics [33–36].

2.2 Multiscale separation within the time-dependent mean-field approximation

The separation of electronic degrees of freedom from those describing heavy classical nuclei and light quantum nuclei can be formally achieved through the time-dependent self-consistent field. An outline of the derivation, using one-dimensional notations, is presented below. The mass of an electron is equal to one atomic mass unit; M and m denote the mass of the classical and quantum particles, respectively. Taking advantage of the mass- and time-scale separation, the following wavefunction ansatz is used

$$\Phi(r_e,x,R,t) \approx \phi(r_e,t;\, x(t),R(t)) \cdot \psi(x,t;\, R(t)) \cdot \chi(R,t) \cdot \exp\left(-\frac{\imath}{\hbar}\int_0^t E(\tau)d\tau\right), \tag{3.1}$$

where r_e, x, and R describe, respectively, the electrons, the light quantum nuclei, and the heavy (quasi)classical nuclei. The wavefunctions depend parametrically on the variables after the semicolon. Inserting this ansatz into the time-dependent Schrödinger equation

$$\imath\hbar\frac{\partial\Phi}{\partial t} = \left\{-\sum_R \frac{\hbar^2}{2M}\Delta_R - \sum_x \frac{\hbar^2}{2m}\Delta_x - \sum_{r_e}\frac{\hbar^2}{2}\Delta_{r_e} + \hat{V}_{int}\right\}\Phi, \tag{3.2}$$

and multiplying on the left by $\langle\chi,\psi|$, $\langle\chi,\phi|$, $\langle\psi,\phi|$, respectively, one obtains the following coupled equations:

$$\imath\hbar\frac{\partial}{\partial t}\phi(r_e,t;\, x(t),R(t)) = \left\{-\sum_{r_e}\frac{\hbar^2}{2}\Delta_{r_e} + V_{r_e-x,R}\right\}\phi(r_e,t;\, x(t),R(t)) \quad \text{electrons} \tag{3.3}$$

$$\imath\hbar\frac{\partial}{\partial t}\psi(x,t;\,R(t))=\left\{-\sum_{x}\frac{\hbar^2}{2m}\Delta_x+V_{x-r_e,R}\right\}\psi(x,t;\,R(t))\quad\text{light quantum nuclei}\tag{3.4}$$

$$\imath\hbar\frac{\partial}{\partial t}\chi(R,t)=\left\{-\sum_{R}\frac{\hbar^2}{2M}\Delta_R+V_{R-r_e,x}\right\}\chi(R,t)\quad\text{heavy (quasi)classical nuclei}\tag{3.5}$$

in which the term V_{X-Y} denotes the mean field interaction of particle X with the system Y. The above separation provides the framework for a "modular" implementation of dynamics in which different methods can be applied to different subsystems.

We describe the NQEs for selected light atoms using the Bohmian Quantum Trajectories framework [37]. This approach naturally includes the effect of anharmonicity of the potential energy surface for the nuclei and is very suited for parallelization on High Performance Computing (HPC) resources. The wave equation for the quantum nuclei (Eq. 3.4) is transformed into a particle trajectory equation by expressing the complex wavefunction $\psi(x,\,t)$ in terms of the real amplitude and phase, as outlined in Section 2.3. The wavefunction $\psi(x,\,t)$ is discretized as an ensemble of trajectories such that the probability density of finding a quantum particle in space is given by $|\psi|^2$ and the trajectory momentum is given by the gradient of the wavefunction phase. The time evolution of quantum trajectories unfolds on the potential energy surface, which contains an additional quantum potential $U(x,\,t)$ besides the classical potential V coming from the electronic structure. That is, the quantum nuclei move according to the combined action of the classical force, $F_{cl}=-\nabla V$, and the quantum force $F_{qm}=-\nabla U$. The quantum potential U describes the nonlocality of the nuclear wavefunction. At any given position in space, U depends on the values of the nuclear wavefunction amplitude and on its second derivatives, $U=-\hbar^2\nabla^2|\psi|/(2m|\psi|)$.

For the classical treatment of the heavy nuclei, the time-dependent wavefunction $\chi(R,\,t)$ is replaced with the classical trajectory whose position and momentum correspond to the expectation values $\langle\chi|R|\chi\rangle\rightarrow R$ and $\langle\chi|-\imath\hbar\frac{\partial}{\partial R}|\chi\rangle\rightarrow P$, respectively. The trajectory momentum is updated according to the force averaged over the probability distribution of the quantum particles according to the Ehrenfest theorem [38]. This is a standard approach underlying ab initio molecular dynamics schemes. For the electrons one can (i) invoke the Born–Oppenheimer approximation and replace the time-dependent electronic wavefunction $\phi(r,t)$ solving the TDSE (Eq. 3.3) with the ground-state wavefunction, or (ii) directly integrate Eq. (3.3). In this chapter we assume that the nuclear dynamics follows on the ground electronic state. The procedures for direct integration of the time-dependent electronic structure are presented in chapter "From ground to excited electronic state dynamics of electron and ion irradiated graphene nanomaterials" by Lingerfelt et al. We primarily employ the DFTB theory for description of the electronic structure because of its optimal balance of accuracy and computational cost. Section 2.5 details a brief description of the DFTB theory.

2.3 Quantum trajectory dynamics

The *quantum or Bohmian trajectory* formulation [39] of the TDSE is discussed below, assuming the same mass m for all degrees of freedom (DOFs) described in Cartesian coordinates. A generalization is given in Ref. [40]. Small bold letters denote vectors and capital bold letters denote matrices, whose dimensionality is N_{dim}. The usual TDSE is then,

$$\hat{H}\psi(\boldsymbol{x},t) = \imath\hbar\frac{\partial}{\partial t}\psi(\boldsymbol{x},t), \ \hat{H} = -\frac{\hbar^2}{2m}\nabla\cdot\nabla + V(\boldsymbol{x}). \tag{3.6}$$

Using the polar form of the wavefunction (the amplitude $A(\boldsymbol{x}, t)$ and the phase $S(\boldsymbol{x}, t)$ are real functions),

$$\psi(\boldsymbol{x},t) = A(\boldsymbol{x},t)\exp\left(\frac{\imath}{\hbar}S(\boldsymbol{x},t)\right), \tag{3.7}$$

and defining the classical ($\boldsymbol{p}$) and the nonclassical ($\boldsymbol{r}$) momentum components,

$$\boldsymbol{p} := \nabla S, \ \boldsymbol{r} := \frac{\nabla A}{A} \tag{3.8}$$

the TDSE leads to the following equations of motion for a trajectory described by the associated quantities ($\boldsymbol{x}_t$, $\boldsymbol{p}_t$, $\boldsymbol{r}_t$, S_t) :

$$\frac{d\boldsymbol{x}_t}{dt} = \frac{\boldsymbol{p}_t}{m} \tag{3.9}$$

$$\frac{d\boldsymbol{p}_t}{dt} = -\nabla(V+U)|_{x=x_t} \tag{3.10}$$

$$\frac{d\boldsymbol{r}_t}{dt} = -\left(\boldsymbol{r}_t\cdot\nabla + \frac{\nabla\cdot\nabla}{2}\right)\frac{\boldsymbol{p}_t}{m} \tag{3.11}$$

$$\frac{dS_t}{dt} = \frac{\boldsymbol{p}_t\cdot\boldsymbol{p}_t}{2m} - (V+U)|_{x=x_t}. \tag{3.12}$$

The subscript t labels attributes of the quantum trajectories (**QT**s), which discretize the initial wavefunction and, as an ensemble, represent $\psi(\boldsymbol{x}, t)$ at all times. The quantum-mechanical features of dynamics enter these classical-like equations of motion via the quantum potential U, expressed in terms of $\boldsymbol{r}$ as,

$$U(\boldsymbol{x},t) = -\frac{\hbar^2}{2m}(\boldsymbol{r}\cdot\boldsymbol{r} + \nabla\cdot\boldsymbol{r}). \tag{3.13}$$

The quantum potential is nonlocal in $\boldsymbol{x}$ and, being proportional to $\hbar^2/m$, suggests a simple formal transition to classical mechanics: $U \to 0$. (The atomic units of $\hbar = 1$ are used henceforth.) The trajectory weight w_t, which is the probability density within the volume element associated with each QT, $\delta\boldsymbol{x}_t = \delta x_1\cdot\delta x_2\ldots\delta x_{N_{dim}}$, is conserved in time [41],

$$w_t = |\psi(\boldsymbol{x}_0)|^2\delta\boldsymbol{x}_0 = |\psi(\boldsymbol{x}_t)|^2\delta\boldsymbol{x}_t. \tag{3.14}$$

Thus, the expectation values of position-dependent operators $\hat{O}$ are readily computed as a sum over all QTs, their number being N_{traj}:

$$\langle\hat{O}\rangle = \int|\psi(\boldsymbol{x},t)|^2 O(\boldsymbol{x})d\boldsymbol{x} = \sum_{k=1}^{N_{traj}} O(\boldsymbol{x}_t^{(k)})w^{(k)}. \tag{3.15}$$

The expectation values are used to analyze dynamics and, also, to construct approximations to the quantum potential U [41, 42] necessary to keep the approach practical [43].

2.4 Approximations to the quantum force

The QM effects in the QT formulation come from the action of the quantum potential U, which is generally singular and reflects the complexity of the time-dependent wavefunction. In the context of nuclear dynamics, however, the mean-field type approximations to the quantum potential are often useful and cost-efficient [41]. For example, for a Gaussian wavefunction, the components of $\boldsymbol{r}$ are linear functions of $\boldsymbol{x}$, U is quadratic and the quantum force is linear in $\boldsymbol{x}$. The "linearized" quantum force (LQF) can be defined variationally (once integration by parts is used) by the Least Squares Fit [44] of the components of $\boldsymbol{r}$, i. e., by the minimization of a functional I,

$$I = \sum_{\alpha} \langle \| (A^{-1}\nabla A)_{\alpha} - \tilde{r}_{\alpha} \|^2 \rangle. \tag{3.16}$$

The subscript α labels the components of vectors, or DOFs of the system, $\alpha = 1 \ldots N_{dim}$. The fitting functions $\tilde{r}_{\alpha} = \sum_i c_{i,\alpha} f_i$ are expansions of the components of the vector $A^{-1}\nabla A$ in a linear basis $\boldsymbol{f}$ of the size $N_{dim} + 1$, $\boldsymbol{f} = \{x_1, \ldots, x_{N_{dim}}, 1\}$. Minimization with respect to the expansion coefficients $\{c_i\}$ leads to a single linear matrix equation [41]. Apart from the cost of solving this matrix equation, the numerical cost of the global LQF scheme scales linearly with the systems size N_{dim} and with the number of trajectories. This approximation is used in the study of adsorption of H-atom on graphene model (next section).

The global approximation to the quantum potential can be defined without the explicit knowledge of the nonclassical momentum $\boldsymbol{r} = (\nabla A)/A$, by invoking integration by parts in Eq. (3.16). However, calculation of $\boldsymbol{r}$ along the trajectories according to Eq. (3.11) enables more accurate spatially semilocal or local approximation schemes [45–47]. Another option for going beyond the global LQF approximation is to use larger, e.g., quadratic in $\boldsymbol{x}$, basis to fit $\boldsymbol{r}$ and $\boldsymbol{p}$, for specific atoms or subsets of strongly coupled DOFs. Note that even if the fitting basis is nonlinear, the LSF procedure itself still reduces to a single linear matrix equation. Yet working with the subsets of the DOFs rather than with all DOFs at once, is necessary for practical reasons of controlling the total basis size; it is also consistent with the typical view of the condensed-phase chemical processes as involving strongly coupled primary "system" modes interacting with the weakly coupled secondary "bath" modes representing the molecular environment. The quantum force approximation is, then, generated through the following two-step LSF procedure [48].

(i) The global LQF. We perform the Least-Squares Fit of $\boldsymbol{r}$ and $\boldsymbol{p}$ within the linear basis $\boldsymbol{f}$, i.e., minimize the global (with respect to the DOFs) functionals,

$$I_r = \sum_{\alpha} \langle \| r_{\alpha} - \tilde{r}_{\alpha} \|^2 \rangle \quad I_p = \sum_{\alpha} \langle \| p_{\alpha} - \tilde{p}_{\alpha} \|^2 \rangle. \tag{3.17}$$

For each DOF r_{α} and p_{α} are expanded in the Cartesian coordinates of *all* atoms. This first step is similar to the LQF of Eq. (3.16) except now we fit the actual values of nonclassical and classical momenta evolved along the trajectories. This step describes the quantum force at the LQF level due to correlation of motion between different nuclei.

(ii) The atom-specific high-order fit. The residual $\boldsymbol{r}$ and $\boldsymbol{p}$ are fitted with the higher order polynomials for a single nucleus at a time, using quadratic or higher order Taylor basis in

DOFs describing one specific nucleus at a time. Denoting these fitting functions with the double tilde the nuclear-specific functionals are:

$$I_r^{nucl} = \sum_{\alpha\, nucl} \langle \| r_\alpha - \widetilde{r}_\alpha - \widetilde{\widetilde{r}}_\alpha \|^2 \rangle \; I_p^{nucl} = \sum_{\alpha\, nucl} \langle \| p_\alpha - \widetilde{p}_\alpha - \widetilde{\widetilde{p}}_\alpha \|^2 \rangle, \tag{3.18}$$

The second step adds more flexibility to the approximate quantum potential to account for a non-Gaussian shape of the evolving wavefunction. The Least Squares Fit involves solving a linear matrix equation for the matrix size equal to that of the fitting basis.

2.5 Approximate DFT electronic structure

A major hindrance in the development and application of quantum dynamics is the prohibitive cost of electronic structure calculations and the scaling of the related algorithms. To illustrate this, we note that 4 min spent on the single calculation of energy and forces translates into 1 month spent on a 10,000 step trajectory. Therefore it is very important to use inexpensive electronic structure methods that are fast, and yet maintain reliable accuracy. Whereas the conventional DFT methods provide a workhorse for a static electronic structure calculations of hundreds to thousands of atoms, yet their relatively high computational cost makes DFT impractical for routine ab initio dynamics of systems consisting more than a few hundred atoms. Semiempirical and approximate DFT methods based on tight-binding parametrization become critical for modeling large systems as first recognized by Godecker [49, 50].

In this regard we discuss one such method called DFTB theory, which is a very promising approach offering broad applicability [51–53]. Although DFTB is approximately 1000 times more expensive than the classical force fields, it is up to 1000 times cheaper than standard density functional theories. Thus DFTB fills the gap between classical force fields and DFT and is an attractive candidate for direct molecular dynamics simulations of bulk and condensed matter systems.

DFTB is an approximate DFT method in which only valence electrons are treated quantum mechanically while all core electrons and nuclei are approximated via pairwise interatomic repulsive potential E_{rep}

$$E = \sum_i 2f_i \langle \phi_i | H_{core} | \phi_i \rangle + \sum_{\substack{A,B \\ A>B}}^{Atoms} \gamma^{AB} \Delta q^A \Delta q^B + \sum_{A>B}^{Atoms} E_{rep}^{AB} \tag{3.19}$$

where f_i is an occupation number (typically 0 or 1) and i runs over all molecular orbitals. The first term describes the interaction of valence electrons with core ions (nuclei and core electrons). The second term is responsible for electron–electron interaction. Symbols Δq^A and γ^{AB} are, respectively, a charge at center A and a chemical hardness-based parameter describing electron–electrons interactions between centers A and B, which depends on the interatomic distance. The last term describes the interaction between core ions obtained from a fit. An important feature of DFTB is a correct Coulomb asymptotic behavior for interaction of charged molecules. This is due to the fact that γ^{AB} behaves as $1/R_{AB}$ for large interatomic distances. DFTB also provides an inexpensive tool for the description of low lying excitation. Higher energy electronic excitations are less reliable. This limitation is inherent to the use of a minimal basis set (Slater-type orbitals). In addition, DFTB cannot describe processes involving core electrons. The discussion of techniques for modeling electronic excitation is a subject of the next chapter "From ground to excited electronic state dynamics of electron and ion irradiated graphene nanomaterials" by Lingerfelt et al.

The above energy expression can be rewritten in the matrix form suitable for high performance implementations

$$E = Tr[H_{core}P] + \frac{1}{2}Tr[G(P)P] + \sum_{A>B}^{Atom} E_{rep}^{AB} \tag{3.20}$$

Symbol P is a reduced one-electron density matrix obtained from molecular orbitals coefficients C, such that $P = CfC^T$. Here, H_{core} and $G(P)$ are atomic orbitals matrices that describe, respectively, the interaction of electrons with core (H_{core}) and electron–electrons interaction ($G(P)$).

The main reason for the low computational cost of DFTB is due to the fact that (a) only valence electrons are considered while core electrons are neglected, (b) a minimal basis set is used (Slater basis), (c) only two-center (pairwise) integrals are used in the calculations. A consequence of (a) and (b) is that for a given molecular system, all matrices (H,G,P) are 5–10 times smaller than in DFT. A consequence of (c) is that the cost for the formation of all matrices is significantly lower than in DFT.

The solution to the electronic structure needs to be obtained iteratively due to the dependence of matrix $G(P)$ on the electronic structure itself. For that purpose a diagonalization is typically used. The most expensive part of the DFTB is diagonalization, which for the Self-Consistent-Charge version (SCC-DFTB) has to be performed approximately 10–20 times per evaluation of the energy and forces of systems with wide HOMO-LUMO gap [51, 52]. Typically for a trajectory type simulation of a molecular system consisting of a few hundred atoms, around 80% of the time is spent on the diagonalization (including orbital transformations), while approximately 15% is spent on the evaluation of forces, and the remaining 5% of time is spent on formation of H_{core}, G and overlap matrices. The overall percentage of time spend on $O(N^3)$ operations becomes even larger with increasing size of molecules. This reflects the fact that the formation of DFTB matrices scales quadratically whereas the diagonalization and BLAS3 operations become more and more dominant due to their cubically scaling computational cost [49].

The combined Quantum Trajectory and Electronic Structure (QTES) based on high throughput implementation of the spin-unpolarized SCC-DFTB with Fermi-Dirac smearing at electronic temperature T_{el} [51–55] is described in Ref. [56]. The calculations are performed on-the-fly for a few thousand trajectories propagated for a several thousand time-steps. The parallel implementation is based on Open Multi-Processing and Message Passing Interface, with evaluations of V and gradients distributed over several hundred cores, and minimal information passed to the head-node where the trajectory attributes for the entire QT ensemble are updated and the output quantities are computed.

3 Simulations

Depending on the type of particles and their energy, the beam irradiations can lead to different physical effects and therefore have different applications. Focused ion and electron beam techniques have been used to engineer defects, for patterning, surface functionalization, and nanopore fabrication [57–62]. Ar atoms cluster beams, ionized for electrostatic acceleration, are an efficient sputterer of organic contamination from graphene surface [63–65]. Beside removal of contamination, the irradiation with Ar cluster ion beams can be utilized to fabricate nanopores but can also lead to rupturing of graphene membranes [61–63, 66, 67]. Unlike Ar beams, H beams can penetrate, scatter or adsorb on a graphene membrane. The adsorption of H atoms at the graphene surface is known as hydrogenation, which has been shown to modify the reactivity and electronic properties of graphene [68–70]. The different behavior of Ar and H beams can be ascribed to their different atomic and ionic sizes. The radius of H atoms is much smaller than that of Ar atoms and thus the ion is much smaller as well. The small size of H and H^+ makes their transmission through graphene possible, leading to much smaller energy barriers, especially in the latter case.

The choice of methodology for description of graphene irradiation depends on the mass and energy of beam particles. Due to their large mass and its chemical inactivity, Ar atoms can be treated as classical particles and the effect of irradiation (interaction) with graphene can be described by employing classical dynamics framework without referring to quantum dynamics machinery. For energetic H atoms, the de Broglie wavelength is short and the classical Newtonian framework is also sufficient to describe its dynamics. For low energy H or H^+ beams, however, NQEs are expected to play an important role and thus require quantum treatment [71–73].

We present below a few examples for simulation of graphene irradiation with Ar and H/D beams of various energies and transmission of H^+/D^+ through graphene to illustrate the modeling techniques discussed earlier. We first discuss the irradiation of graphene by an Ar cluster ion beam. Different mechanisms for Ar interactions with supported and suspended graphene are observed, which can be exploited for removal of contaminations or fabrication of nanopores. We then discuss scattering and interaction of an energetic H beam with graphene, focusing on adsorption, reflection, and transmission phenomena. After that we turn to the low energy regime and focus on modeling adsorption of H/D on a graphene flake. The differences between classical and quantum treatment are discussed to expose the role of NQEs. In the final example the transmission of low energy H^+/D^+ through membranes in a liquid instead of gas phase environment is analyzed and discussed for graphene and its isoelectronic hBN analagoue. As these examples illustrate, employing approximate DFT electronic structure methods such as DFTB, while inexpensive, requires benchmarking against more accurate methods and appropriate adjustments of interaction potentials in the critical interaction regions.

3.1 Processing graphene with an Ar cluster ion beam

An Ar cluster ion beam is a positively charged aggregate of Ar atoms, namely Ar_n^+, where n ranges from a few dozens to several thousands atoms, which can be used for nanoscale processing of material surfaces. The ionized clusters are electrostatically accelerated to tens of keV energy before striking the materials surface. It has been demonstrated that irradiation with an Ar_n^+ beam is an effective tool to remove organic contamination from the graphene surface. The surface of graphene synthesized in chemical vapor deposition followed by postprocessing is usually covered with more than a nanometer thick layer of contamination, often from polymethylmethacrylate (PMMA) that is used for transfer and lifting. The distribution of contamination on a graphene surface is nonuniform and can vary from clean areas to areas covered with ~10 nm thick contaminants. In general during the irradiation of graphene with Ar_n^+, the sputtering of contaminant and graphene damage can occur concurrently, the latter of which can lead to formation of nanopores [62, 65–67].

Here we analyze the effect of irradiation of graphene with Ar_n^+ using DFTB based classical molecular dynamics. Two types of graphene samples were modeled: suspended and supported. Both types are considered for nanoscale electronics device applications. The computational model consisted of a 5×5 nm^2 graphene flake with approximately 1000 C atoms and an ion cluster with 100 Ar atoms. For modeling the supported graphene, a second graphene layer was placed underneath in the graphite orientation and was frozen during the MD simulations to mimic the effect of support. For modeling the suspended graphene, only one graphene layer was used, which was frozen on the edges during the MD simulations to mimic the effect of suspension. The dangling bonds of the graphene flake's edges were passivated with H. The computational model consisted of over 2100 atoms in total for supported graphene and about half as many for suspended graphene. Employing DFTB for ab initio molecular dynamics is dictated by a balance between accuracy (electronic structure methods are preferable for description of processes involving bond breaking and formation) and computational cost (MD simulations with DFT for 1000+ atoms are typically too expensive).

The main challenge in the computational simulations of the graphene irradiation with Ar_n^+ is the description of potentials for the cluster and its interaction with the graphene surface. This problem is threefold: (1) The standard sets of Slater–Koster parameters used in DFTB simulations do not include noble gas elements. Thus, any choice of DFTB parameters set requires extending it by including Ar–Ar and Ar–C parameters. (2) The character of Ar–Ar bonding and interaction with graphene is not covalent but weak van der Waals type interactions. As such it is notoriously difficult to model using DFT methods. Resorting to more accurate many-body methods such as coupled clusters for validation or adjustment is desired. (3) The overall positive charge of an Ar_n^+ means that not all Ar atoms in the cluster are "made equal". At least one Ar atom is a cation and needs to be treated differently. Contrary to the Ar–Ar case, the Ar–Ar^+ interaction is two orders of magnitude stronger and is the primarily attractive force that holds the cluster together.

The interaction energy profiles for Ar–Ar and Ar–Ar^+ from different electronic structure methods are shown in Fig. 3.2. Highly accurate and experimentally validated Aziz potential for neutral Ar–Ar dimer is also shown for comparison. The CCSD(T) potential for Ar–Ar shows a binding energy amounting to 0.011 eV at 3.8 Å separation and agrees very well with the spectroscopic Aziz potential [74–76] as excepted. Also with CCSD(T), the binding energy for Ar–Ar^+ amounts to 1.26 eV at interatomic separation of 2.4 Å. Interestingly, the binding energy by CAM-B3LYP with dispersion corrections included is very close to the Aziz data. However, it significantly overestimates the binding energy for the Ar–Ar^+ dimer as indicated by the discrepancy with respect to the coupled cluster curve.

For the current DFTB simulations we used the MIO parameter set to describe C–C and C–H interactions [77]. The CCSD(T) theory, which intrinsically includes dispersion interaction, along with 6-311++G(3df,3pd) basis, was used for the repulsion term (E_{rep}) of DFTB with a 1.6 nm cutoff [78, 79]. The Slater–Koster parameters (Hamiltonian and overlap matrices)

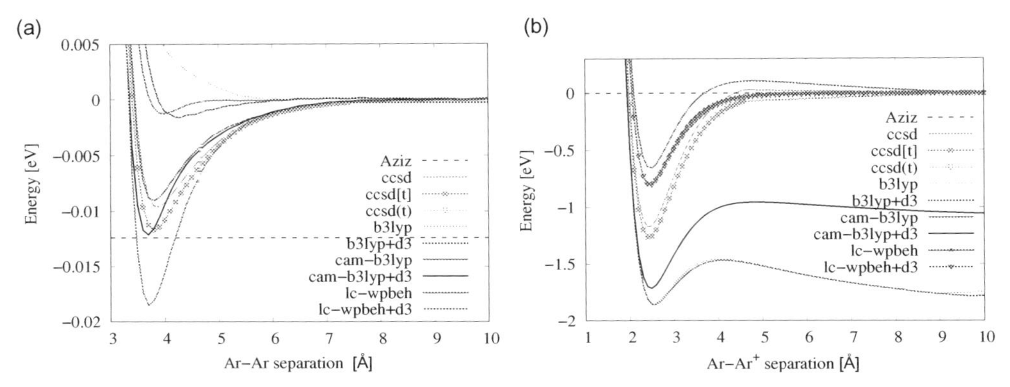

FIG. 3.2 Interaction energy of neutral Ar–Ar (A) and cationic Ar–Ar$^+$ (B) as a function of interatomic separation from different electronic structure methods. Dispersion-corrected DFT results are labeled as "+d3." Experimentally measured interaction energy for neutral Ar–Ar case is shown as horizontal *dashed lines*. Energy is in E_h and interatomic distance is in Å.

for Ar–Ar, Ar–C and Ar–H were prepared from higher level, all-electron DFT with long-range corrected (LC)-wPBEh functional calculations and a 6-311++G(3df,3pd) basis set. The new Ar parameter set included *s* and *p* valence orbitals.

The DFTB simulations of graphene irradiation with an Ar_n^+ beam were performed within the Born–Oppenheimer approximation and with Fermi-Dirac smearing (T_{el}=1000 K). The electronic temperature and Fermi-Dirac smearing effectively improves SCF convergence for processes in which a significant number of bonds are being broken and/or formed. The time step for molecular dynamics was set to 0.5 fs. The initial energy of the Ar_n^+ beam in the simulations was varied from 2.5 to 25 eV per Ar atom. This corresponds to a total beam energy ranging from 2.5 to 25 keV.

The representative snapshots from direct dynamics of graphene irradiation with Ar_n^+ using DFTB are shown in Fig. 3.3. The supported and suspended graphene layers show different behavior under Ar_n^+ beam irradiation. The irradiation of the suspended graphene induced vacancies and nanopores, with a threshold of ~15 eV per Ar atom for rupturing of the suspended graphene. However, no significant damage to supported graphene was observed with even higher beam energies. The mechanism for the Ar_n^+ induced damage of the suspended graphene involves significant deformation of the graphene layer prior to its rupture, which is initiated by C–C bond breaking after the impact of the Ar_n^+ beam. For supported graphene, this would require a much higher beam energy. Effectively the support underneath graphene membrane facilitates the redistribution and dissipation of kinetic energy deposited by Ar_n^+ beam during the impact as indicated by the outgoing concentric wave propagating away from impact point as visible in Fig. 3.3A.

3.2 Graphene irradiation with an energetic H beam

Next we discuss the results of classical molecular dynamics of graphene bombarded by a H beam for impact energies up to 200 eV. Generally speaking, hydrogen beams can be

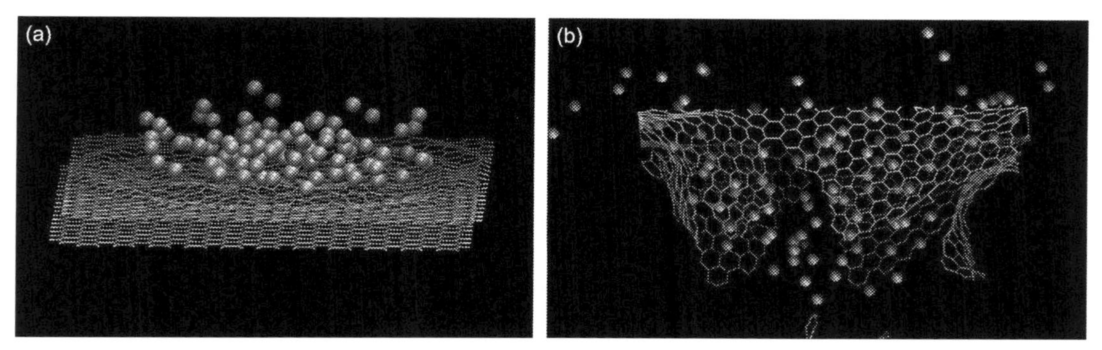

FIG. 3.3 Ab initio MD simulations using DFTB of (A) a supported graphene irradiated with an Ar^{+}_{100} cluster of 2.5 keV beam energy (25 eV per Ar atom) captured at 0.55 ps showing no surface damage and of (B) a suspended graphene irradiated with an Ar^{+}_{100} cluster of 1.5 keV beam energy (15 eV per Ar atom) captured at 1.0 ps showing rupturing of graphene. The Ar^{+} that carries the charge is shown in *blue* (gray in print version).

provided in the form of both H_2 molecules and H atoms [80, 81], the latter of which is for our concern in this chapter. In this energy regime the incident hydrogen behaves classically and NQE can be neglected. Understanding the effect of interaction of energetic hydrogen with graphene is important for understanding of the damage of plasma facing carbon tile in the ITER fusion reactor [6, 82, 83]. Also, graphene-based electronic systems in space vehicles might be sensitive to the damage caused by cosmic radiation containing a wide spectrum of particles, a significant component of which would be light atoms from solar wind [6]. The defects caused by energetic light particles include lattice defects, creation of vacancies, as well as chemical changes.

To simulate effects of irradiation on graphene one can apply direct molecular dynamics methods in which electronic structure is treated explicitly using quantum mechanics, while the motion of the nuclei is described using classical molecular dynamics. This allows to accurately describe bond breaking and formation due to the interaction of graphene with projectile hydrogen and related chemical changes. Such an approach is however limited by the computational cost of electronic structure theory. Employing DFTB instead of DFT allows alleviating this limitation in the system sizes, time scales, and statistics that can be obtained.

Here, the initial structure of 3×3 nm^2 graphene sheet with 336 C atoms was obtained from Nose–Hoover thermostated DFTB molecular dynamics at 300 K. To include the effect of statistical uncertainty in the irradiation, for each impact energy 1000 independent trajectory simulations were performed with randomly chosen initial positions of the incident H atom above the surface of the graphene sheet. We note that the usual DFTB parameters are not suitable to describe short range interatomic interactions as those distances are not relevant for the expected chemistry. To allow for the high-energy impact, the repulsive part of DFTB parameters (PBC-0-3 set) was fitted to the binary Ziegler–Biersack–Littmark repulsive potentials [84, 85].

Fig. 3.4 compares the DFTB potential energies of hydrogen–graphene and hydrogen–coronene as a function of H position above the graphene/coronene plane. The coronene potentials show bonding that is roughly 1 eV weaker and a potential barrier at the hexagon center that is 1 eV higher, reflecting the changes in electronic structure between hydrogen-terminated and periodic sp^2 carbon. Despite these differences, the forms of the H-graphene

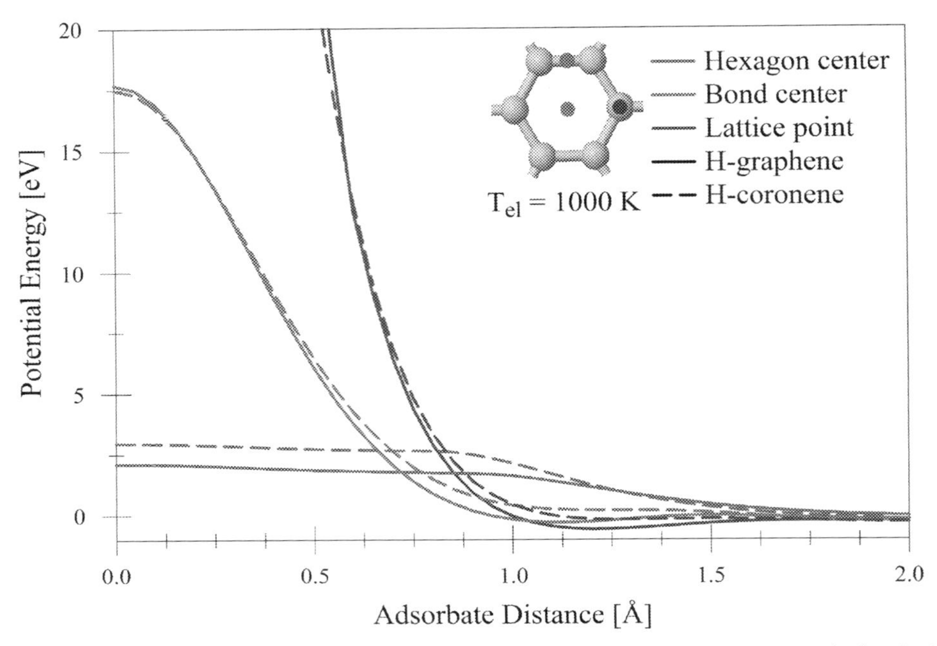

FIG. 3.4 Potential energies of the hydrogen–coronene and hydrogen–graphene interaction calculated with DFTB method as a function of distance for different impact positions. *Reprinted with permission from R.C. Ehemann, J. Dadras, P.R.C. Kent, J. Jakowski, P.S. Krstic, Detection of hydrogen using graphene, Nano Res. Lett. 7 (2012) 198, https://doi.org/10.1186/1556-276X-7-198.*

and H-coronene interactions are very similar, highlighting that short range potentials are critical for high-energy impact. Furthermore, the DFTB results for the H-coronene system agree qualitatively well with the DFT calculations in Ref. [86]. These observations combined together suggest that the DFTB parameters are acceptable alternative of more expensive DFT potentials for the study of graphene.

The DFTB simulations of irradiation show three possible outcomes: adsorption, reflection, and transmission of incident hydrogen, without observing sputtering. The adsorption is observed for incident energies not exceeding 1 eV, as shown in the left panel, Fig. 3.5A. The reflection of H dominates in a mid-energy range between 1 and 10 eV with the probability peak at approximately 2 eV. The threshold for transmission of H through graphene should be greater than 2 eV. Indeed, the energy barrier for H transfer across graphene is about 3–4 eV from DFT calculations [87, 88]. For high energies up to 200 eV, the event is mainly transmission with insignificant reflection.

By examining the position within the hexagon where incident hydrogens are adsorbed, reflected, or transmitted (see right panel, Fig. 3.5B), one can infer the nature of the interaction potential. Maximum adsorption clearly indicates the position of lattice atoms. Probability clustering around the carbon atoms also shows their in-plane lattice vibrations. Reflection can occur at every position in the graphene hexagon (see 1 eV), but the clustering for incident energies of 5 eV indicates a CC bonding network. The transmission of H is most probable near the center of hexagons.

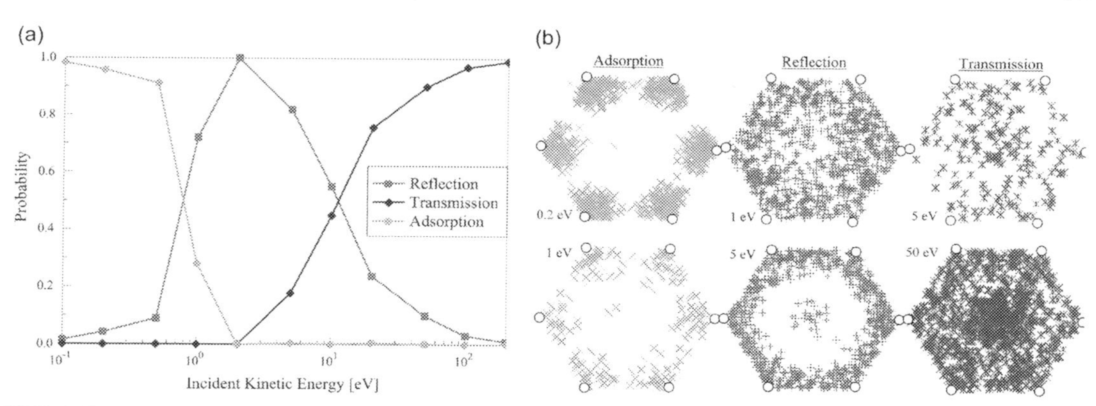

FIG. 3.5 (A) Adsorption, reflection, and transmission probabilities for H irradiation of graphene as a function of incident energy. (B) Positions of adsorption, reflection, and transmission events for the quantum-classical calculations for selected incident energies. Position of carbon atoms are shown as *white circles. Reprinted with permission from R.C. Ehemann, J. Dadras, P.R.C. Kent, J. Jakowski, P.S. Krstic, Detection of hydrogen using graphene, Nano Res. Lett. 7 (2012) 198, https://doi.org/10.1186/1556-276X-7-198.*

3.3 Classical vs quantum simulations of H and D adsorption on graphene

Next we turn to the modeling of H/D adsorption on graphene in the low energy regime. As shown above, during graphene irradiation with atomic hydrogen, reflection and transmission are the prevalent processes at high kinetic energy of the incident atoms, while the regime of low incident energies (below 1 eV) is dominated by the hydrogen adsorption. During low-energy motion of hydrogen, the NQEs such as the zero-point energy (ZPE) and tunneling become more prominent and may influence chemical reactivity [10, 11, 89, 90]. Therefore, it is essential to account for the quantum mechanical nature of low-energy hydrogen, and to accurately reproduce the potential energy landscape, in particular the energy barriers. To be able to describe this regime, we have tuned the DFTB potential using electronic temperature as tuning parameter so that the features of the PES important for adsorption of H and D on graphene are reasonably well-reproduced as shown in Fig. 3.6. See Ref. [56] for details on the procedure used to adjust the DFTB parameters.

We use the approximate quantum trajectory dynamics, summarized in Section 2.3, based on the evolution of an ensemble of interdependent trajectories to describe the quantum nature of incident "light" hydrogen, combined with classical Newtonian dynamics of remaining "heavy" atoms, and with the DFTB description of electronic structure. The quantum-mechanical effects are introduced via the quantum potential, determined by the time evolution of the nuclear wavefunction. The corresponding quantum force, added to the usual classical force determined by the electronic structure of a system, formally generates all NQEs within the trajectory description of the wavefunction dynamics. The linearized quantum force used here—derived from the simplest globally defined approximation to the quantum potential—is exact for a Gaussian-type wavefunctions, and it captures the leading NQEs, e.g., the ZPE, energy of wavefunction localization, moderate tunneling and wavefunction bifurcation [41].

As noted in the previous section, the most reactive condition for hydrogen adsorption is when the incoming hydrogen collides with the lattice center (carbon atom) as opposed to the

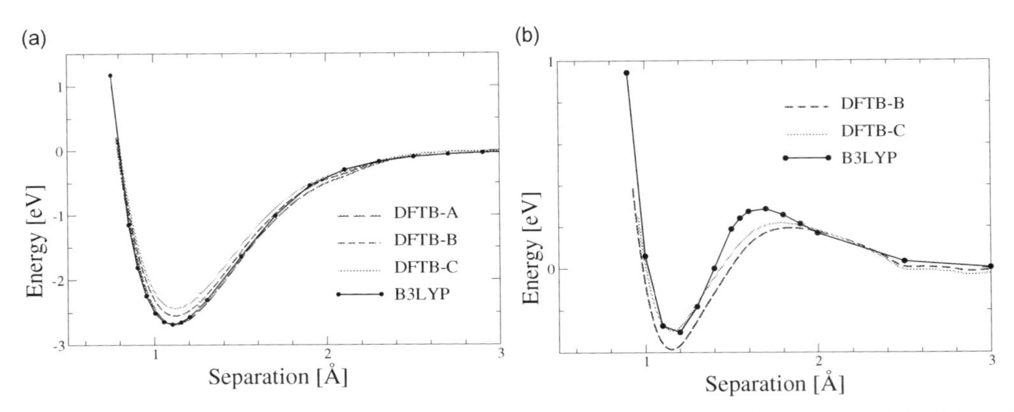

FIG. 3.6 Comparison of the DFTB and DFT/B3LYP energy for interaction of H with a graphene flake $C_{37}H_{15}$ along the normal to the graphene surface above C atom (lattice center) as a function of separation between H and C. (A) The structure of a graphene flake is allowed to fully relax as the incident H approaches the C center of the graphene surface (slow motion limit). (B) The structure of graphene is completely frozen during the potential energy scan (fast motion of H limit). For DFTB the potential energy was adjusted using electronic temperature as a parameter. The labels (A)–(C) correspond to different values of the tuning parameter. *Adapted with permission from L. Wang, J. Jakowski, S. Garashchuk, Adsorption of a hydrogen atom on a graphene ake examined with quantum trajectory/electronic structure dynamics, J. Phys. Chem. C 118 (2014) 16175–16187, https://doi.org/10.1021/jp503261k; copyright 2014 American Chemical Society.*

center of the CC bond or with the hexagon center, especially at low incident energy. In general, the carbon bonding network dynamically responds to the impact of incident hydrogen. The carbon lattice can absorb some fraction of the kinetic energy of incident hydrogen in the course of dynamics and change hybridization from sp^2 to sp^3 depending on the kinetic energy of hydrogen.

Overall, the potential experienced by incoming hydrogen will fall between two limiting cases: (a) vibrationally "adiabatic" collision in which the carbon bonding network can fully relax and hybridization can change from sp^2 to sp^3 in response to incoming atom, and (b) vibrationally "diabatic" collision with a frozen carbon network, when the collision time is too short for the carbon network to adapt to the approaching hydrogen atom. The corresponding potential energy surfaces (PES) for a graphene flake $C_{37}H_{15}$ calculated from three sets of adjusted DFTB parameters and DFT are plotted in Fig. 3.6. In the case of the adiabatic process, which is the typical evolution of a system along the 'reaction path', the relaxation of the carbon bonding network in graphene results in the rehybridization of the C atom from sp^2 to sp^3, leading to the barrierless formation of a covalent CH bond sitting in a deep potential well of ~2.5 eV. In contrast, the PES describing the diabatic process, i.e., the positions of the carbon atoms do not change during the collision process, reveals that the incoming H experiences a low-energy barrier on the level of 0.2 eV followed by a shallow well on the level of only 0.4 eV. The actual potential experienced by incoming hydrogen depends on its kinetic energy and falls in between these two limiting situations.

The dynamics simulations show that hydrogen adsorption occurs in four consecutive steps: (1) hydrogen passes over the barrier, (2) inelastic collision of hydrogen with the planar sp^2 graphene, which leads to the transfer of a portion of its kinetic energy to graphene, (3) a physisorption stage in which the H atom becomes trapped in the shallow well of potential energy after partial deposition of its kinetic energy with C in a sp^2 configuration, and (4) conversion of physisorption into chemisorption through the relaxation of the graphene lattice and formation of a CH bond with C in the sp^3 configuration. The last step of graphene lattice relaxation is the slowest one. Several vibrational C–H oscillation cycles were observed in MD simulations before graphene lattice reorganized and corresponding force constants (and frequencies) for C–H vibrations changed from the value corresponding to sp^2 hybridization to sp^3 ones [56].

We now discuss NQE and the effects of the quantum corrections on the dynamics of H observed with the ensemble of quantum trajectory dynamics and on the evolution of the graphene flake (classical part of the system). First, turning off the quantum potential within the quantum trajectory simulations turns the ensemble of quantum trajectories into an ensemble of classical trajectories. This allows to directly estimate the effects of neglecting quantum correlations between trajectories by turning *on* and *off* the quantum potential. Next, the (quasi)classical part of the system, that is the carbon network lattice, experiences a force from the quantum part of a system that is averaged over the ensemble of trajectories. The results in Fig. 3.7A are obtained using the Ehrenfest-type treatment, where first we calculate an ensemble averaged expectation value for the hydrogen position and then use this position to calculate forces acting on graphene. The results of (quasi)classical-like [29] description of graphene are presented in Fig. 3.7B.

The simulations with the classical treatment of H, i.e., with quantum potential turned off, show sharp change in adsorption probability that corresponds to a well-defined energy window for adsorption. This can be understood intuitively in the following way. For the very small incident energies that are below the adsorption window, all trajectories are reflected from the barrier energy. For the incident energies falling within the adsorption window the incident hydrogen has enough kinetic energy to pass above the barrier and to collide with

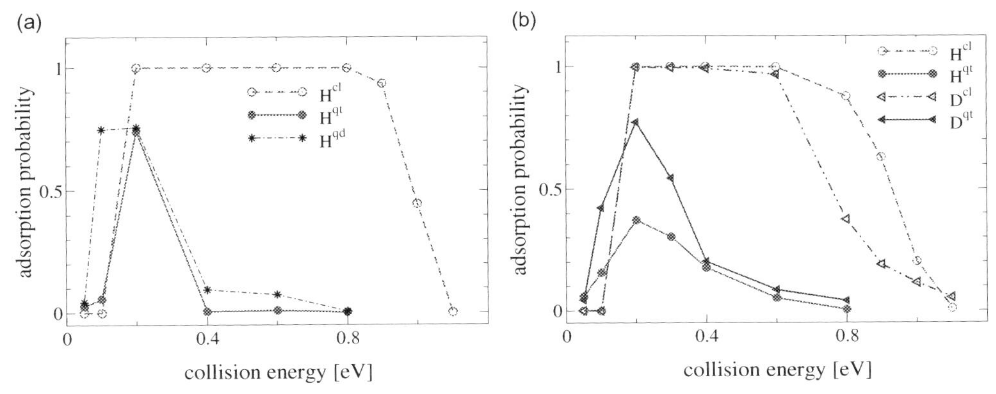

FIG. 3.7 Probabilities for adsorption of H and D atoms on graphene as a function of the collision energy. Superscripts "qt" and "cl" refer to the probabilities from the dynamics with and without the quantum correction on the force. Dynamics with the graphene atoms initially at equilibrium (single QT ensemble) and with the ZPE distribution (multiple QT ensembles) are shown in panels (A) and (B), respectively. Results of the reduced-dimensionality exact quantum dynamics of H in a time-dependent potential (panel (A), details in text) are labeled "qd." *Adapted with permission from S. Garashchuk, J. Jakowski, L.Wang, B.G. Sumpter, Quantum trajectoryelectronic structure approach for exploring nuclear effects in the dynamics of nanomaterials, J. Chem. Theory Comput. 9 (2013) 5221–5235, https://doi.org/10.1021/ct4006147; copyright 2013 American Chemical Society.*

the repulsive wall at short range. The collision leads to H transferring some of its kinetic energy to graphene network and becoming trapped in the potential well between the short range repulsive wall and the barrier. Finally, for the kinetic energies above the adsorption window, the incident H passes above the barrier and transfers some of its energy to the graphene flake, but the change in its kinetic energy is too small for H to become trapped in the potential well, resulting instead in reflection.

The main NQE observed in this system is that the quantum potential, i.e., the localization energy of the initial wavefunction, effectively spreads the momenta of individual quantum trajectories as they exchange energy, whereas in the classical case all trajectories start out with the same momenta and do not exchange energy. Thus, their adsorption probability is nearly "binary" resulting in the adsorption "window" of [0.2, 1.0] eV. In the QT simulation, even for the collision energies below 0.2 eV, the QTs at the leading edge of the wavepacket gets the energy boost from the QT ensemble, thus certain fraction of the QTs has sufficient energy to cross the barrier and become adsorbed at the lattice center. This behavior can be viewed as shallow to moderate tunneling, when compared to the exact quantum simulation on a grid performed in one-dimension along the collision coordinate, using the time-dependent potential from the QT simulation. In the exact quantum dynamics simulation, the adsorption probability at 0.1 eV is higher than for the QT dynamics, while the classical dynamics at this energy gives zero adsorption.

At higher collision energies, we attribute the reduced adsorption probability in the QT simulation to a significant fraction of trajectories being too "energetic" for effective energy exchange with the graphene, something that happens in the classical simulation at energies above 1 eV. We do note, however, that the classical and QT dynamics unfolds on effectively different potential energy surfaces since the classical force acting on the carbon atoms is averaged over the ensembles of trajectories, which have different positions with and without the quantum correction on dynamics.

To relax the Ehrenfest treatment of the graphene flake, we have also introduced sampling of the carbon positions according to the ZPE of the C–C stretch. The adsorption probability, averaged over multiple QTs ensembles, is shown in Fig. 3.7. These ensembles are independent of each other; and we find that 11 and 14 ensembles were adequate for the average probabilities computed for H and D, respectively. The overall trend in probabilities obtained from the dynamics with and without the quantum correction is similar to that discussed above. Comparing the isotopes, both H and D show a well-defined adsorption peak at 0.4 eV; the higher adsorption probability for D is explained by the relatively longer interaction times of D (which for the same kinetic energy moves slower than H) with the graphene flake, allowing for more efficient transfer of the collision energy to the graphene flake. At the collision energy of 0.2 eV, the ratio of adsorption probabilities is close to 3, which is not observed in the classical simulations, an effect that could be tested experimentally. Finally we note that the difference between the classical and QT simulations involving D is reduced compared to that for H, an expected dependence on the mass of quantum particles.

3.4 Transmission of H^+ and D^+ through graphene and hBN

Finally we discuss the transmission of protons and deuterons (H^+ and D^+) through graphene and its isoelectronic analogue hBN. Understanding transmission of small molecules and species through graphene and other atomically thin two-dimensional materials is of great importance for potential applications as membranes suitable for sieving gases, liquids, and separation of hydrogen isotopes. Here we focus on the role of NQEs and isotopic substitution effects on the transmission rates [91]. As discussed above, when the kinetic energy of neutral H is on the level of 1 eV or less, the only observed processes are adsorption and reflection of H. Transmission of H through graphene can be observed when its kinetic energy is on the level of a couple of eV, especially for collision through the hexagon center. However, the situation is very different for transmission of charged species in a liquid environment.

It has been experimentally demonstrated that mono-layer of graphene and mono- and bi-layer of hBN are permeable to H^+ at room temperature, but are not permeable to H [71]. The experimental device consisted of graphene and hBN membranes immersed in Nafion, which is a proton conducting polymer [92]. The dependence of proton conductivity as a function of temperature was measured. The results were fit with an Arrhenius expression $\exp(-E_a/k_B T)$ from which activation energies, E_a, for proton transmission were estimated to be 0.3 eV for hBN and 0.78 eV for graphene. These activation energies are systematically lower than the theoretical barriers in the ranges of [0.7, 1.0] eV and [1.2, 1.5] eV for mono-layer hBN and graphene, respectively [71, 87, 91, 93–98]. It has also been shown that the transmission of a proton is 10 times greater than that of a deuteron, with the difference attributed to the zero-point energy [99].

The proton and deuteron transfer in the experimental setting involves a network of hydrogen bondings. The simplest model that accounts for energetic effects of hydrogen bonding relies on two water molecules flanking the two sides of the membrane to serve as proton donor and acceptor. In such case the overall process can be written as:

$$H_3O^+ - M - H_2O \rightarrow H_2O - M - H_3O^+ \tag{3.21}$$

where M is a graphene or hBN membrane. The resulting potential energy profile for proton transmission through the membrane is characterized by two energy minima and two barriers. Fig. 3.8 shows the potential energy for hydrogen transmission through borazine ($B_3H_3N_3$), the smallest model for hBN, as a function of distance from the membrane's surface and corresponding molecular structures. The potential energy on the other side of the membrane is simply a mirror image. At the minimum I proton is adsorbed at the surface of the membrane (borazine's hexagon here) whereas at minimum II the proton is involved in hydrogen bonding with nearby water forming hydronium. To remove a proton from hydronium, external work is required to break the hydrogen bond in hydronium which is represented by barrier II. Similarly passing through the center of a hexagon ring of the membrane requires work (barrier I) to "loosen" the bonding network in the membrane to open a transmission channel.

Here we describe the simulation of proton and deuteron transfer through graphene and hBN membranes. In both cases the models of the membrane are constructed as optimized graphene and hBN flakes consisting of 19 fused hexagonal rings [91]. The potential energy surfaces for graphene and hBN are obtained from coupled cluster theory and are shown in Fig. 3.9A. As can be seen, the barrier I for the graphene flake is 0.5 eV higher than for hBN. This is consistent with higher permeability and lower activation energy for a hBN membrane than for a graphene membrane. The membrane itself was frozen during simulations. The simulation of proton and deuteron transmission was achieved by employing an exact one-dimensional quantum dynamics model along the normal direction to the membrane surface. The quantum dynamical model was based on wavepacket correlation-function

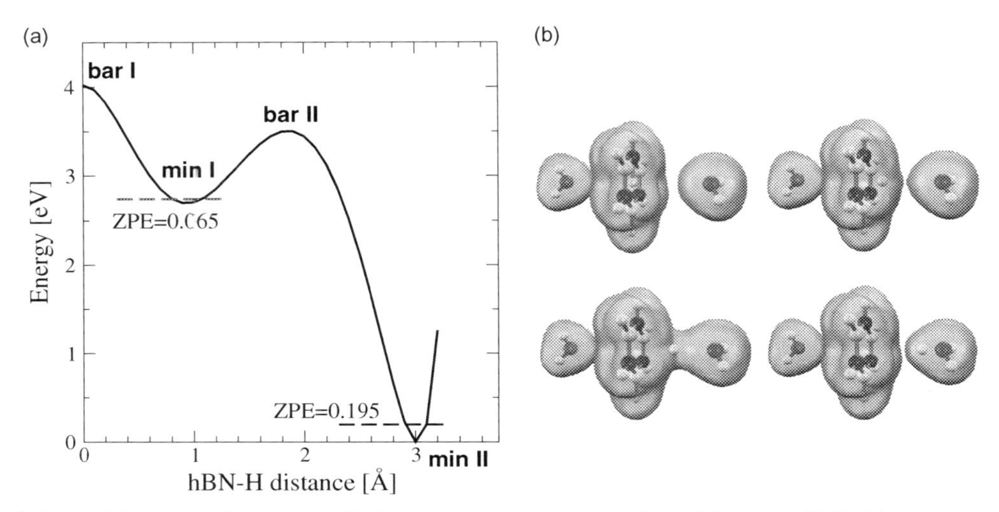

FIG. 3.8 (A) The potential energy profile for the proton transmission through borazine ($B_3H_3N_3$) computed at the CCSD/6-31G** level shown as a function of r. The ZPE-corrected energies at the minima are indicated with *dashes*. Only the ZPE along the proton transfer coordinate is considered. Hydrogen bonding with environmental water is characterized by two barriers, labeled as "bar I" at $r = 0$ Å and "bar II" at $r = 2$ Å, and two minima labeled as "min I" at $r = 1$ Å and "min II" at $r = 3$ Å, where r is the proton-ring distance. (B) The respective electronic densities, computed at the MP2/cc-PVTZ level, are shown for the oxygen-ring distance fixed at $R = 4.0$ Å. *Reprinted with permission from N.T. Ekanayake, J. Huang, J. Jakowski, B.G. Sumpter, S. Garashchuk, Relevance of the nuclear quantum effects on the proton/deuteron transmission through hexagonal boron nitride and graphene monolayers, J. Phys. Chem. C 121 (2017) 24335–24344, https://doi.org/10.1021/acs.jpcc.7b08152; copyright 2017 American Chemical Society.*

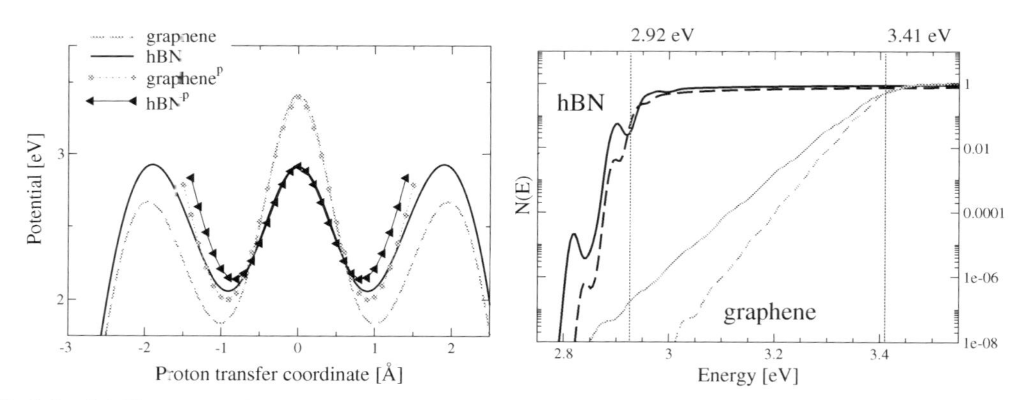

FIG. 3.9 (A) The potential energy profile as a function of the proton–membrane separation. The membrane is modeled with 19 rings with two water molecules (proton donor and acceptor) on both sides of the membrane. Similar gas phase curves (without the water molecules) are shown for comparison and labeled with "p" superscript. (B) The energy-dependent transmission probabilities for hBN (*black*) and graphene (*green*, gray in print version) interacting with the proton (*solid line*) and deuteron (*dashed line*). The barrier heights are marked with vertical left *red* line (gray in print version) for hBN at 2.92 eV and *blue*, right line (gray in print version) for graphene at 3.41 eV. *Reprinted with permission from N.T. Ekanayake, J. Huang, J. Jakowski, B.G. Sumpter, S. Garashchuk, Relevance of the nuclear quantum effects on the proton/deuteron transmission through hexagonal bron nitride and graphene monolayers, J. Phys. Chem. C 121 (2017) 24335–24344, https://doi.org/10.1021/acs.jpcc.7b08152; copyright 2017 American Chemical Society.*

formulation of S-matrix scattering theory [100]. The reactant and product states correspond to H_3O^+ or H_2DO^+ on different sides of membrane and are described by Gaussian wavepacket localized at minimum II. Noticeably, transmission of proton or deuteron through a membrane requires passing through three energy barriers: twice through barrier II (on both sides of membrane) and once through barrier I (at the center of membrane). The proton and deuteron transmission probabilities $N(E)$ for a graphene and hBN membrane as a function of kinetic energy, E, are shown in Fig. 3.9B. Simulations show that the proton is more likely to transfer through the membrane than the deuteron. The resonant maxima of transmission probabilities for hBN are attributed to its triple barrier. That is, the nonmonotonic transmission probabilities are due to shallow tunneling and above-barrier reflection when the wavepacket energy is comparable with the barrier height. Contrary to this the transmission through graphene is determined by a single barrier I which dominates the entire energy profile. Overall, more classical transmission and smaller proton/deuteron isotope effects for hBN are expected than for graphene after thermal averaging.

Next, from the energy-dependent transmission probabilities, $N(E)$, the quantum thermal rate constants $k(T)$ are obtained as an average over Boltzmann factor $k^q(T) = (2\pi Q(T))^{-1} \int_0^\infty N(E) \cdot \exp(-E/k_B T) dE$, where $Q(T)$ is the translational partition function. Similarly, classical thermal rates are related to the barrier height V_b as $k^{cl}(T) = k_B T/(2\pi Q(T)) \exp(-V_b/k_B T)$, which is directly related to experimentally measured activation energy E_a through the Arrhenius reaction rate expression $k^A(T) = A \exp(-E_a/k_B T)$. Finally, the tunneling factor $\kappa(T) = k^q(T)/k^{cl}(T)$ is obtained from the ratio of quantum and classical rate constants. Our estimates for the room temperature kinetic isotope effect (KIE) are 3–4 for hBN and 20–30 for graphene, whereas the experimental values are closer to 10 for both systems. We note that 1D models tend to exaggerate quantum effects while full ab initio dynamics involving dynamical membrane relaxation along with a more realistic 3D quantum treatment of protons such as quantum trajectories is expected to give results closer to experiment. Nevertheless, 1D quantum models for quantum treatment of selected protons provide an inexpensive first approximation for estimation of NQEs and isotopic substitution effects.

4 Summary

There are significant experimental and theoretical interests in application of beams of atomic and ionic clusters and species for graphene engineering. In this chapter, we have presented techniques for direct dynamics modeling of irradiation of graphene with beams of atomic and ionic species. The discussed techniques include both classical and quantum trajectory methods in conjunction with DFTB. Nuclear quantum effects and isotopic substitution effects are also discussed. Depending on the type of particles in the beam and the beam energy, the irradiation can lead to different physical effects. The specific choice of methodology and its validity for description of graphene irradiation depend on the mass and energy of beam particles. All simulations presented here assume that electronic excitations can be neglected during the nuclear motion and that the ground electronic state sufficiently describes the potential energy during the dynamics. This is not always the case. In the next chapter "From ground to excited electronic state dynamics of electron and ion irradiated graphene nanomaterials" by Lingerfelt et al. we discuss the effect of electronic excitation in the beam–matter interactions.

Acknowledgments

This work was conducted at the Center for Nanophase Materials Sciences of the Oak Ridge National Laboratory, a U.S. Department of Energy Office of Science User Facility. This work used the Extreme Science and Engineering Discovery Environment (XSEDE), which is supported by National Science Foundation grant No. ACI-1548562 (allocation TG-DMR110037) and resources of the Oak Ridge Leadership Computing Facility (OLCF) and of the Compute and Data Environment for Science (CADES) at the Oak Ridge National Laboratory, which is supported by the Office of Science of the U.S. Department of Energy under Contract No. DE-AC05-00OR22725. SG acknowledges partial support by the National Science Foundation under Grants CHE-1955768 and CHE-1565985. Any Opinions, findings and conclusions or recommendations expressed in this material are those of the authors and do not necessarily reflect those of the National Science Foundation.

References

[1] V. Georgakilas, M. Otyepka, A.B. Bourlinos, V. Chandra, N. Kim, K.C. Kemp, P. Hobza, R. Zboril, K.S. Kim, Functionalization of graphene: covalent and non-covalent approaches, derivatives and applications, Chem. Rev. 112 (2012) 6156–6214, https://doi.org/10.1021/cr3000412.
[2] Y.-W. Son, M.L. Cohen, S.G. Louie, Energy gaps in graphene nanoribbons, Pys. Rev. Lett. 97 (2006) 216803, https://doi.org/10.1103/PhysRevLett.97.216803.
[3] V. Barone, O. Hod, G.E. Scuseria, Electronic structure and stability of semiconducting graphene nanoribbons, Nano Lett. 6 (2006) 2748–2754, https://doi.org/10.1021/nl0617033.
[4] V. Barone, J.E. Peralta, Magnetic boron nitride nanoribbons with tunable electronic properties, Nano Lett. 8 (2008) 2210–2214, https://doi.org/10.1021/nl080745j.
[5] A. Lopez-Bezanilla, J. Huang, H. Terrones, B.G. Sumpter, Boron nitride nanoribbons become metallic, Nano Lett. 11 (2011) 3267–3273, https://doi.org/10.1021/nl201616h.
[6] R.C. Ehemann, J. Dadras, P.R.C. Kent, J. Jakowski, P.S. Krstic, Detection of hydrogen using graphene, Nano Res. Lett. 7 (2012) 198, https://doi.org/10.1186/1556-276X-7-198.
[7] I. Deretzis, G. Fiori, G. Iannaccone, G. Piccitto, A. La Magna, Quantum transport modeling of defected graphene nanoribbons, Physica E 44 (2012) 981–984, https://doi.org/10.1016/j.physe.2010.06.024.
[8] N. Gorjizadeh, Y. Kawazoe, Chemical functionalization of graphene nanoribbons, J. Nanomater. 20 (2010) 1–7, https://doi.org/10.1155/2010/513501.
[9] K. Wakabayashi, Y. Takane, M. Yamamoto, M. Sigrist, Electronic transport properties of graphene nanoribbons, N. J. Phys. 11 (2009) 095016, https://doi.org/10.1088/1367-2630/11/9/095016.
[10] M. Karplus, R.N. Porter, R.D. Sharma, Exchange reactions with activation energy. I. Simple barrier potential for (H, H_2), J. Chem. Phys. 43 (1965) 3259–3287.
[11] X. Zhu, P. Lopes, A.D. MacKerell, Recent developments and applications of the CHARMM force fields, Wiley Interdiscip. Rev. Comput. Mol. Sci. 2 (2012) 167–185.
[12] J.C. Light, T. Carrington Jr, Discrete variable representations and their utilization, Adv. Chem. Phys. 114 (2000) 263–310.
[13] F. Huarte-Larranaga, U. Manthe, Quantum dynamics of the $CH_4 + H \rightarrow CH_3 + H_2$ reaction: full-dimensional and reduced dimensionality rate constant calculations, J. Phys. Chem. A 105 (2001) 2522–2529.
[14] D.V. Shalashilin, I. Burghardt, Gaussian-based techniques for quantum propagation from the time-dependent variational principle: formulation in terms of trajectories of coupled classical and quantum variable, J. Chem. Phys. 129 (2008) 084104.
[15] H.D. Meyer, U. Manthe, L.S. Cederbaum, The multi-configurational time-dependent hartree approach, Chem. Phys. Lett. 165 (1990) 73–78.
[16] H.B. Wang, M. Thoss, Multilayer formulation of the multiconfiguration time-dependent Hartree theory, J. Chem. Phys. 119 (2003) 1289–1299.

[17] I. Burghardt, H.-D. Meyer, L.S. Cederbaum, Approaches to the approximate treatment of complex molecular systems by the multiconfiguration time-dependent hartree method, J. Chem. Phys. 111 (1999) 2927–2939.
[18] M. Ben-Nun, J. Quenneville, T.J. Martinez, *ab initio* multiple spawning: photochemistry from first principles quantum molecular dynamics, J. Chem. Phys. 104 (2001) 5161–5175.
[19] D.V. Shalashilin, M.S. Child, Time dependent quantum propagation in phase space, J. Chem. Phys. 113 (2000) 10028–10036.
[20] R. Martinazzo, M. Nest, P. Saalfrank, G.F. Tantardini, A local coherent-state approximation to system-bath quantum dynamics, J. Chem. Phys. 125 (2006) 194102.
[21] Y.H. Wu, V.S. Batista, Matching-pursuit for simulations of quantum processes, J. Chem. Phys. 118 (2003) 6720–6724.
[22] S.Y. Kim, S. Hammes-Schiffer, Hybrid quantum/classical molecular dynamics for a proton transfer reaction coupled to a dissipative bath, J. Chem. Phys. 124 (2006) 244102.
[23] O. Prezhdo, V.V. Kisil, Mixing quantum and classical mechanics, Phys. Rev. A 56 (1997) 162–175.
[24] J. Gao, D.G. Truhlar, Quantum mechanical methods for enzyme kinetics, Ann. Rev. Phys. Chem. 53 (2002) 467–505.
[25] G. Naray-Szabo, A. Warshel, Computational Approaches to Biochemical Reactivity, vol. 19, Kluwer Academic Publishers, New York, NY, 1997.
[26] O.V. Prezhdo, C. Brooksby, Quantum backreaction through the Bohmian particle, Phys. Rev. Lett. 86 (2001) 3215–3219.
[27] S.S. Iyengar, J. Jakowski, Quantum wave packet *ab initio* molecualr dynamics, J. Chem. Phys. 122 (2005) 114105.
[28] W.H. Miller, The semiclassical initial value representation: a potentially practical way for adding quantum effects to classical molecular dynamics simulations, J. Phys. Chem. A 105 (2001) 2942–2955.
[29] G.C. Schatz, J.M. Bowman, A. Kuppermann, Exact quantum, quasiclassical, and semiclassical reaction probabilities for collinear $F + D_2 \rightarrow FD + D$ reaction, J. Chem. Phys. 63 (1975) 685–696.
[30] F. Paesani, G.A. Voth, The properties of water: insights from quantum simulations, J. Phys. Chem. B 113 (2009) 5702–5719.
[31] S. Habershon, D.E. Manolopoulos, Zero point energy leakage in condensed phase dynamics: an assessment of quantum simulation methods for liquid water, J. Chem. Phys. 131 (2009) 244302.
[32] D. Marx, M.E. Tuckerman, G.J. Martyna, Quantum dynamics via adiabatic *ab initio* centroid molecular dynamics, Comput. Phys. Commun. 118 (1999) 166–184.
[33] J. Jakowski, K. Morokuma, Liouville-von neumann molecular dynamics, J. Chem. Phys. 130 (2009) 224106.
[34] O.V. Prezhdo, P.J. Rossky, Mean-field molecular dynamics with surface hopping, J. Chem. Phys. 107 (1997) 825.
[35] H.M. Jaeger, S. Fischer, O.V. Prezhdo, Decoherence-induced surface hopping, J. Chem. Phys. 137 (2012) 22A545.
[36] M.J. Bedard-Hearn, R.E. Larsen, B.J. Schwartz, Mean-field dynamics with stochastic decoherence MF-SD: a new algorithm for nonadiabatic mixed quantum/classical molecular-dynamics simulations with nuclear-induced decoherence, J. Chem. Phys. 123 (2005) 234106.
[37] S. Garashchuk, J. Jakowski, L. Wang, B.G. Sumpter, Quantum trajectory-electronic structure approach for exploring nuclear effects in the dynamics of nanomaterials, J. Chem. Theory Comput. 9 (2013) 5221–5235, https://doi.org/10.1021/ct4006147.
[38] P. Ehrenfest, Bemerkung über die angenäherte Gültigkeit der klassischen Mechanik innerhalb der Quantenmechanik, Z. Phys. 45 (1927) 455–457, https://doi.org/10.1007/BF01329203.
[39] D. Bohm, A suggested interpretation of the quantum theory in terms of "hidden" variables, I and II, Phys. Rev. 85 (1952) 166–193.
[40] V.A. Rassolov, S. Garashchuk, G.C. Schatz, Quantum trajectory dynamics in arbitrary coordinates, J. Phys. Chem. A 110 (2006) 5530–5536.
[41] S. Garashchuk, V.A. Rassolov, Energy conserving approximations to the quantum potential: dynamics with linearized quantum force, J. Chem. Phys. 120 (2004) 1181–1190.
[42] S. Garashchuk, V. Rassolov, O. Prezhdo, Semiclassical Bohmian dynamics, in: Reviews in Computational Chemistry, vol. 27, Wiley, 2011, pp. 111–210.
[43] V.A. Rassolov, S. Garashchuk, Computational complexity in quantum chemistry, Chem. Phys. Lett. 464 (2008) 262–264.
[44] W.H. Press, B.P. Flannery, S.A. Teukolsky, W.T. Vetterling, Numerical Recipes: The Art of Scientific Computing, third ed., Cambridge University Press, Cambridge, 2007.

[45] V.A. Rassolov, S. Garashchuk, Bohmian dynamics on subspaces using linearized quantum force, J. Chem. Phys. 120 (2004) 6815–6825.
[46] S. Garashchuk, M.V. Volkov, Incorporation of quantum effects for selected degrees of freedom into the trajectory-based dynamics using spatial domains, J. Chem. Phys. 137 (2012) 074115.
[47] S. Garashchuk, V. Rassolov, Quantum trajectory dynamics based on local approximations to the quantum potential and force, J. Chem. Theory Comput. 15 (2019) 3906–3916, https://doi.org/10.1021/acs.jctc.9b00027.
[48] B. Gu, R.J. Hinde, V.A. Rassolov, S. Garashchuk, Estimation of the ground state energy of an atomic solid by employing quantum trajectory dynamics with friction, J. Chem. Theory Comput. 11 (2015) 2891–2899, https://doi.org/10.1021/ct501176m.
[49] S. Godecker, Linear scaling electronic structure methods, Rev. Mod. Phys. 71 (1999) 1085.
[50] C. Bannwarth, S. Ehlert, S. Grimme, GFN2-xTB–an accurate and broadly parametrized self-consistent tight-binding quantum chemical method with multipole electrostatics and density-dependent dispersion contributions, J. Chem. Theory Comput. 15 (2019) 1652–1671, https://doi.org/10.1021/acs.jctc.8b01176.
[51] G. Zheng, M. Lundberg, J. Jakowski, T. Vreven, M. Frisch, K. Morokuma, Implementation and benchmark tests of the DFTB method and its application in the ONIOM method, Int. J. Quantum Chem. 109 (2009) 1841–1854, https://doi.org/10.1002/qua.22002.
[52] M. Elstner, D. Porezag, G. Jungnickel, J. Elsner, M. Haugk, T. Frauenheim, S. Suhai, G. Seifert, Self-consistent-charge density-functional tight-binding method for simulations of complex materials properties, Phys. Rev. B 58 (11) (1998) 7260–7268.
[53] D. Porezag, T. Fraunheim, T. Kohler, G. Seifert, R. Kaschner, Construction of tight-binding-like potentials on the basis of density-functional theory: application to carbon, Phys. Rev. B 51 (1995) 12947–12957, https://doi.org/10.1103/PhysRevB.51.12947.
[54] J. Mazzuca, S. Garashchuk, J. Jakowski, Description of proton transfer in soybean lipoxygenase-1 employing approximate quantum trajectory dynamics, Chem. Phys. Lett. 542 (2012) 153–158, https://doi.org/10.1016/j.cplett.2012.06.019.
[55] J. Jakowski, B. Hadri, S.J. Stuart, P. Krstic, S. Irle, D. Nugawela, S. Garashchuk, Optimization of density functional tight-binding and classical reactive molecular dynamics for high-throughput simulations of carbon materials, XSEDE'12, Conference Proceedings. ACM. 36 (2012) 1–7, https://doi.org/10.1145/2335755.2335832.
[56] L. Wang, J. Jakowski, S. Garashchuk, Adsorption of a hydrogen Atom on a graphene flake examined with quantum trajectory/electronic structure dynamics, J. Phys. Chem. C 118 (2014) 16175–16187, https://doi.org/10.1021/jp503261k.
[57] D.C. Bell, M.C. Lemme, L.A. Stern, J.R. Williams, C.M. Marcus, Precision cutting and patterning of graphene with helium ions, Nanotechnology 20 (2009) 455301.
[58] A. Dey, A. Chroneos, N.S. Braithwaite, R.P. Gandhiraman, S. Krishnamurth, Plasma engineering of graphene, Appl. Phys. Rev. 3 (2016) 021301.
[59] V. Iberi, I. Vlassiouk, X.-G. Zhang, B. Matola, A. Linn, D.C. Joy, A.J. Rondinone, Maskless lithography and in situ visualization of conductivity of graphene using helium ion microscopy, Sci. Rep. 5 (2015) 11952, https://doi.org/10.1038/srep11952.
[60] J. Kotakoski, C. Brand, Y. Lilach, O. Cheshnovsky, C. Mangler, M. Arndt, J.C. Meyer, Toward two-dimensional all-carbon heterostructures via ion beam patterning of sngle-layer graphene, Nano Lett. 15 (2015) 5944–5949, https://doi.org/10.1021/acs.nanolett.5b02063.
[61] C.J. Russo, J.A. Golovchenko, Atom-by-atom nucleation and growth of graphene nanopores, Proc. Natl. Acad. Sci. U.S.A. 109 (2012) 5953–5957, https://doi.org/10.1073/pnas.1119827109.
[62] N. Toyoda, I. Yamada, Gas cluster ion Beam equipment and applications for surface processing, IEEE Trans. Plasma Sci. 36 (2008) 1471–1488.
[63] S. Kim, A.V. Ievlev, J. Jakowski, I.V. Vlassiouk, X. Sang, C. Brown, O. Dyck, R.R. Unocic, S.V. Kalinin, A. Belianinov, B.G. Sumpter, S. Jesse, O.S. Ovchinnikova, Multi-purposed Ar gas cluster ion beam processing for graphene engineering, Carbon 131 (2018) 142–148.
[64] B.J. Tyler, B. Brennan, H. Stec, T. Patel, L. Hao, I.S. Gilmore, A.J. Pollard, Removal of organic contamination from graphene with a controllable mass-selected Argon gas cluster Ion Beam, J. Phys. Chem. C 119 (2015) 17836–17841, https://doi.org/10.1021/acs.jpcc.5b03144.
[65] Z. Zabihi, H. Araghi, Formation of nanopore in a suspended graphene sheet with argon cluster bombardment: a molecular dynamics simulation study, Nuclear Instrum. Methods Phys. Res. Section B: Beam Interactions Mater. Atoms 343 (2015) 48–51, https://doi.org/10.1016/j.nimb.2014.11.022.

[66] S. Zhao, J. Xue, L. Liang, Y. Wang, S. Yan, Drilling nanopores in graphene with clusters: a molecular dynamics study, J. Phys. Chem. C 116 (2012) 11776–11782, https://doi.org/10.1021/jp3023293.
[67] K. Yoon, A. Rahnamoun, J.L. Swett, V. Iberi, D.A. Cullen, I.V. Vlassiouk, A. Belianinov, S. Jesse, X. Sang, O.S. Ovchinnikovai, A.J. Rondinone, R.R. Unocic, A.C. van Duin, Atomistic-scale simulations of defect formation in graphene under Noble gas ion irradiation, ACS Nano 10 (2016) 8376–8384.
[68] J.L. Achtyl, R.R. Unocic, L. Xu, Y. Cai, M. Raju, W. Zhang, R.L. Sacci, I.V. Vlassiouk, P.F. Fulvio, P. Ganesh, D.J. Wesolowski, S. Dai, A.C.T. van Duin, M. Neurock, F.M. Geiger, Aqueous proton transfer across single-layer graphene, Nat. Commun. 6 (2015) 6539, https://doi.org/10.1038/ncomms7539.
[69] F. Spath, J. Gebhardt, F. Dull, U. Bauer, P. Bachmann, C. Gleichweit, A. Gorling, H. Steinruck, C. Papp, Hydrogenation and hydrogen intercalation of hexagonal boron nitride on Ni(111): reactivity and electronic structure, 2D Materials 4 (2017) 035026.
[70] S. Tang, Z. Cao, Structural and electronic properties of the fully hydrogenated boron nitride sheets and nanoribbons: insight from first-principles calculations, Chem. Phys. Lett. 488 (2010) 67–72, https://doi.org/10.1016/j.cplett.2010.01.073.
[71] S. Hu, M. Lozada-Hidalgo, F.C. Wang, A. Mishchenko, F. Schedin, R.R. Nair, E.W. Hill, D.W. Boukhvalov, M.I. Katsnelson, R.A.W. Dryfe, I.V. Grigorieva, H.A. Wu, A.K. Geim, Proton transport through one-atom-thick crystals, Nature 516 (2014) 227–230, https://doi.org/10.1038/nature14015.
[72] O. Leenaerts, B. Partoens, F.M. Peeters, Graphene: a perfect nanoballoon, Appl. Phys. Lett. 93 (2008) 193107, https://doi.org/10.1063/1.3021413.
[73] W. Fang, J. Chen, Y. Feng, X.-Z. Li, A. Michaelides, The quantum nature of hydrogen, Int. Rev. Phys. Chem. 38 (2019) 35–61, https://doi.org/10.1080/0144235X.2019.1558623.
[74] R.A. Aziz, A highly accurate interatomic potential for argon, J. Chem. Phys. 99 (1993) 4518–4525, https://doi.org/10.1063/1.466051.
[75] A.A. Buchachenko, J. Jakowski, G. Chalasinski, M.M. Szczesniak, S.M. Cybulski, *ab initio* based study of the aro$^-$ photoelectron spectra: selectivity of spin-orbit transitions, J. Chem. Phys. 112 (2000) 5852–5865, https://doi.org/10.1063/1.481186.
[76] J. Jakowski, G. Chalasinski, J. Gallegos, M.W. Severson, M.M. Szczesniak, Characterization of Ar_nO^- clusters from *ab initio* and diffusion Monte Carlo calculations, J. Chem. Phys. 118 (2003) 2748–2759, https://doi.org/10.1063/1.1531110.
[77] M. Gaus, Q. Cui, M. Elstner, DFTB3: extension of the self-consistent-charge density-functional tight-binding method (SCC-DFTB), J. Chem. Theory Comput. 7 (2011) 931–948, https://doi.org/10.1021/ct100684s.
[78] Y. Shao, Z. Gan, E. Epifanovsky, A.T.B. Gilbert, M. Wormit, J. Kussmann, A.W. Lange, A. Behn, J. Deng, X. Feng, D. Ghosh, M. Goldey, P.R. Horn, L.D. Jacobson, I. Kaliman, R.Z. Khaliullin, T. Kuś, A. Landau, J. Liu, E.I. Proynov, Y.M. Rhee, R.M. Richard, M.A. Rohrdanz, R.P. Steele, E.J. Sundstrom, H.L.W. III, P.M. Zimmerman, D. Zuev, B. Albrecht, E. Alguire, B. Austin, G.J.O. Beran, Y.A. Bernard, E. Berquist, K. Brandhorst, K.B. Bravaya, S.T. Brown, D. Casanova, C.-M. Chang, Y. Chen, S.H. Chien, K.D. Closser, D.L. Crittenden, M. Diedenhofen, R.A.D. Jr, H. Do, A.D. Dutoi, R.G. Edgar, S. Fatehi, L. Fusti-Molnar, A. Ghysels, A. Golubeva-Zadorozhnaya, J. Gomes, M.W. Hanson-Heine, P.H. Harbach, A.W. Hauser, E.G. Hohenstein, Z.C. Holden, T.-C. Jagau, H. Ji, B. Kaduk, K. Khistyaev, J. Kim, J. Kim, R.A. King, P. Klunzinger, D. Kosenkov, T. Kowalczyk, C.M. Krauter, K.U. Lao, A.D. Laurent, K.V. Lawler, S.V. Levchenko, C.Y. Lin, F. Liu, E. Livshits, R.C. Lochan, A. Luenser, P. Manohar, S.F. Manzer, S.-P. Mao, N. Mardirossian, A.V. Marenich, S.A. Maurer, N.J. Mayhall, E. Neuscamman, C.M. Oana, R. Olivares-Amaya, D.P. O'Neill, J.A. Parkhill, T.M. Perrine, R. Peverati, A. Prociuk, D.R. Rehn, E. Rosta, N.J. Russ, S.M. Sharada, S. Sharma, D.W. Small, A. Sodt, T. Stein, D. Stück, Y.-C. Su, A.J. Thom, T. Tsuchimochi, V. Vanovschi, L. Vogt, O. Vydrov, T. Wang, M.A. Watson, J. Wenzel, A. White, C.F. Williams, J. Yang, S. Yeganeh, S.R. Yost, Z.-Q. You, I.Y. Zhang, X. Zhang, Y. Zhao, B.R. Brooks, G.K. Chan, D.M. Chipman, C.J. Cramer, W.A.G. III, M.S. Gordon, W.J. Hehre, A. Klamt, H.F.S. III, M.W. Schmidt, C.D. Sherrill, D.G. Truhlar, A. Warshel, X. Xu, A. Aspuru-Guzik, R. Baer, A.T. Bell, N.A. Besley, J.-D. Chai, A. Dreuw, B.D. Dunietz, T.R. Furlani, S.R. Gwaltney, C.-P. Hsu, Y. Jung, J. Kong, D.S. Lambrecht, W. Liang, C. Ochsenfeld, V.A. Rassolov, L.V. Slipchenko, J.E. Subotnik, T.V. Voorhis, J.M. Herbert, A.I. Krylov, P.M. Gill, M. Head-Gordon, Advances in molecular quantum chemistry contained in the Q-Chem 4 program package, Mol. Phys. 113 (2015) 184–215, https://doi.org/10.1080/00268976.2014.952696.
[79] M. Valiev, E.J. Bylaska, N. Govind, K. Kowalski, T.P. Straatsma, H.J.J. van Dam, D. Wang, J. Nieplocha, E. Apra, T.L. Windus, W.A de Jong, NWChem: a comprehensive and scalable open-source solution for large scale

molecular simulations, Comput. Phys. Commun. 181 (2010) 1477–1489, https://doi.org/10.1016/j.cpc.2010.04.018.

[80] G. Ebel, R. Krohne, H. Meyer, U. Buck, R. Schinke, T. Seelemann, P. Andresen, Rotationally inelastic scattering of NH_3 with H_2: molecular-beam experiments and quantum calculations, J. Chem. Phys. 93 (1990) 6419–6432, https://doi.org/10 1063/1.458958.

[81] B. van Zyl, N.G. Utterback, R.C. Amme, Generation of a fast atomic hydrogen beam, Rev. Sci. Instrum. 47 (1976) 814–819, https://doi.org/10.1063/1.1134758.

[82] H. Nakamura, A. Takayama, A. Ito, Molecular dynamics simulation of hydrogen isotope injection into graphene, Contributions Plasma Phys. 48 (2008) 265–269, https://doi.org/10.1002/ctpp.200810046.

[83] A. Ito, H. Nakamura, Molecular dynamics simulation of bombardment of hydrogen atoms on graphite surface, Commun. Comput. Phys. 4 (2008) 592–610.

[84] J.F. Ziegler, J.P. Biersack, The Stopping and Range of Ions in Matter, Springer, Boston, MA, 1985.

[85] J.F. Ziegler, M.D. Ziegler, J.P. Biersack, SRIM—the stopping and range of ions in matter (2010), Nucl. Instrum. Methods Phys. Res. B: Beam Interact. Mater. At. 268 (2010) 1818–1823, https://doi.org/10.1016/j.nimb.2010.02.091. 19th International Conference on Ion Beam Analysis.

[86] L. Jeloaica, V. Sidis, DFT investigation of the adsorption of atomic hydrogen on a cluster-model graphite surface, Chem. Phys. Lett. 300 (1999) 157–162, https://doi.org/10.1016/S0009-2614(98)01337-2.

[87] M. Miao, M.B. Nardelli, Q. Wang, Y. Liu, First principles study of the permeability of graphene to hydrogen atoms, Phys. Chem. Chem. Phys. 15 (2013) 16132–16137, https://doi.org/10.1039/c3cp52318g.

[88] L. Tsetseris, S.T. Pantelides, Graphene: an impermeable or selectively permeable membrane for atomic species? Carbon 67 (2014) 58–63, https://doi.org/10.1016/j.carbon.2013.09.055.

[89] M.J. Knapp, K. Rickert, J.P. Klinman, Temperature-dependent isotope effects in soybean lipoxygenase-1: correlating hydrogen tunneling with protein dynamics, J. Am. Chem. Soc. 24 (2002) 3865–3874.

[90] D.R. Killelea, V.L. Campbell, N.S. Shuman, A.L. Utz, Bond-selective control of a heterogeneously catalyzed reaction, Science 319 (2008) 790–793.

[91] N.T. Ekanayake, J. Huang, J. Jakowski, B.G. Sumpter, S. Garashchuk, Relevance of the nuclear quantum effects on the proton/deuteron transmission through hexagonal bron nitride and graphene monolayers, J. Phys. Chem. C 121 (2017) 24335–24344, https://doi.org/10.1021/acs.jpcc.7b08152.

[92] K.A. Mauritz, R.B. Moore, State of understanding of Nafion, Chem. Rev. 104 (2004) 4535–4585, https://doi.org/10.1021/cr0207123.

[93] W.L. Wang, E. Kaxiras, Graphene hydrate: theoretical prediction of a new insulating form of graphene, N. J. Phys. 12 (2010) 125012, https://doi.org/10.1088/1367-2630/12/12/125012.

[94] Q. Zhang, M. Ju, L. Chen, X.C. Zeng, Differential permeability of proton isotopes through graphene and graphene analogue monolayer, J. Phys. Chem. Lett. 7 (2016), https://doi.org/10.1021/acs.jpclett.6b01507. 3395–3340.

[95] M. Seel, R. Pandey, Proton and hydrogen transport through two-dimensional monolayers, 2D Materials 3 (2016) 025004, https://doi.org/10.1088/2053-1583/3/2/025004.

[96] J.M.H. Kroes, A. Fasolino, M.I. Katsnelson, Density functional based simulations of proton permeation of graphene and hexagonal boron nitride, Phys. Chem. Chem. Phys. 19 (2017) 5813–5817, https://doi.org/10.1039/C6CP08923B.

[97] Y.-B. Xin, Q. Hu, D.-H. Niu, X.-H. Zheng, H.-L. Shi, M. Wang, Z.-S. Xiao, A.-P. Huang, Z.-B. Zhang, Research progress of hydrogen tunneling in two-dimensional materials, Acta Phys. Sinica 66 (2017) 056601, https://doi.org/10.7498/aps.66.056601.

[98] I. Poltavsky, L. Zheng, M. Mortazavi, A. Tkatchenko, Quantum tunneling of thermal protons through pristine graphene, J. Chem. Phys. 148 (2018) 204707, https://doi.org/10.1063/1.5024317.

[99] M. Lozada-Hidalgo, S. Hu, O. Marshall, A. Mishchenko, A.N. Grigorenko, R.A.W. Dryfe, B. Radha, I.V. Grigorieva, A.K. Geim, Sieving hydrogen isotopes through two-dimensional crystals, Science 351 (2016) 68–70, https://doi.org/10.1126/science.aac9726.

[100] R. Kosloff, Time-dependent quantum-mechanical methods for molecular dynamics, J. Phys. Chem. 92 (1988) 2087–2100.

CHAPTER

4

From ground to excited electronic state dynamics of electron and ion irradiated graphene nanomaterials☆

David Lingerfelt[a,*], *Panchapakesan Ganesh*[a], *Bobby G. Sumpter*[a], *and Jacek Jakowski*[a,b,*]

[a]Center for Nanophase Materials Sciences, Oak Ridge National Laboratory, Oak Ridge, TN, United States [b]Computational Sciences and Engineering Division, Oak Ridge National Laboratory, Oak Ridge, TN, United States

*Corresponding authors: E-mail: lingerfeltdb@ornl.gov; jakowskij@ornl.gov

1 Introduction

Graphene exhibits extreme mechanical strength [1], presents high electron and hole mobilities [2], and supports exotic quantum mechanical phenomena (e.g., charge carriers which exhibit the linear dispersion relations of massless Dirac fermions at low energies) [3–6]. Owing to its unique set of properties, graphene is considered a promising platform for the development of information processing devices in the "Beyond Moore's Law" era [6–8], which is expected to rely on a combination of classical (digital and analogue), neuro-inspired, and quantum computing modalities [9]. As a semimetal with a very low density of states at its Fermi level, bulk graphene is not appropriate in its pristine form for traditional

☆Notice: This chapter has been authored by UT-Battelle, LLC under Contract No. DE-AC05-00OR22725 with the U.S. Department of Energy. The United States Government retains and the publisher, by accepting the article for publication, acknowledges that the United States Government retains a nonexclusive, paid-up, irrevocable, world-wide license to publish or reproduce the published form of this chapter, or allow others to do so, for United States Government purposes. The Department of Energy will provide public access to these results of federally sponsored research in accordance with the DOE Public Access Plan (http://energy.gov/downloads/doe-public-access-plan).

https://doi.org/10.1016/B978-0-12-819514-7.00003-8

semiconductor applications. A large body of work has emerged as a result, which is centered around the modification of graphene to induce a band gap. Numerous studies have exposed that a band gap can be "opened" in graphene through interactions with certain substrates [10], formation of graphene nanoribbon structures [11], polycrystallinity [12], and inclusion of heteroatom dopants [13, 14].

Beyond the potential for functionalized graphene to supplant silicon in traditional digital computing devices, the unique electronic structure of graphene nanomaterials can also be harnessed to make functionalized graphene materials amenable for certain quantum information processing applications. For instance, the edges of graphene flakes support topologically protected states that may find applications as qubits with long coherence times [15], and the strong nonlinearity of interactions between quantum-confined plasmon excitations in graphene nanoribbons may provide a route to high-fidelity two qubit gates for universal photonic quantum computation [16]. Mechanically strained graphene has also been predicted to develop nonlinear anomalous Hall current [17], and it is argued—based on including small phenomenological perturbations in the tight binding Hamiltonian for graphene—that other topologically nontrivial phases (for example, quantum spin and quantum anomalous Hall phases) may be realizable, which would further extend the reach of graphene into the realm of quantum materials for information processing applications [18]. Intermediate to the quantum and classical computing regimes, various methods of doping and straining graphene have also shown promise for spin- [19] and valley-tronic [20] applications.

Functioning of next generation electronic applications and quantum information devices critically relies on manipulation of excited states in a controllable fashion. Harnessing excited states is also a promising avenue for future chemistries. Electronic excitations can alter the potential energy landscape for chemical processes allowing reactions to proceed along routes inaccessible at the ground state [21]. Excitations to electronic states in which charge density is significantly displaced relative to the ground state density often decay efficiently through transfer of energy to local vibrations. This is caused by strong coupling between the ground and excited states induced by the motions of nuclei residing in the areas of displaced charge density. Excitation of localized defect states introduced into the spectrum of condensed phase materials like graphene can modulate the reactivity of the material in the vicinity of the defect [22]. This opens the door to new graphene functionalization chemistries through selective electronic excitation. Understanding how to tailor spectra of electronic states via a particular method of functionalization opens opportunities for the new technological applications in photochemistry, nanophotonics, nonequilibrium charge transport processes, and quantum information processing.

Electronic excitation can be enacted through laser irradiation and electrostatic biasing, but also through interaction with the evanescent electric and magnetic fields associated with beams of swiftly moving charged particles like those utilized in electron and ion beam microscopes [23]. The electric Coulomb fields from the charged particles comprising these beams decay in intensity with the distance from the charged particles. This—in conjunction with the ability to focus beams of charged particles into small (in some cases, subpicometer) spot sizes—has enabled the probing of materials' electronic structure with unprecedented spatial resolution. This is in stark contrast to the spot sizes achievable using lasers, where the oscillating electromagnetic fields cannot be focused beyond one half of their wavelength (i.e., the diffraction limit). Instruments relying on focused electron and ion beams are uniquely capable of exciting highly localized electronic and vibrational states; capabilities which hold transformative potential for atomically precise in situ modifications of two-dimensional materials like graphene [24–26]. However, early experimental measures for the cross sections of these processes show that many millions of electrons worth of beam current are required in order to bring about a single reaction step for, e.g., the diffusion of threefold coordinated silicon atom [24, 27], suggesting that there is significant room for optimization of beam-induced atomic manipulation once its underlying mechanism is more fully understood.

In pursuit of a first principles description of electron and ion beam-induced reactions in materials, researchers in this field have turned to ab initio molecular dynamics simulations of materials [28–31]. Chapter "From classical to quantum dynamics of atomic and ionic species interacting with graphene and its analogue" by Garashchuk et al. includes several examples of such studies [32–35]. However, in these ab initio molecular dynamics simulations electronic excitations are ignored and electronically adiabatic evolution is assumed, which can lead to vibrational evolution that may deviate from that which is experimentally observed. For example, breakdown of the Born–Oppenheimer approximation can result when nuclei are imparted with multiple electron volts of kinetic energy.

In general, computational materials science relies on the density functional theory as the workhorse electronic structure theory. Different basis functions are used to express the electronic states, including atom-centered Gaussian functions, plane waves, grids and numerical basis sets. The most appropriate choice depends on the problem at hand. Plane waves are well suited for studies of periodic systems such as crystals and pristine 2D materials, whereas localized functions are robust in description of molecular systems and clusters. While pristine bulk graphene is well-represented as an infinitely extended two-dimensional lattice of planar threefold coordinated carbon atoms, the inclusion of defects into this lattice breaks this periodicity. Ideally, to model graphene nanostructures containing impurities, numerous calculations of large-but-finite clusters of graphene with specific defect configurations should generally be considered. The properties of nonperiodic systems like these are more quickly convergent when the electronic degrees of freedom are expressed in a basis of localized functions (e.g., atom-centered Gaussian functions) than when delocalized basis functions (e.-g., plane waves) are used. In the same way that the Dirac cone physics which emerges for pristine graphene is most straightforwardly addressed using electronic structure methods formulated for periodic solids in the momentum-space representation, edge- and defect-localized states are more easily expressed in "real" (or position) space. The quantity of interest in this chapter (the electronic response to electron/ion beams) further motivates our preference for the position-space representation here, since the application of an external point-source electric field lifts any spatial periodicity that the Hamiltonian may have otherwise exhibited.

A description of the emerging computational and theoretical tools and techniques for studying electron/ion beam-induced manipulation of graphene will be the focus of this chapter. While the previous chapter (Chapter "From classical to quantum dynamics of atomic and ionic species interacting with graphene and its analogue" by Garashchuk et al.) focuses on the classical and quantum dynamics of nuclei on the ground electronic state and nuclear quantum effects, in this chapter we discuss methods for modeling effects of beam-induced electronic excitations, with the content organized as follows. We first consider formal aspects of the selection rules for excitations induced by external charged particles. We introduce time-dependent electronic structure theory methods for solving the explicit electronic dynamics resulting from an arbitrary perturbation directly in the time domain, and for evaluating the same quantities perturbatively in the frequency domain through application of linear response theory. We go on to show how the position dependence of electronic excitations induced by the fields associated with external charged particles can be accessed (in an internally consistent fashion) in both of these formalisms. Finally, we demonstrate how these tools can be used to explain some experimentally observed features of beam-induced reactions promoted in modern aberration corrected scanning transmission electron microscopes, and provide some perspective on opportunities for expanding upon the current state-of-the-art approaches.

2 Theory

2.1 Selection rules for electronic excitations induced by point-source electric fields

The swift charged particles that comprise high-energy beams exhibit small de Broglie wavelengths, so the fields that they emanate can (to a reasonable approximation) be substituted for those of a point charge. At the same time, electrons in materials (except for the core electrons of heavier elements) exhibit velocities much smaller than the speed of light, so the scalar (electric) potential dominates in the description of their interaction with the beam electrons. As such, it is reasonable to omit the vector (magnetic) potential from the matter–field interaction Hamiltonian when modeling valence excitations. The Lorentz-invariant expression for the scalar potential of a charged particle undergoing constant velocity motion is given by the Liénard–Wiechert (LW) potential [36]. In the limit where the charged particle's velocity approaches zero, the LW scalar potential reproduces the classical electrostatic (Coulomb) potential. In the opposite limit where the charged particle's speed approaches that of light, its interaction with the material becomes increasingly brief, and contributes to the material's electronic Hamiltonian with a delta function envelope in time (i.e., an impulse perturbation). The LW potential also captures the relativistic retardation effects which lead to decreased electric field intensity emanated by such a moving particle along its direction of propagation (and proportionally increased intensity in the transverse directions) [37]. In the ultra-relativistic limit, where the particle's velocity matches the speed of light exactly, the emanated electric field would be nonzero only in the plane perpendicular to its direction of motion.

In considering the response of effectively two-dimensional systems like graphene to the passage of an ultra-relativistic electron, the scalar potential can be qualitatively approximated by an electrostatic potential impulse activated at the instant that the beam electron enters the plane of the material (Fig. 4.1). This impulse approximation for the perturbing potential experienced by the material is most appropriate when the beam electrons are fast enough to vacate the region of strong interaction before any significant redistribution of population in the excited state manifold can occur.

The electrostatic potential energy of a material exposed to an external particle with charge q_{ext} located at position $\mathbf{r}_{ext}$ can be formulated as an operator in the space of the material's electronic coordinates $\mathbf{r}$ through canonical quantization of the scalar potential expression (Eq. 4.1).

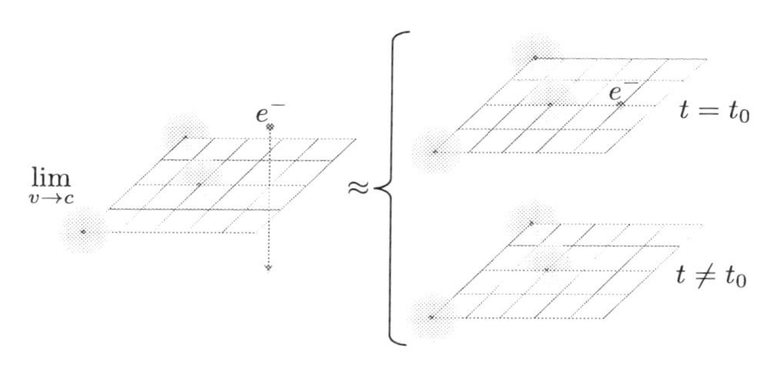

FIG. 4.1 A cartoon of the electrostatic impulse approximation to the electric potential of an ultra-relativistic electron, where the system experiences an impulsive electrostatic potential at the time that the point charge would be passing through the plane of the material (and zero potential at other times).

$$\widehat{V}(q_{ext}, \mathbf{r}_{ext}) = \frac{q_{ext}}{|\mathbf{r} - \mathbf{r}_{ext}|}, \tag{4.1}$$

Selection rules for electronic excitations are determined by the off-diagonal matrix elements of this matter–field interaction operator, evaluated between initial and final states, $M_{0\to k} = \langle \Psi_0 | \widehat{V} | \Psi_k \rangle$, with the transition rates from state Ψ_0 to state Ψ_k due to the perturbing potential $\widehat{V}$ given by Fermi's golden rule to be proportional to the squared modulus of $M_{0\to k}$. Note that in the case of perturbation by an external point charge, these transition rates will depend explicitly on the position of the point charge.

When the source of the electric potential ($\mathbf{r}_{ext}$ in Eq. (4.1)) is well-separated spatially from the charge distribution of the material, it is convenient to expand the perturbing potential $\widehat{V}$ in the electric multipole moments of the material's charge density around the origin (which we take to be located at the material's center of charge) [38–43]:

$$\widehat{V} = q(\phi)_0 - \hat{\mu}_\alpha (E_\alpha)_0 - \frac{1}{3}\widehat{\Theta}_{\alpha\beta}(E_{\alpha\beta})_0 + \cdots \tag{4.2}$$

In Eq. (4.2), Greek character subscripts index Cartesian components, and a zero subscript indicates that a quantity is evaluated at the origin. ϕ denotes the scalar potential from an external point charge, E denotes the electric field associated with this potential, and $E_{\alpha\beta}$ is shorthand notation indicating the gradient of the electric field ($\nabla_\alpha E_\beta$). $\hat{\mu}$ and $\widehat{\Theta}$ are the electric dipole and quadrupole moment operators acting in the electronic coordinates of the material, and q is the material's electric monopole moment (i.e., the net charge.) Higher order terms in this expansion go as the products of higher electric multipole moments of the material's charge distribution and corresponding higher derivatives of the external electric field. Note that Einstein's summation convention has been employed, in which repeated indices are summed over.

The first term on the right-hand side of Eq. (4.2) is a constant that modulates the energies of the electronic states but yields zero transition moment strength between them due to their orthogonality ($\langle \Psi_j | q(\phi)_0 | \Psi_k \rangle = 0$, for all $j \neq k$). The second term describes the coupling of the dipole moment of the material's charge distribution to an external electric field. This is the leading order term in the light–matter interaction Hamiltonian which allows transitions between, e.g., atomic states differing by one quantum of orbital angular momentum (i.e. $\Delta L = \pm 1$). The third term captures the coupling between the quadrupole moment of the material's charge distribution and the gradient of an external electric field, which allows the $\Delta L = 0, \pm 2$ transitions which are observed very weakly in UV–visible optical absorption spectra.

The relative contribution from each term in Eq. (4.2) to $M_{0\rightarrow k}$ at a given point charge position is determined through the position dependence of the magnitude of the electric field and its spatial derivatives. $E_{\alpha\beta}$ decays an order of magnitude faster than E itself with respect to distance from the point charge (and $E_{\alpha\beta\gamma}$ an order faster than that), indicating that terms beyond the dipolar one can contribute appreciably only when the material's charge distribution is close to (or overlapping) the perturbing point charge's position. Each of the terms in Eq. (4.2) vanish asymptotically as the point charge is positioned at greater distances from the material. As this limit is approached, the quadrupolar and higher order multipolar terms quickly vanish, and electric dipole selection rules emerge for point charge-induced transitions beyond a threshold separation distance between point charge and material.

In the event that the perturbing charge is located well inside of the material's charge distribution, the expansion of the scalar potential about the position of the perturbing charge itself is more quickly convergent. In this case, the dependence of selection rules on the position of the perturbing point charge can be most straightforwardly gleaned through the application of group theory. The direct product of the representations of the electric potential from the perturbing point charge, initial, and final electronic states must form a basis for the totally symmetric representation of the material's point group in order for the transition to be symmetry-allowed [44]. The task of resolving the symmetry-dictated position dependence of the selection rules then reduces to identifying all of the point charge positions which give an electric potential that transforms according to an irreducible representation of the material's point group. Consider, for example, the illustrative case of a perturbing point charge positioned at the exact center of a benzene ring, such that its associated electric potential transforms as the totally symmetric representation. In this case, the only allowed transitions are those between states belonging to the same irreducible representation.

2.2 Time-dependent electronic structure theory and electronic excitations

Most theoretical approaches for the calculation of excited states of materials derive from the time-dependent Schrödinger equation (TDSE) and can therefore be categorized as time-dependent electronic structure theories. Excited state energies and transition probabilities between states can be accessed from the TDSE either through time-domain simulation of the electronic dynamics, or via response theory approaches solved in the frequency domain. In the time domain, or "real-time" (RT) approach, excitation energies and transition probabilities are resolved by integrating the TDSE for a system initialized in an electronic energy eigenstate (typically the ground state) that is thrust out of stationarity by an impulsive electric field perturbation. Fourier transformation of different operators' expectation values collected during the resulting dynamics yields information on the spectrum of states which become populated as a result of the perturbation. For sufficiently weak perturbations, the same information can be gleaned directly in the frequency domain without actually integrating the TDSE through a perturbative treatment of the response of the reference state.

Like its ground state counterpart, time-dependent Kohn–Sham density functional theory (TD-DFT) exhibits the highest accuracy-to-expense ratios of all excited state electronic structure approaches for large systems. TD-DFT has long been a prominent method for computational investigations of the optical excitation of materials, and it is relatively straightforward to extend the existing efficient TD-DFT implementations found in most modern computational chemistry packages to treat the problem of electronic excitations induced in nanomaterials by the passage of fast charged particles as in the case of electron (or ion) impact spectroscopy [45]. When the beam of external charged particles encroaches upon or penetrates the region of bound electron density of a target material, the series expansion of the interaction potentials (Eq. 4.2) is very slow to converge. First principles methods designed to explicitly capture the effects of the spatially inhomogeneous electric fields near point charges are needed to describe the dipole disallowed transitions (i.e., those which cannot be promoted by electric fields that are homogenous over the volume of the materials) which are known to be enacted by electron impact [46, 47] but are essentially absent in the analogous optical absorption spectra.

Historically, theoretical treatments of the dynamics of swift charged particles incident upon a material have been formulated (in the momentum representation) as a scattering problem in which a transfer of momentum between the incident particle and charged particles comprising the materials connects incoming and outgoing momentum eigenstates of the incident particle and electronic excitations within the materials [23, 48–52]. However, since our goal here is to understand the electronic response of functionalized graphene to the passage of a single incident charged particle of high initial momentum (i.e., small deBroglie wavelength), it is reasonable to assume that no multiple scattering takes place and that a nonquantum mechanical treatment of the incident particle is acceptable. It is convenient in this case to take a materials-centric vantage of this inelastic scattering event, and focus exclusively on the electronically excited states which become populated as a function of the initial conditions of the incident particle (charge, position of impact/point of closest approach to the molecule, etc.)

2.3 The TD-DFT formalism

As TD-DFT is now a well established method, we only provide a brief overview of the formalism and necessary working equations here. We refer readers to references [53] and [54] for a complete derivation and review of the method. Atomic units have been used throughout this chapter unless otherwise noted, in which $\hbar = 4\pi\epsilon_0 = e = m_e = 1$. We present TD-DFT working equations assuming pure functionals here for brevity, since the extension to global hybrid functionals which include a fraction of the exact Hartree–Fock exchange is trivial and detailed elsewhere [55, 56]. In the Kohn–Sham (KS) density functional theory, the many-electron Schrödinger equation is recast as an effective one-electron problem for a fictitious system of noninteracting electrons (described by KS orbitals, $\{\phi_i\}$) evolving under an external potential, such that the density ($\rho(\mathbf{r})$) of the fully interacting system is reproduced by the noninteracting KS orbitals [57, 58].

In KS-DFT [57, 58], the time evolution of the electronic density is completely captured by that of the occupied KS orbitals. The equation of motion for the KS orbitals is given in Eq. (4.3) [59, 60].

$$i\frac{\partial \phi_i(t)}{\partial t} = \widehat{F}\phi_i(t) \tag{4.3}$$

The KS operator, $\widehat{F}$, is

$$\widehat{F} = \hat{h} + \int d\mathbf{r}' \frac{\rho(\mathbf{r}')}{|\mathbf{r} - \mathbf{r}'|} + \frac{\delta E_{\mathrm{xc}}[\rho(\mathbf{r})]}{\delta \rho(\mathbf{r})} \tag{4.4}$$

where

$$\hat{h} = -\frac{1}{2}\nabla^2 - \sum_{k=1}^{m} \frac{Z_k}{|\mathbf{r} - \mathbf{R}_k|} \tag{4.5}$$

and m is the number of nuclei in the system, $\mathbf{R}_k$ and Z_k are the coordinates and charge of the kth nucleus, $\mathbf{r}$ are electronic coordinates, and E_{xc} is the exchange-correlation energy functional. In the spin-restricted treatment, the total density is given by:

$$\rho(\mathbf{r}) = \sum_i f_i \cdot |\phi_i(\mathbf{r})|^2, \tag{4.6}$$

where the summation in Eq. (4.6) runs over all KS orbitals and the f_i denotes the occupation number. The KS orbitals can be expanded in a set of N_b nonorthogonal basis functions (in this chapter, atom-centered Gaussian functions), $\{\chi_\mu\}$.

$$\phi_i(t) = \sum_{\mu}^{N_b} C_{\mu i}(t)\,\chi_\mu \tag{4.7}$$

Unless explicitly noted otherwise, Greek characters (μ, ν, λ, σ) index atomic orbital basis functions, lowercase Roman characters (i, j/a, b) index (occupied/virtual) KS orbitals of the KS-DFT ground state, and capital Roman characters (IJ) will index the (all electron) energy eigenstates.

The one particle reduced density matrix (1RDM) $\mathbf{P}$ can be expressed in this basis through the orbital expansion coefficients from Eq. (4.7):

$$P_{\mu\nu}(t) = \sum_{i}^{N_b} C^*_{\mu i}(t) \cdot f_i \cdot C_{\nu i}(t) \tag{4.8}$$

The KS matrix elements are given in the Gaussian function basis by Eq. (4.9).

$$F_{\mu\nu}(t) = h_{\mu\nu} + \sum_{\lambda\sigma} P_{\lambda\sigma}(t)(\mu\nu|\lambda\sigma) + v_{\mathrm{xc},\mu\nu} \tag{4.9}$$

where:

$$(\mu\nu|\lambda\sigma) = \int \mathbf{dr dr}'\, \chi^*_\mu(\mathbf{r})\chi_\nu(\mathbf{r}) \frac{1}{|\mathbf{r} - \mathbf{r}'|} \chi^*_\lambda(\mathbf{r}')\chi_\sigma(\mathbf{r}') \tag{4.10}$$

$$v_{\mathrm{xc},\mu\nu} = \int \mathbf{dr}\, \chi^*_\mu(\mathbf{r}) \frac{\delta E_{\mathrm{xc}}}{\delta \rho(\mathbf{r})} \chi_\nu(\mathbf{r}). \tag{4.11}$$

The 1RDM and KS matrix can be reexpressed in an orthonormal basis $\{\chi'_\mu\}$ by way of, e.g., the symmetric Löwdin transformation,

$$\mathbf{P}' = \mathbf{S}^{1/2}\mathbf{P}\mathbf{S}^{1/2}$$
$$\mathbf{F}' = \mathbf{S}^{-1/2}\mathbf{F}\mathbf{S}^{-1/2},$$

where $S_{\mu\nu} = \int d\mathbf{r}\chi_{\mu}^{*}(\mathbf{r})\chi_{\nu}(\mathbf{r})$.

The TD-KS equation can be written in Liouville–von Neumann form as an equation of motion for the 1RDM in the orthogonalized basis:

$$i\frac{\partial \mathbf{P'(t)}}{\partial t} = [\mathbf{F'}(t), \mathbf{P'}(t)] \tag{4.12}$$

Expectation values of one-body operators and the electronic energy are calculated according to Eqs. (4.13) and (4.14) (where $A_{\mu\nu}$ are matrix elements of an arbitrary one-electron operator, $\hat{A}$, in the nonorthogonal Gaussian function basis)

$$\langle \hat{A} \rangle = \sum_{\mu\nu} P_{\nu\mu} A_{\mu\nu} \tag{4.13}$$

$$\langle E_{ele} \rangle = \langle \hat{h} \rangle + \frac{1}{2}\sum_{\mu\nu}\sum_{\lambda\sigma} P_{\mu\nu} P_{\lambda\sigma}(\mu\nu|\lambda\sigma) + E_{\mathrm{xc}}[\rho(\mathbf{r})] \tag{4.14}$$

2.4 "Real-time" TD-DFT

The electronic equation of motion Eq. (4.12) has the following general solution:

$$\mathbf{P}'(t) = \mathbf{U}(t_0, t)\mathbf{P}'(t_0)\mathbf{U}^{\dagger}(t_0, t) \tag{4.15}$$

The time evolution operator, $\mathbf{U}$ is a time-ordered complex exponential (denoted in Eq. (4.16) by $\mathcal{T}\exp$) of the KS matrix integrated over the interval (t_0, t).

$$\mathbf{U}(t_0, t) = \mathcal{T}\exp\left(-i\int_{t_0}^{t} dt'\, \mathbf{F}(t')\right) \tag{4.16}$$

Since the KS matrix is itself time-dependent (through its dependence on the time-dependent 1RDM) it generally does not commute with itself at different times. Numerous low-order numerical integration techniques for solving the initial value problem for Eq. (4.12) have been developed and implemented, though only those which respect the symplectic structure of the equations of motion are suitable for longer-time simulations [61–64].

The population of the Ith energy eigenstate, ρ_I, that results from application of a given perturbation can be accessed through Fourier transformation of autocorrelation functions of the expectation values for operators collected during the simulation according to the relation shown in Eq. (4.17) [65–67].

$$\mathfrak{F}[\langle A(0)A(t)\rangle](\omega) = \sum_{IJ} \rho_I |\langle \Psi_I|\hat{A}|\Psi_J\rangle|^2 \delta(\omega - \omega_{IJ}) \tag{4.17}$$

By evaluating the autocorrelation functions for different electric and magnetic multipole operators, populations of different excited states can be resolved. For instance, when $\hat{A}$ is taken to be the electric dipole moment operator, $\widehat{\mu}$, the spectrum reports only on the transitions which have nonzero transition dipole strength to the ground state (i.e., dipole-allowed transitions). One could similarly evaluate the electric quadrupole moment autocorrelation function, $\widehat{\theta}$, to select out only the quadrupole-allowed transitions (and so on for higher order multipole moment operators). Finally, the autocorrelation function of the expectation value of the scalar potential itself, $\widehat{V}(q,\mathbf{r}_q)$, produces a spectrum that reports on all of the transitions that are promoted by the external charge impulse.

2.5 Linear response TD-DFT

Starting from Eq. (4.12) and solving (via first-order time-dependent perturbation theory) for the response of a system initially in its ground state to a monochromatic perturbation yields a symplectic eigenvalue problem known as the linear response (LR) TD-DFT equation, with solutions comprised of excitation energies, $\{\omega_I\}$, and one particle transition densities $\{\mathbf{X}^I,\mathbf{Y}^I\}$ between ground and excited states [54, 56, 68–71]:

$$\begin{bmatrix} \mathbf{A} & \mathbf{B} \\ \mathbf{B}^* & \mathbf{A}^* \end{bmatrix}\begin{bmatrix} \mathbf{X} \\ \mathbf{Y} \end{bmatrix} = \omega \begin{bmatrix} \mathbf{I} & \mathbf{0} \\ \mathbf{0} & -\mathbf{I} \end{bmatrix}\begin{bmatrix} \mathbf{X} \\ \mathbf{Y} \end{bmatrix} \tag{4.18}$$

$$\begin{aligned} A_{ia,jb} &= \delta_{ij}\delta_{ab}(F_{aa} - F_{ii}) + (ia|jb) + (ia|f_{\mathrm{xc}}|jb) \\ B_{ia,jb} &= (ia|bj) + (ia|f_{\mathrm{xc}}|bj), \end{aligned}$$

where:

$$(ia|f_{\mathrm{xc}}|jb) = \int \mathbf{drdr'}\phi_i^*(\mathbf{r})\phi_a(\mathbf{r})\frac{\delta^2 E_{\mathrm{xc}}}{\delta\rho(\mathbf{r})\delta\rho(\mathbf{r'})}\phi_b^*(\mathbf{r'})\phi_j(\mathbf{r'}), \tag{4.19}$$

and the KS matrix and two electron integrals were transformed to the canonical KS orbital basis using the orbital coefficients, $\mathbf{C}^0$ (Eq. 4.20).

$$\begin{aligned} (ia|jb) &= \sum_{\mu\nu\lambda\sigma} C^0_{\mu i} C^0_{\nu a} C^0_{\lambda j} C^0_{\sigma b}(\mu\nu|\lambda\sigma) \\ F_{ii} &= \sum_{\mu\nu} C^{0*}_{\mu i} F_{\mu\nu} C^0_{\nu i} \end{aligned} \tag{4.20}$$

The ground to excited state transition moments for one-body operators can be evaluated in the LR formalism according to Eq. (4.21) [70, 72, 73]. For the coupling between ground and excited states induced by an external point charge, the matrix elements of interest are those of $\widehat{V}$ from Eq. (4.1), and transition rates between the ground and excited states can the be approached from the vantage of the (state-to-state) Fermi's golden rule expression [74] given in Eq. (4.22). The functional dependence of the transition potentials between electronic energy eigenstates on $\mathbf{r}_{ext}$ can be explored by reforming $\mathbf{V}(\mathbf{r}_{ext})$ for different $\mathbf{r}_{ext}$ and computing its trace with the transition densities for a given excited state according to Eq. (4.21).

$$\langle\Psi_0|\widehat{V}(\mathbf{r}_{ext})|\Psi_I\rangle = \sum_{ia} V_{ia}(\mathbf{r}_{ext})X^I_{ia} + V_{ai}(\mathbf{r}_{ext})Y^I_{ia} \tag{4.21}$$

$$w_{0I}(\mathbf{r}_{ext}) = 2\pi\left|\langle\Psi_0|\widehat{V}(\mathbf{r}_{ext})|\Psi_I\rangle\right|^2 \tag{4.22}$$

Eq. (4.18) can be solved for different excited states across a range of nuclear coordinates in order to resolve the excited state potential energy surfaces. With access to the excited state 1RDM matrices, the analytic energy gradients of excited states can be evaluated, and critical points (equilibrium geometries for reactants and products, as well as any transition states and stable intermediates encountered along the reaction coordinate) can be obtained using the same geometry optimization methods as for finding critical points on the ground state potential energy surface [75]. Once these structures have been identified, minimum energy reaction pathways between them can be optimized by application of the nudged elastic band approach [76]. This combination of geometry and reaction coordinate optimization represents a powerful workflow for mapping excited state reaction pathways that has been widely employed in the field of computational photochemistry [77]. The same approach can be adopted to understand excited state reactions that can be induced through convergent electron/ion beam exposure.

3 Simulations

3.1 Internal consistency of the LR- and RT-TD-DFT treatments of beam-induced excitations

In the limit where the perturbation is sufficiently weak to preclude any nonlinear response, excited state populations extracted from RT-TD-DFT for a given impulsive perturbation should mirror the transition rates/probabilities calculated from LR-TD-DFT for the same perturbation. This correspondence has been previously demonstrated for excitations induced by weak homogenous electric field perturbations [78], but since the strength of the perturbation experienced by a material due to interaction with an external point charge is strongly dependent on the position of the perturbing external charge, it is not clear that the same correspondence will also be borne out for point-source electric field type perturbations applied at a given position.

To gauge the degree of agreement between these two methods, we have tested the position dependence of the electronic response of a simple atomic toy system (carbide ion, C^{4-}) to external charge ($q_{ext} = 1$) impulse perturbations. Across the entire range of positions spanning distance between 0 and 5 angstroms from the system's center of charge, the Fermi's golden rule transition rates and the excited state populations extracted from the RT-TD-DFT simulations show nearly perfect linear proportionality (Fig. 4.2). For modeling the material's response to more highly charged species (e.g., those associated with beams of bare nuclei) RT-TD-DFT is more appropriate than LR-TD-DFT due to its ability to capture potential nonlinear response of the density to the stronger perturbation.

3.2 Excited state structural transformations of point defects in graphene nanomaterials

In this section, we summarize a study in which the methods outlined earlier in this chapter were employed to investigate the effect of electronic excitation on the structural dynamics of graphene quantum dots (GQDs) of different edge morphologies that were substitutionally

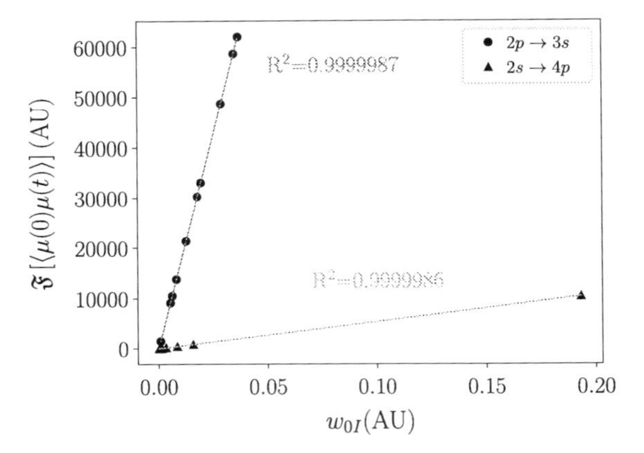

FIG. 4.2 Intensities of frequency domain dipole autocorrelation function (Eq. (4.17), y axis) and Fermi's golden rule transition rates (Eq. (4.22), x axis) for two transitions of the carbide ion, evaluated for a range of point charge (q_{ext}=1) positions. Lines of best fit exhibit near-unity correlation coefficients, indicating the internal consistency of the two approaches through proportionality of the LR-TD-DFT transition rates and RT-TD-DFT excited state populations.

doped with phosphorous or silicon [79]. Both P and Si can directly replace the threefold coordinated carbons in hexagonal graphene lattice to form the Si-C_3 and P-C_3 defects. However, due to dissimilar sizes of carbon and the period three dopants, the equilibrium structures of the Si-C_3 and P-C_3 defects exhibit a pyramidal, out-of-plane distortion at the defect site that has been directly confirmed for the Si-C_3 defect through electron tomography [27], and is supported by atomically resolved electron energy loss spectra corroborated by first principles modeling of the nonplanar P-C_3 defect [80–82].

All calculations were carried out using a locally modified version of the NWChem quantum chemistry package [83, 84]. For this study, the real-time TD-DFT module implemented in NWchem was utilized to propagate the 1RDM [55, 61, 62, 85]. Bader charge analysis [86, 87], based on the theory of atoms in molecules [88], was used to partition the molecular species into atomic volumes in order to characterize the displacement of electron density in the excited states relative to the ground state.

Inversion of these pyramidal defects is dictated by symmetry to proceed through a planar transition state. The lowest energy reaction profiles plotted in Fig. 4.3 correspond to the ground state PES along the optimized minimum energy reaction coordinate for defect inversion. Note that only half of the symmetric reaction profile for inversion is plotted here for brevity. The barrier for P defect inversion in the ground state (1.82 eV for zigzag, 1.21 eV for armchair GQD) is significantly larger than that for the Si defect (0.35 eV for zigzag, 0.13 eV for armchair GQD). The defects' exhibit similar sizes (tabulated covalent radii of singly bonded Si and P are 1.16 and 1.11 Å) [89], so one would not anticipate significant differences in steric hindrance along the inversion coordinate for Si and P. Instead, nontrivial differences in the electronic structure of the P- and Si-doped GQDs are to blame for the large disparity in potential barriers for their inversion.

Analysis of the LR-TD-DFT transition density reveals that the lowest-lying excited electronic states of each of the doped GQDs exhibit significant transition density contributed by defect-localized orbitals. Their energies are plotted along the inversion coordinate in Fig. 4.3. The first excited states of the doped GQDs are energetically well-separated from the next lowest excited states, and exhibit decrease in the potential barrier for inversion along

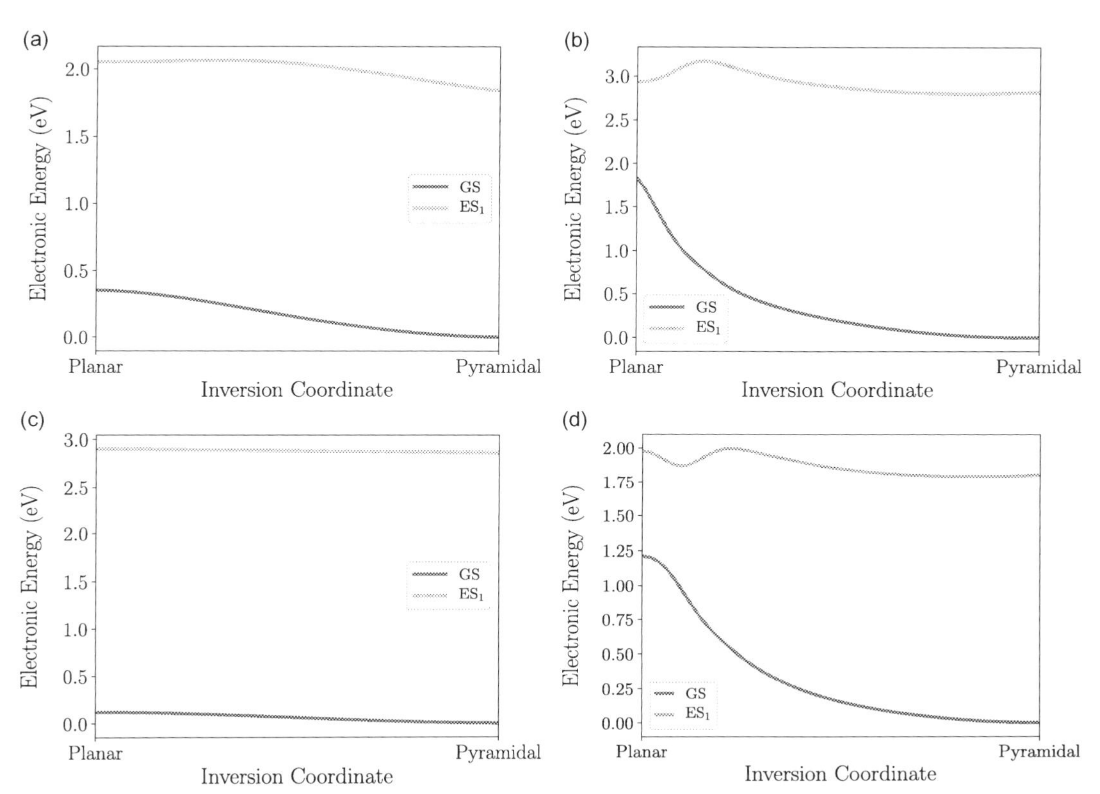

FIG. 4.3 Potential energy surfaces for the ground and first excited states of the Si (A) and P (B) doped zigzag GQDs, and Si (C) and P (D) armchair GQDs along the optimized ground state inversion coordinate calculated at the B3LYP/6-311g(d) level of theory. *Data from figure 4 of D.B. Lingerfelt, P. Ganesh, J. Jakowski, B.G. Sumpter, Electronically nonadiabatic structural transformations promoted by electron beams, Adv. Funct. Mater. (2019) 1901901, replotted here*

TABLE 4.1 Oscillator strengths (dimensionless) for photoexcitation to the low-lying excited states of GQDs with Si-C_3 and P-C_3 defects.

	Armchair GQD		Zigzag GQD	
	Si	**P**	**Si**	**P**
ES_1	0.03529	0.00065	0.00109	0.00017

the minimum energy inversion coordinate calculated in the electronic ground state. This indicates that the cross sections for inversion induced by electron beam impact may be enhanced through inelastic electron scattering, or concomitant application of electromagnetic radiation resonant with the excited state energy gap. Oscillator strengths for photoexcitation calculated in the electric dipole (length gauge) approximation are tabulated in Table 4.1. Each of the transitions are at least weakly allowed, suggesting that photoexcitation may in some cases ease the manipulation of defects in graphene with electron beams.

A visual analysis of the density differences between ground and excited states suggests that the first excited states of the Si- and P-doped species usher electron density away from the defect and its three bonding partners, and disperse it around the remainder of the molecule. The lowest-lying excited states of the P-doped GQDs exhibit a large decrease in the inversion barrier, and also show large contributions to their transition density from excitations which deplete the population of the HOMO, which is carbon–carbon bonding in character in the vicinity of the defect and effectively rigidifies the P-C_3 defect and prevents the facile inversion observed for the Si-C_3 defect in the ground state [79].

The net amount of electron density depleted at the defect site upon excitation (including bonding electron density contributed by the dopant in terms of the of the theory of atoms in molecules) can be gauged through Bader's state-specific charge partitioning scheme. Bader charges for the defect atom were determined using ground and excited state charge densities for the lowest-lying excited states at the (ground state) transition state geometries of the doped GQDs (Table 4.2) and show correlation between the extent of charge transferred away from the defect site and the lowering of the electronic barrier for the inversion process.

Fermi's golden rule transition rates from the ground to excited states induced by an external charged particle were evaluated on a grid to map out the spatial dependence of the point charge-induced population rates of the states exhibiting lower barriers for inversion. The dependence of these rates on the position of the point charge was evaluated at the ground state equilibrium geometries of the GQDs, and is visualized as isosurfaces for the low-lying excited states associated with lowered barriers for inversion in Figs. 4.4 and 4.5. Not only are the states associated with decreased inversion barriers relative to the ground state, they are found to exhibit significant transition rates when the point charge is placed near the dopant atom itself. This suggests that a convergent electron beam may simultaneously be capable of enacting a transition to an electronic state that exhibits a lower barrier for a particular structural transformation and transferring momentum to the material's nuclei to drive the reaction over the diminished barrier. It is crucial in this case that changes in the PES along the inversion coordinate upon electronic excitation are considered when investigating the cross sections for electron beam-induced dynamics of defects in materials.

While a large potential barrier for inversion of the pyramidal P-C_3 defect forbids this process from being promoted thermally, the barrier for Si-C_3 inversion is nearly on the order of k_BT at room temperature in the ground state. However, P-C_3 inversion is made significantly more favorable in low-lying defect centered excited states. The amount that the inversion barrier is reduced in the excited state correlates with the charge density depleted in the region of the defect and its bonds to neighboring atoms. For P-C_3, the low-lying states with reduced inversion barriers deplete the population of an orbital that is responsible for rigidifying

TABLE 4.2 Differences in electron density (relative to the ground state) integrated over the atomic volumes of the P and Si defects as determined by the Bader (atoms-in-molecules) partitioning scheme applied to the ground and lowest energy excited states at the transition state geometry.

	Armchair GQD		Zigzag GQD	
	Si	**P**	**Si**	**P**
ES_1	−0.070578	−0.017319	−0.096267	−0.219782

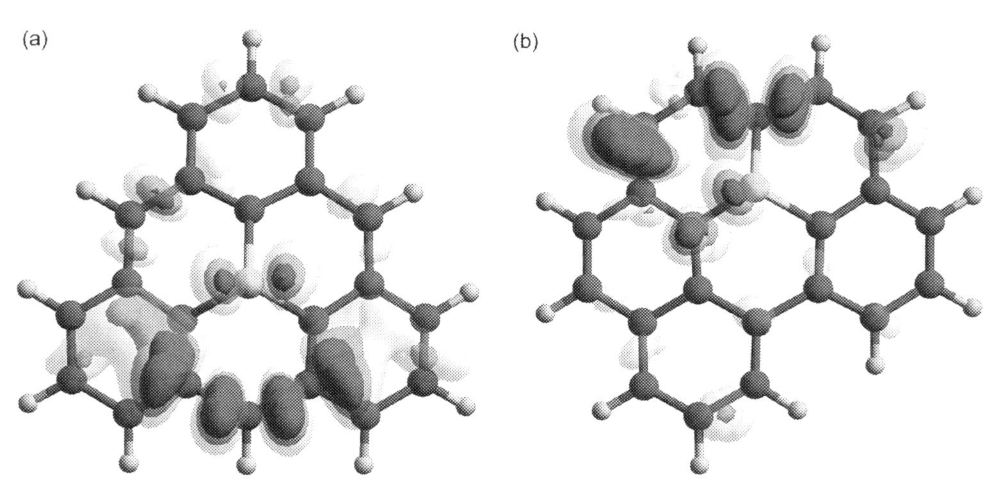

FIG. 4.4 Overlaid isosurfaces of the ground to first excited state transition rates (with isovalues corresponding to lightest/middle/darkest isosurface listed below in atomic units) for the first excited state of phosphorous (indicated by *green sphere*, gray in print version) doped zigzag (A) and armchair (B) GQDs evaluated at their ground state equilibrium geometry. *Data from figure 7 of D.B. Lingerfelt, P. Ganesh, J. Jakowski, B.G. Sumpter, Electronically nonadiabatic structural transformations promoted by electron beams, Adv. Funct. Mater. (2019) 1901901, replotted here.*

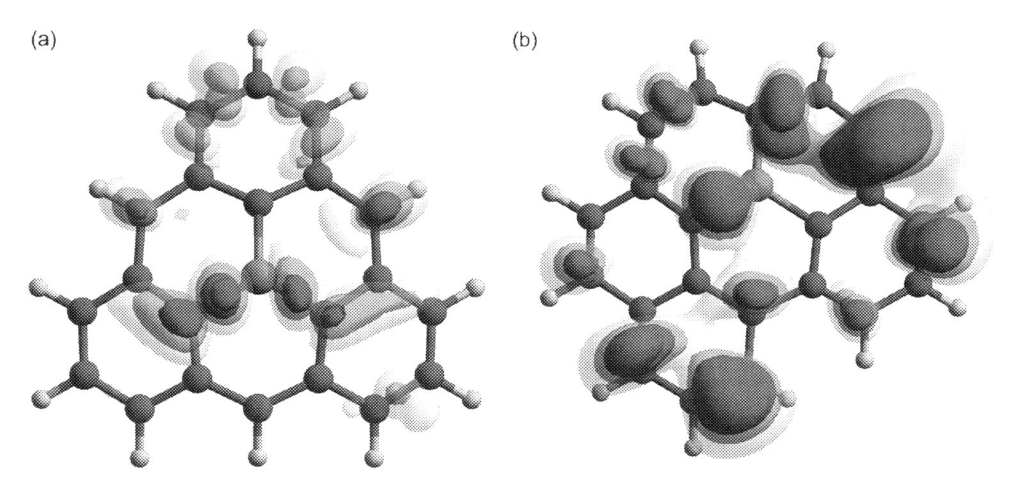

FIG. 4.5 Overlaid isosurfaces of the ground to first excited state transition rates (with isovalues corresponding to lightest/middle/darkest isosurface listed below in atomic units) for the first excited state of silicon (indicated by *teal sphere*, gray in print version) doped zigzag (A) and armchair (B) GQDs evaluated at their ground state equilibrium geometry. *Data from figure 8 of D.B. Lingerfelt, P. Ganesh, J. Jakowski, B.G. Sumpter, Electronically nonadiabatic structural transformations promoted by electron beams, Adv. Funct. Mater. (2019) 1901901, replotted here.*

the P-C_3 defect in the ground state, leading to stabilization of the planar transition state for defect inversion in the excited states. The defect-localized states associated with diminished barriers for inversion can be populated directly by the application of a point charge perturbation at certain positions near the defect, suggesting a nontrivial role for electronic excitations in facilitating beam-induced defect chemistries.

3.3 Future directions for simulating materials under electron/ion irradiation

First principles modeling of the nonequilibrium electronic response of nano- to meso-scaled materials to perturbation by localized charges has been facilitated as described in this chapter through the use of TD-DFT. However, the electronic structure of certain important graphene nanomaterials (zigzag-edged graphene nanoribbons for instance) is characterized by strong static correlation, for which the single Slater determinant description of electronic states invoked in DFT fails categorically. The time evolution of these strongly correlated systems can be tractably approached using Hubbard model Hamiltonians with parameters calculated from first principles [90], but the formalisms outlined in this chapter can also be applied in time-dependent multireference electronic structure methods such as multiconfigurational self-consistent field theory.

While excited state potential energy surface mapping as described in this chapter provides information about the lowest energy pathway for a given reaction, often times an excess of energy is deposited in the material through the beam interaction, in which case other higher energy pathways can become equally important to consider. In order to probe the mechanisms of structural reorganization induced in materials by finite energy electron beams, one must explicitly model the electronic and vibrational dynamics, coupled through the electron–nuclear attraction, such that high-energy vibrations excited thorough knock-on processes can lead to electronic excitation and relaxation from excited electronic states can precipitate large-amplitude vibrations and isomerization. Mixed quantum-classical simulation techniques originally introduced to treat the coupled electronic and vibrational evolution of materials subject to optical excitation are currently being adopted as a means to this end [91, 92]. Mixed quantum-classical approaches can be categorized based on their treatment of dynamical correlations between electronic and ionic motions. Simulation methods which permit bifurcations of the classical trajectories in regions of strong nonadiabatic coupling (e.g., trajectory surface hopping and multiple spawning type approaches) systematically account for these correlations, and those which produce a single trajectory evaluated using a population-weighted average potential energy surface (Ehrenfest dynamics) cannot [93]. It has been shown that accounting for this dynamical correlation is necessary to satisfy the principle of detailed balance, and therefore to approach asymptotically correct probability distributions from longer-time simulations [94]. Thus, in order to accurately predict the outcome of a given electron/ion beam perturbation, mean-field approaches are likely to be insufficient.

The description of the perturbation experienced by the system from the passage of a swift electron can also be improved so as to be more widely applicable. While the approximate ultra-relativistic limit for the interaction potential put forward in this chapter is a useful first development in this area for two-dimensional systems, representing the scalar potential of the beam by an electrostatic (Coulomb) potential impulse is not appropriate for three-dimensional materials even in the ultra-relativistic limit. Multiple scattering and channeling effects—i.e., the diversion of scattered electrons through high-probability pathways through the material—will become important in this scenario. Beyond the deficiency of the impulsive treatment of the interaction potential, the Lorentz-invariant retarded potential would ideally be employed in order to capture the anisotropy of the electric field that imbues the electronic transition probabilities with beam energy dependence. The vector potential could also be included in the electronic Hamiltonian through the generalized momentum operator to more accurately describe interactions with higher kinetic energy electrons in materials occupying core and deeper valence levels.

4 Summary

In this chapter, we have presented two complimentary methods to solve for the electronic response of materials to the point-source electric fields emanated by the charged particles comprising convergent beams employed in scanning electron/ion microscopes. The instantaneous excitation rates due to the presence of an external point charge were evaluated within the framework of time-dependent density functional theory in the frequency domain through application of linear response theory. Likewise, excited state populations resulting from application of a point-source electric field impulse were accessed directly in the time domain through numerical integration of the electronic equations of motion. The internal consistency of these two approaches was demonstrated through the proportionality observed between the response theory transition rates and the excited state populations resulting from identical perturbations. This approach was applied—in combination with excited state reaction pathway optimization—in a study that highlights the importance of considering the effects of electronic excitations induced by electron beams on the kinetics of reactions involving point defects in graphene nanostructures. Finally, we outlined some future directions for development that would enhance both the accuracy and general applicability of the methods put forth in this chapter.

Acknowledgments

This work was performed at the Center for Nanophase Materials Sciences, a U.S. Department of Energy Office of Science User Facility, and used resources of the Compute and Data Environment for Science (CADES) and of the Oak Ridge Leadership Computing Facility (OLCF) at the Oak Ridge National Laboratory, which is supported by the Office of Science of the U.S. Department of Energy under Contract No. DE-AC05-00OR22725 and of Extreme Science and Engineering Discovery Environment (XSEDE), which is supported by National Science Foundation grant no. ACI-1548562 (allocation TG-DMR110037).

References

[1] C. Lee, X. Wei, J.W. Kysar, J. Hone, Measurement of the elastic properties and intrinsic strength of monolayer graphene, Science 321 (5887) (2008) 385–388.
[2] S.V. Morozov, K.S. Novoselov, M.I. Katsnelson, F. Schedin, D.C. Elias, J.A. Jaszczak, A.K. Geim, Giant intrinsic carrier mobilities in graphene and its bilayer, Phys. Rev. Lett. 100 (1) (2008) 016602.
[3] Y. Cao, V. Fatemi, S. Fang, K. Watanabe, T. Taniguchi, E. Kaxiras, P. Jarillo-Herrero, Unconventional superconductivity in magic-angle graphene superlattices, Nature 556 (7699) (2018) 43.
[4] N. Tombros, A. Veligura, J. Junesch, M.H.D. Guimar aes, I.J. Vera-Marun, H.T. Jonkman, B.J. Van Wees, Quantized conductance of a suspended graphene nanoconstriction, Nat. Phys. 7 (9) (2011) 697.
[5] K.S. Novoselov, A.K. Geim, S.V. Morozov, D. Jiang, M.I. Katsnelson, I.V. Grigorieva, S.V. Dubonos, A.A. Firsov, Two-dimensional gas of massless Dirac fermions in graphene, Nature 438 (7065) (2005) 197.
[6] K.S. Novoselov, A.K. Geim, S.V. Morozov, D. Jiang, Y. Zhang, S.V. Dubonos, I.V. Grigorieva, A.A. Firsov, Electric field effect in atomically thin carbon films, Science 306 (5696) (2004) 666–669.
[7] K. Kim, J.-Y. Choi, T. Kim, S.-H. Cho, H.-J. Chung, A role for graphene in silicon-based semiconductor devices, Nature 479 (7373) (2011) 338–344.
[8] A.K. Geim, K.S. Novoselov, The rise of graphene, in: Nanoscience and Technology: A Collection of Reviews From Nature Journals, World Scientific, 2010, pp. 11–19.

[9] J.M. Shalf, R. Leland, Computing beyond moore's law, Computer 48 (12) (2015) 14–23.
[10] S.Y. Zhou, G.-H. Gweon, A.V. Fedorov, P.N. First, W.A. De Heer, D.-H. Lee, F. Guinea, A.H.C. Neto, A. Lanzara, Substrate-induced bandgap opening in epitaxial graphene, Nat. Mater. 6 (10) (2007) 770–775.
[11] V. Barone, O. Hod, G.E. Scuseria, Electronic structure and stability of semiconducting graphene nanoribbons, Nano Lett. 6 (12) (2006) 2748–2754.
[12] J. Zhang, J. Gao, L. Liu, J. Zhao, Electronic and transport gaps of graphene opened by grain boundaries, J. Appl. Phys. 112 (5) (2012) 053713.
[13] X. Fan, Z. Shen, A.Q. Liu, J.-L. Kuo, Band gap opening of graphene by doping small boron nitride domains, Nanoscale 4 (6) (2012) 2157–2165.
[14] M. Shahrokhi, C. Leonard, Tuning the band gap and optical spectra of silicon-doped graphene: many-body effects and excitonic states, J. Alloys Compd. 693 (2017) 1185–1196.
[15] R.B. Laughlin, Quantized Hall conductivity in two dimensions, Phys. Rev. B 23 (10) (1981) 5632.
[16] I.A. Calafell, J.D. Cox, M. Radonjić, J.R.M. Saavedra, F.J.G de Abajo, L.A. Rozema, P. Walther, Quantum computing with graphene plasmons, Npj Quantum Inf. 5 (1) (2019) 1–7.
[17] I. Sodemann, L. Fu, Quantum nonlinear Hall effect induced by Berry curvature dipole in time-reversal invariant materials, Phys. Rev. Lett. 115 (21) (2015) 216806.
[18] J.L. Lado, N. García-Martínez, J. Fernández-Rossier, Edge states in graphene-like systems, Synth. Met. 210 (2015) 56–67.
[19] S.M. Avdoshenko, I.N. Ioffe, G. Cuniberti, L. Dunsch, A.A. Popov, Organometallic complexes of graphene: toward atomic spintronics using a graphene web, ACS Nano 5 (12) (2011) 9939–9949.
[20] C. Yesilyurt, S. Ghee Tan, G. Liang, M.B.A. Jalil, Perfect valley filter in strained graphene with single barrier region, AIP Adv. 6 (5) (2016) 056303.
[21] N.J. Turro, V. Ramamurthy, V. Ramamurthy, J.C. Scaiano, Principles of Molecular Photochemistry: An Introduction, University Science Books, 2009.
[22] B.G. Levine, M.P. Esch, B.S. Fales, D.T. Hardwick, W.-T. Peng, Y. Shu, Conical intersections at the nanoscale: molecular ideas for materials, Annu. Rev. Phys. Chem. 70 (2019) 21–43.
[23] R.F. Egerton, Electron Energy-Loss Spectroscopy in the Electron Microscope, Springer US, 2011, ISBN: 9781441995834.
[24] T. Susi, J. Kotakoski, D. Kepaptsoglou, C. Mangler, T.C. Lovejoy, O.L. Krivanek, R. Zan, U. Bangert, P. Ayala, J.C. Meyer, Q. Ramasse, Silicon-carbon bond inversions driven by 60-keV electrons in graphene, Phys. Rev. Lett. 113 (11) (2014) 115501.
[25] O. Dyck, S. Kim, S.V. Kalinin, S. Jesse, Placing single atoms in graphene with a scanning transmission electron microscope, Appl. Phys. Lett. 111 (11) (2017) 113104.
[26] O. Dyck, C. Zhang, P.D. Rack, J.D. Fowlkes, B. Sumpter, A.R. Lupini, S.V. Kalinin, S. Jesse, Electron-beam introduction of heteroatomic Pt-Si structures in graphene, Carbon 161 (2020) 750–757.
[27] C. Hofer, V. Skakalova, M.R.A. Monazam, C. Mangler, J. Kotakoski, T. Susi, J.C. Meyer, Direct visualization of the 3D structure of silicon impurities in graphene, Appl. Phys. Lett. 114 (5) (2019) 053102.
[28] H.-P. Komsa, J. Kotakoski, S. Kurasch, O. Lehtinen, U. Kaiser, A.V. Krasheninnikov, Two-dimensional transition metal dichalcogenides under electron irradiation: defect production and doping, Phys. Rev. Lett. 109 (3) (2012) 035503, https://doi.org/10.1103/PhysRevLett.109.035503.
[29] J. Kotakoski, C.H. Jin, O. Lehtinen, K. Suenaga, A.V. Krasheninnikov, Electron Knock-on damage in hexagonal boron nitride monolayers, Phys. Rev. B 82 (11) (2010) 113404, https://doi.org/10.1103/PhysRevB.82.113404.
[30] J. Kotakoski, D. Santos-Cottin, A.V. Krasheninnikov, Stability of graphene edges under electron beam: equilibrium energetics versus dynamic effects, ACS Nano 6 (1) (2011) 671–676.
[31] T. Susi, J. Kotakoski, R. Arenal, S. Kurasch, H. Jiang, V. Skakalova, O. Stephan, A.V. Krasheninnikov, E.I. Kauppinen, U. Kaiser, J.C. Meyer, Atomistic description of electron beam damage in nitrogen-doped graphene and single-walled carbon nanotubes, ACS Nano 6 (10) (2012) 8837–8846, https://doi.org/10.1021/nn303944f.
[32] R.C. Ehemann, J. Dadras, P.R.C. Kent, J. Jakowski, P.S. Krstic, Detection of hydrogen using graphene, Nano Res. Lett. 7 (2012) 198, https://doi.org/10.1186/1556-276X-7-198.
[33] S. Kim, A.V. Ievlev, J. Jakowski, I.V. Vlassiouk, X. Sang, C. Brown, O. Dyck, R.R. Unocic, S.V. Kalinin, A. Belianinov, B.G. Sumpter, S. Jesse, O.S. Ovchinnikova, Multi-purposed Ar gas cluster ion beam processing for graphene engineering, Carbon 131 (2018) 142–148.

[34] S. Garashchuk, J. Jakowski, L. Wang, B.G. Sumpter, Quantum trajectory-electronic structure approach for exploring nuclear effects in the dynamics of nanomaterials, J. Chem. Theory Comput. 9 (2013) 5221–5235, https://doi.org/10.1021/ct4006147.
[35] L. Wang, J. Jakowski, S. Garashchuk, Adsorption of a hydrogen atom on a graphene flake examined with quantum trajectory/electronic structure dynamics, J. Phys. Chem. C 118 (2014) 16175–16187, https://doi.org/10.1021/jp503261k.
[36] J.D. Jackson, Classical Electrodynamics, Wiley, 1975, ISBN: 9780471431329.
[37] R.P. Feynman, R.B. Leighton, M.L. Sands, The Feynman Lectures on Physics: Mainly Electromagnetism and Matter, Addison-Wesley Publishing Company, 1965, ISBN: 9780201021172.
[38] A. Stone, The Theory of Intermolecular Forces, Second ed., Oxford University Press, Oxford, 2013.
[39] R.N. Zare, Angular Momentum: Understanding Spatial Aspects in Chemistry and Physics, Second ed., Wiley-Interscience, New York, 1991.
[40] G.B. Arfken, H.J. Weber, Mathematical Methods for Physicists, Fourth ed., Academic Press, San Diego, CA, 1995.
[41] J. Applequist, A multipole interaction theory of electric polarization of atomic and molecular assemblies, J. Chem. Phys. 83 (18) (1985) 809–826.
[42] L. Jansen, Tensor formalism for coulomb interactions and asymptotic properties of multipole expansions, Phys. Rev. 110 (3) (1958) 661–669.
[43] L.D. Barron, Molecular Light Scattering and Optical Activity, Cambridge University Press, 2009, ISBN: 9781139453417.
[44] F.A. Cotton, Chemical Applications of Group Theory, John Wiley & Sons, 2003.
[45] D.B. Lingerfelt, P. Ganesh, J. Jakowski, B.G. Sumpter, Understanding beam-induced electronic excitations in materials, J. Chem. Theory Comput. 16 (2) (2020) 1200–1214, https://doi.org/10.1021/acs.jctc.9b00792.
[46] I.V. Hertel, K.J. Ross, Octopole-allowed transitions in the electron energy-loss spectra of potassium and rubidium, J. Chem. Phys. 50 (1) (1969) 536–537, https://doi.org/10.1063/1.1670833.
[47] E.N. Lassettre, A. Skerbele, V.D. Meyer, Quadrupole-allowed transitions in the electron-impact Spectrum of N_2, J. Chem. Phys. 45 (9) (1966) 3214–3226, https://doi.org/10.1063/1.1728096.
[48] H. Bethe, Zur Theorie des Durchgangs schneller Korpuskularstrahlen durch Materie, Ann. Phys. (Berl.) 397 (3) (1930) 325–400.
[49] M. Inokuti, B. Bederson, Bethe's contributions to atomic and molecular physics, Phys. Scr. 73 (2) (2006) C98.
[50] M. Inokuti, Inelastic collisions of fast charged particles with atoms and molecules—the Bethe theory revisited, Rev. Mod. Phys. 43 (3) (1971) 297.
[51] R.D. Leapman, P. Rez, D.F. Mayers, K, L, and M Shell Generalized Oscillator Strengths and Ionization Cross Sections for Fast Electron Collisions, J. Chem. Phys. 72 (2) (1980) 1232–1243.
[52] N.J. Carron, An Introduction to the Passage of Energetic Particles Through Matter, CRC Press, 2006.
[53] M.A.L. Marques, E.K.U. Gross, Time-dependent density functional theory, Annu. Rev. Phys. Chem. 55 (1) (2004) 427–455, https://doi.org/10.1146/annurev.physchem.55.091602.094449.
[54] A. Dreuw, M. Head-Gordon, Single-reference ab initio methods for the calculation of excited states of large molecules, Chem. Rev. 105 (11) (2005) 4009–4037.
[55] K. Lopata, N. Govind, Modeling fast electron dynamics with real-time time-dependent density functional theory: application to small molecules and chromophores, J. Chem. Theory Comput. 7 (5) (2011) 1344–1355.
[56] F. Furche, R. Ahlrichs, Adiabatic time-dependent density functional methods for excited state properties, J. Chem. Phys. 117 (16) (2002) 7433–7447.
[57] P. Hohenberg, W. Kohn, Inhomogeneous electron gas, Phys. Rev. 136 (3B) (1964) B864.
[58] W. Kohn, L.J. Sham, Self-consistent equations including exchange and correlation effects, Phys. Rev. 140 (4A) (1965) A1133.
[59] P.A.M. Dirac, Note on exchange phenomena in the thomas atom, in: Mathematical Proceedings of the Cambridge Philosophical Society, 26, Cambridge University Press, 1930, pp. 376–385. vol.
[60] E. Runge, E.K.U. Gross, Density-functional theory for time-dependent systems, Phys. Rev. Lett. 52 (12) (1984) 997–1000, https://doi.org/10.1103/PhysRevLett.52.997.
[61] J. Jakowski, K. Morokuma, Liouville-von Neumann molecular dynamics, J. Chem. Phys. 130 (22) (2009) 224106.
[62] S. Blanes, F. Casas, J.A. Oteo, J. Ros, The Magnus expansion and some of its applications, Phys. Rep. 470 (5-6) (2009) 151–238.

[63] A. Schleife, E.W. Draeger, Y. Kanai, A.A. Correa, Plane-wave pseudopotential implementation of explicit integrators for time-dependent Kohn-Sham equations in large-scale simulations, J. Chem. Phys. 137 (22) (2012) 22A546.
[64] A. Castro, M.A.L. Marques, A. Rubio, Propagators for the time-dependent Kohn-Sham equations, J. Chem. Phys. 121 (8) (2004) 3425–3433.
[65] D.W. Oxtoby, Vibrational population relaxation in liquids, Adv. Chem. Phys. (1981) 487–519.
[66] S.A. Egorov, K.F. Everitt, J.L. Skinner, Quantum dynamics and vibrational relaxation, J. Phys. Chem. A 103 (47) (1999) 9494–9499.
[67] B.J. Berne, G.D. Harp, On the calculation of time correlation functions, Adv. Chem. Phys. 17 (1970) 63–227.
[68] P. Jørgensen, J. Simons, Second Quantization-Based Methods in Quantum Chemistry, Second ed., Academic Press, New York, 1981.
[69] J. Olsen, H.J.A. Jensen, P. Jørgensen, Solution of the large matrix equations which occur in response theory, J. Comput. Phys. 74 (2) (1988) 265–282.
[70] P. Ring, P. Schuck, The Nuclear Many-Body Problem, Springer, 2004, ISBN: 9783540212065.
[71] M.E. Casida, Time-dependent density functional response theory for molecules, in: Recent Advances In Density Functional Methods: (Part I), World Scientific, 1995, pp. 155–192.
[72] A.E. Hansen, T.D. Bouman, Hypervirial relations as constraints in calculations of electronic excitation properties: the random phase approximation in configuration interaction language, Mol. Phys. 37 (6) (1979) 1713–1724.
[73] J. Linderberg, Y. Öhrn, Propagators in Quantum Chemistry, Wiley, 2004, ISBN: 9780471662570.
[74] G.C. Schatz, M.A. Ratner, Quantum Mechanics in Chemistry, Dover Publications, 2002, ISBN: 9780486420035.
[75] H.B. Schlegel, Geometry optimization, Wiley Interdiscip. Rev. Comput. Mol. Sci. 1 (5) (2011) 790–809.
[76] H. Jónsson, G. Mills, K.W. Jacobsen, Nudged elastic band method for finding minimum energy paths of transitions, in: G.C. Bruce J Berne, D.F. Coker (Eds.), Classical and Quantum Dynamics in Condensed Phase Simulations, World Scientific, 1998, https://doi.org/10.1142/9789812839664_0016.
[77] M.A. Robb, M. Garavelli, M. Olivucci, F. Bernardi, A computational strategy for organic photochemistry, Rev. Comput. Chem. 15 (2000) 87–146.
[78] K. Lopata, N. Govind, Modeling fast electron dynamics with real-time time-dependent density functional theory: application to small molecules and chromophores, J. Chem. Theory Comput. 7 (5) (2011) 1344–1355.
[79] D.B. Lingerfelt, P. Ganesh, J. Jakowski, B.G. Sumpter, Electronically nonadiabatic structural transformations promoted by electron beams, Adv. Funct. Mater. (2019) 1901901.
[80] H.-m. Wang, H.-x. Wang, Y. Chen, Y.-j. Liu, J.-x. Zhao, Q.-h. Cai, X.-z. Wang, Phosphorus-doped graphene and (8, 0) carbon nanotube: structural, electronic, magnetic properties, and chemical reactivity, Appl. Surf. Sci. 273 (2013) 302–309.
[81] E. Cruz-Silva, F. Lopez-Urias, E. Munoz-Sandoval, B.G. Sumpter, H. Terrones, J.-C. Charlier, V. Meunier, M. Terrones, Electronic transport and mechanical properties of phosphorus- and phosphorus-nitrogen-doped carbon nanotubes, ACS Nano 3 (7) (2009) 1913–1921.
[82] T. Susi, T.P. Hardcastle, H. Hofsäss, A. Mittelberger, T.J. Pennycook, C. Mangler, R. Drummond-Brydson, A.J. Scott, J.C. Meyer, J. Kotakoski, Single-atom spectroscopy of phosphorus dopants implanted into graphene, 2D Mater. 4 (2) (2017) 021013.
[83] D.W. Silverstein, N. Govind, H.J.J. van Dam, L. Jensen, Simulating one-photon absorption and resonance raman scattering spectra using analytical excited state energy gradients within time-dependent density functional theory, J. Chem. Theory Comput. 9 (12) (2013) 5490–5503.
[84] M. Valiev, E.J. Bylaska, N. Govind, K. Kowalski, T.P. Straatsma, H.J.J.V. Dam, D. Wang, J. Nieplocha, E. Apra, T.-L. Windus, W.A de Jong, NWChem: A comprehensive and scalable open-source solution for large scale molecular simulations, Comput. Phys. Commun. 181 (9) (2010) 1477–1489.
[85] W. Magnus, On the exponential solution of differential equations for a linear operator, Commun. Pure Appl. Math. 7 (4) (1954) 649–673.
[86] G. Henkelman, A. Arnaldsson, H. Jónsson, A fast and robust algorithm for Bader decomposition of charge density, Comput. Mater. Sci. 36 (3) (2006) 354–360.
[87] W. Tang, E. Sanville, G. Henkelman, A grid-based Bader analysis algorithm without lattice bias, J. Phys. Condens. Matter 21 (8) (2009) 084204.
[88] R.F.W. Bader, Atoms in molecules, Acc. Chem. Res. 18 (1) (1985) 9–15.

[89] P. Pyykkö, M. Atsumi, Molecular single-bond covalent radii for elements 1-118, Chem. Eur. J 15 (1) (2009) 186–197.

[90] N. Schlünzen, S. Hermanns, M. Bonitz, C. Verdozzi, Dynamics of strongly correlated fermions: ab initio results for two and three dimensions, Phys. Rev. B 93 (3) (2016) 035107.

[91] S. Kretschmer, T. Lehnert, U. Kaiser, A.V. Krasheninnikov, Formation of defects in two-dimensional MoS2 in the transmission electron microscope at electron energies below the knock-on threshold: the role of electronic excitations, Nano Lett. 20 (4) (2020) 2865–2870.

[92] A.V. Krasheninnikov, Y. Miyamoto, D. Tománek, Role of electronic excitations in ion collisions with carbon nanostructures, Phys. Rev. Lett. 99 (1) (2007) 016104.

[93] R. Crespo-Otero, M. Barbatti, Recent advances and perspectives on nonadiabatic mixed quantum-classical dynamics, Chem. Rev. 118 (15) (2018) 7026–7068.

[94] P.V. Parandekar, J.C. Tully, Mixed quantum-classical equilibrium, J. Chem. Phys. 122 (9) (2005) 094102.

5

Molecule-graphene and molecule-carbon surface binding energies from molecular mechanics

Thomas R. Rybolt, Jae H. Son, Ronald S. Holt, and Connor W. Frye

Department of Chemistry and Physics, University of Tennessee at Chattanooga, Chattanooga, TN, United States

1 Introduction

There are a variety of interesting and useful applications that are dependent upon the attraction of molecules to surfaces. Some of these adsorption applications include odor removal, environmental remediation in air or water, environmental monitoring, drug storage and release, energetic gas adsorption, separation technologies, gas–solid chromatography, and sensors for chemical or biochemical monitoring. Many of these applications involve forming noncovalent attractions between a molecule or molecules and a surface and are thus associated with reversible physical adsorption. This type of molecule-surface attraction is in contrast to the formation of surface chemical bonds that are associated with chemical adsorption.

Historically, nonporous and porous carbons have been central to many aspects of adsorption. In recent decades, the discovery and development of new allotropes of carbon, including carbon nanotubes and graphene, have given rise to many new studies of both theoretical and applied significance. Both pristine and modified graphene have played an important part in this work. Since the isolation of graphene in 2004 by Geim, Novoselov, and co-workers, there has been a sustained and growing interest in its unique properties and possibilities [1–14]. Studies have focused on the many fascinating properties of the material itself as well as altered forms that have been modified by functionalization through covalent and

Theoretical and Computational Chemistry, Volume 21
https://doi.org/10.1016/B978-0-12-819514-7.00002-6

noncovalent interactions [2]. Studies include such diverse applications as antibiotic removal in water [3]; hydrogen adsorption [4]; and environmental monitoring and protection related to heavy metal ions, organic pollutants, and toxic gases [5].

There are a wide variety of studies that involve the interaction of molecules with graphene or modified graphene surfaces. A sense of this variety is given by noting that studies have included the role of adatoms in graphene fracture mechanics [6]; the effects of CO and NO gas adsorption on pristine, vacancy, and doped graphene [7]; the use of graphene oxide and carbon nanotubes membranes for metal ion removal [8]; the interaction of the enzyme lactate dehydrogenase on pristine and functionalized graphene [9]; the binding of dendrimers to graphene [10]; the adsorption of a peptide on graphene [11]; models for the adsorption of organic compounds on graphene [12]; the use of metal organic framework (MOF) and graphene nanosheet hybrids for the adsorption of carbon dioxide, volatile organic compounds (VOC), hydrogen, and methane [13]; and computation of the ability of subnanometer pores in graphene to function as a membrane for the separation of hydrogen and methane gases [14].

One key aspect of importance is the molecule-surface interaction energy. For many situations, this binding energy is largely dependent on van der Waals (vdW) forces and/or hydrogen bonding. For example, molecular mechanics has been used to estimate the binding energies of nucleobases on graphene [15]. Also matching experimental and calculated binding energies for various molecules on graphene layered pore models allowed information about the porosity of the nanoporous powder Carbosieve S-III (Supelco) to be extracted [16]. Experimental binding energies were obtained from gas–solid chromatography. The background for the experiments used to extract binding energies from gas–solid chromatography and second gas–solid virial coefficients over a range of temperatures is given elsewhere [17].

Molecular mechanics has proven to be useful in providing correlations of experimental and calculated molecule-carbon surface binding energies. Based on these correlations, useful predictions of unknown molecule-carbon surface binding energies can be estimated. For this special purpose and in this work, we focus here on the use of force field calculations using MM2 and MM3 parameters to calculate neutral organic molecule binding energies for graphene, multilayer graphene, carbon nanotubes, and surface-modified graphene.

Consider the specific example of a graphene-based sensor that depends on molecular adsorption. Graphene-based electronic devices have been developed that can act as sensors for specific molecules or for environmental monitoring of pH, temperature, electric current, light, and moisture [18]. A device containing a graphene sensor needs to able to interact with molecules by adsorption, convert the amount of adsorbed molecules to a signal such as a change in resistivity, and have a way to convert the raw data to useful measurements with minimum interference [19]. Chemical sensors have been developed using graphene and surface-functionalized graphene [20–22]. For example, an ozone treatment converted a pristine graphene sensor with a detection limit of about 10 ppm for NO_2 to a sensor with a limit of 1.3 ppb [23].

Graphene has been shown to be a workable sensor material, and sensors have been produced by various means [24], including photolithography [25]. Unmodified graphene has been used as a chemical vapor sensor [26]. Graphene's electrical properties are sensitive to temperature, light, and humidity as well as adsorbed molecules. For use as a sensor, the environmental variables must be held constant, their effects minimized, or their influence determined by, for example, by principle component analysis [27].

NO_2 concentrations in the 10–150 ppb range have been monitored with a graphene sensor, and by operating this sensor at an elevated temperature, interference from other gases was held to a minimum [28]. Another approach was to construct a sensor with multiple surfaces in an array [29]. Poly(methyl methacrylate) and graphene composite laminates were prepared and used to detect volatile organic compounds [30]. A hydrogenated graphene sensor was used as a NO_2 sensor at room temperature [24].

In addition to helping model physical devices such as sensors, computational studies have been used to consider molecule-surface interactions. Density functional theory (DFT) calculations have explored molecule-graphene and molecule-modified graphene surface interactions [31–34]. The binding energies of small molecules like H_2O, NH_3, CO, NO_2, and NO on graphene have been examined [31]. DFT was used to calculate the adsorption of CO_2 gas on graphene with different doped surfaces [34]. Surface hydroxyl groups were found to promote the adsorption of NH_3 on the graphene oxide [32].

There are a variety of common explosive molecules, and detection of them is of continued interest [19,35–37]. Although methods exist for the detection of explosives, greater sensitivity and miniaturization of devices would be useful to detect explosive molecules in extremely low concentrations [38,39]. Graphene-based sensors capable of detecting low concentrations of hazardous gases or vapor-phase explosive molecules have been examined [19,35]. An ideal graphene sensor should have a special adsorption affinity for explosive molecules.

A common feature of a variety of explosive molecules is the NO_2 functional group. The NO_2 nitro group is a common explosophore among explosive molecules [36,37,40]. An explosophore is a part of the molecule that makes a compound explosive [38,39]. Our recent study relates to modeling a graphene nanosensor pore to enhance the attraction of explosive molecules with similarities in structure [41].

Force field calculations show that significant noncovalent interactions can exist between a graphene surface and a molecule [42]. These forces are primarily due to van der Waals (vdW) attractions and have been experimentally demonstrated and effectively modeled [42]. However, only DFT calculations that included a dispersion correction were observed to provide accurate binding energies [43]. One of the applications we illustrate is the use of molecular mechanics to study the enhancement of the adsorption of explosive molecules on a modified graphene bilayer surface [41]. Hydrogen bonding was used as a way to increase the molecule-surface interaction energy.

Before discussing the specific application of a bilayer hydroxylated pore to enhance the attraction of explosive molecules, we first discuss the methods of obtaining molecule-surface binding energies, the general use of MM2 and MM3 parameters to develop useful correlations with experimental binding energy values, and the use of force field calculations to predict binding energies. We used the software Scigress (Fujitsu) for all our MM2 and MM3 calculations [41–43].

2 Experimental

Before any predictions of the molecule-surface interaction energies are made, it is essential to test the model against some actual experimental values. For this purpose, experimental values on a graphitic surface are ideal [42].

Experiments using thermal programmed desorption (TPD) or gas–solid chromatography (GSC) can provide molecule-surface interaction energies. TPD uses multilayer or monolayer desorption and so modeling needs to include molecule-molecule interactions as well as molecule-surface interactions. However, GSC retention times that are extrapolated to the Henry's law region of low coverage yield the interaction energies of isolated molecules on a surface [17].

In various published experiments, the retention times from GSC provided adsorption Henry's law constants (K_H) over a range of temperatures. This data from the literature has been compiled elsewhere [42]. For a given molecule-surface system, a plot of the natural logarithm of K_H versus the reciprocal of the temperature gives a slope of E^*/R, where R is the gas constant and E^* is the experimental molecule-surface binding energy (adsorption interaction energy). Larger E^* values are associated with stronger molecule-surface interactions. Additional experimental and theoretical details are available elsewhere [17].

A number of prior published studies using graphitized thermal carbon black (GTCB) provided a suitable uniform graphitic adsorbent to generate experimental molecule-graphite interactions, E^*(graphite) [42]. In our prior work, a total of 118 different organic molecules with a variety of structures and functionality were gathered from the literature that used GSC on graphitic surfaces to find experimental binding energies [42]. These values were suitable to compare to molecule-graphite binding energies determined from force field calculations.

Experimental binding energies or gas–solid interaction energies, E^*, are reported in units of Kelvin (based on E^*/R) or for E^* directly as eV, meV, kJ/mol, or kcal/mol. Useful conversion factors include 1 kcal/mol $= 4.336411 \times 10^{-2}$ eV $= 43.36411$ meV $= 4.184$ kJ/mol $= 503.217$ K. In our work, the values of the binding energy are typically expressed in kcal/mol.

3 Theory

Molecular mechanics force field calculations are based on a sum of covalent and noncovalent energies [44,45]. Each type of energy has a given functional form and empirical parameters based on the specific atoms involved that match some available experimental data such as X-ray diffraction data giving preferred molecular conformation and three-dimensional atomic locations. Molecular structures are optimized by finding the atomic arrangement that minimizes steric energy [46].

In our prior work, MM2 and MM3 molecular mechanics force field parameters (Scigress, Fujitsu) have been used to model a variety of organic molecules, carbon surfaces, and their molecule-surface interactions [47–51]. Energies were calculated with MM2 or MM3 force field parameters [44,45]. The MM2 covalent bond-related energy contributions include stretch, angle, dihedral, and improper torsion. The MM2 noncovalent energy contributions include electrostatics, hydrogen bonding, and van der Waals. Contributions to the MM3 steric energy include bond stretching, bond angles, dihedral angles, improper torsion, torsion stretching, bend, van der Waals interactions, electrostatics, and hydrogen bonding.

In comparing structures based on two or more parts, the stabilization of the structure relative to its component parts can be determined. Our interest is in noncovalent forces such as van der Waals (vdW) forces or hydrogen bonding that may contribute to hold a structure together. A structure's binding energy (ΔE) may be written as the sum of the individual steric energies ($E_a + E_b + E_c + ...$) minus the steric energy of the combined structure ($E_{structure}$) made of the component parts (a, b, c, ...), where

$$\Delta E = (E_a + E_b + E_c + \ldots) - E_{structure}. \quad (5.1)$$

As written above, the change in energy is associated with the energy input required to disassemble the structure made of the collection of parts. For a simple system of an isolated molecule and a surface, then Eq. (5.1) may be written as

$$\Delta E = (E_m + E_s) - E_{ms} \quad (5.2)$$

where E_m is the energy of the isolated molecule, E_s is the energy of the isolated surface, and E_{ms} is the energy the molecule-surface pair held together by noncovalent forces. Thus, as written in Eq. (5.2), ΔE represents the energy of desorption. Adsorption is characterized by energy stabilization. Desorption is characterized by positive values of ΔE since noncovalent forces must be overcome to separate the parts. If a molecule on a surface is held by physical adsorption, then there is no covalent bond formation. In this work, ΔE values are reported as positive desorption values and thus represent the energy required to move the molecule away from the surface.

An interesting example of this approach is provided by an analysis of the interactions of carbon nanotubes. In prior work, a single-walled (5,5) CNT with an internuclei diameter of 0.67 nm and an internuclei length of 4.46 nm (18 rings) was chosen as a representative CNT to model using MM3 parameters [43]. This isolated (5,5) CNT had a steric energy of 613.0 kcal/mol. A pair of parallel CNTs in contact with each other had a steric energy of 1142.9 kcal/mol. Eq. (5.1) was utilized to calculate the binding energy of these two parallel CNTs, where ΔE (kcal/mol) = (613.0 + 613.0) 1142.9 or 83.1 kcal/mol. The lowering of energy when the two CNTs come in contact is due to vdW forces that cause the CNTs to be attracted to each other. The internuclei separation of two attracted CNTs at the closest approach is analogous to vdW separation of graphene layers in graphite of about 0.34–0.36 nm.

The ΔE for the less favorable perpendicular structure was found to be 17.7 kcal/mol. As expected, the parallel orientation is favored since its stabilization of 83.1 kcal/mol is greater than the 17.7 kcal/mol for the perpendicular CNTs (Fig. 5.1). A previously reported interaction energy for two (5,5) CNTs in contact with each other at right angles was 0.761–0.785 eV or 17.6–18.1 kcal/mol [52]. This result is in agreement with our reported value of 17.7 kcal/mol [43].

As an extension of the above example, molecular mechanics was used to study the noncovalent interactions between single-walled carbon nanotubes and molecular linkers. Groups of nanotubes tend to form tight, parallel bundles (||||). Molecular linkers were introduced into our models to stabilize nanostructures with carbon nanotubes held in perpendicular orientations. Molecular mechanics makes it possible to estimate the strength of noncovalent interactions holding these structures together and to calculate the overall binding energy of the structures.

Inspired by the buckycatcher molecule, a set of linkers were designed and built around a 1,3,5,7-cyclooctatetraene tether with two corannulene containing pincers that extend in opposite directions from the central cyclooctatetraene portion. Each pincer consists of a pair of "arms." These molecular linkers were modified so that the "hand" portions of each pair of "arms" could close together to grab and hold two carbon nanotubes in a perpendicular arrangement [43].

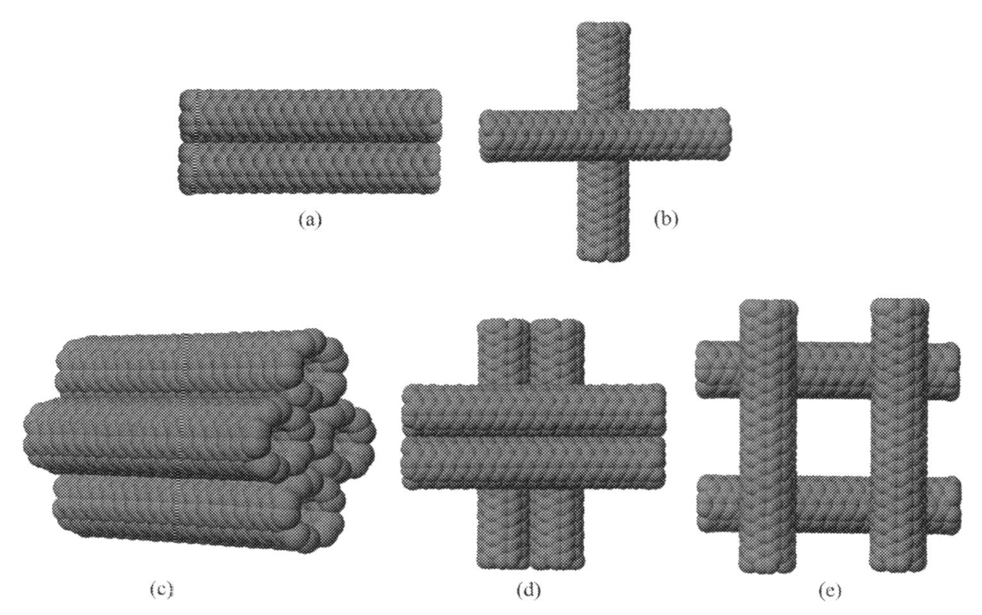

FIG. 5.1 Varied orientations of carbon nanotubes: (A) two parallel, (B) two perpendicular, (C) bundle of four, (D) double perpendicular of four, and (E) hashtag of four [43].

To illustrate the possibility of more complicated and open perpendicular CNTs structures, our primary goal was to create a model of a nanohashtag (#) CNT conformation more stable than any parallel CNT arrangements with bound linker molecules forming clumps of CNTs and linkers in non-hashtag arrangements. This goal was achieved using a molecular linker ($C_{280}H_{96}$) that utilizes van der Waals interactions to two perpendicular oriented CNTs (Fig. 5.2). Hydrogen bonding was then added between linker molecules to augment the stability of the hashtag structure. In the hashtag structure with hydrogen bonding, four (5,5) CNTs each of length 4.46 nm (18 rings) with an internuclei diameter of 0.67 nm and four linkers ($C_{276}H_{92}N_8O_8$) stabilized the hashtag so that the average binding energy per pincer was 118 kcal/mol [43].

While the work described above represents a more unusual application of Eq. (5.1), it does illustrate the ability of molecular mechanics to do faster calculations for systems that may involve thousands of atoms. Molecular mechanics has proven to be useful to obtain correlations and estimates of molecule-surface binding energies. Surfaces such as CNTs or graphene layers may be uniform over an extended region. However, porous solids may have intricate variations of structure.

For complicated porous structures, representations of pores can be modified until a set of experimental values of binding energy, E^*, is found to most closely match a set of computed values of binding energy, ΔE. In general, the model is not an exact physical representation of the surface except in simple systems like graphite or graphene or well-defined carbon nanotubes. However, a model surface that produces the right calculated ΔE values for a series of experimental E^* values can still provide a useful representation of molecule-surface interaction energies and allow for further predictions of unknown values.

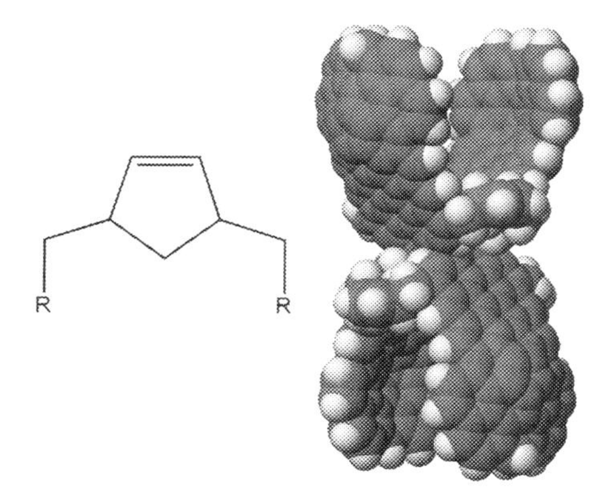

FIG. 5.2 The molecular linker molecule with cyclopentene bridges attached to hold each pair or pincer "arms" in the open conformation to be able to receive a CNT. Each "R" represents the point of attachment of the bridge to the linker arms [43].

4 Analysis and results

4.1 Comparisons of methods

Before reporting on results obtained from the analysis of the 118 E^* values referred to in the Experimental section, it will be useful to make a comparison of some other experimental, molecular mechanics, and density functional theory (DFT) methods. Molecule-surface binding energies have been evaluated by various means both experimental and computational. It is useful to compare molecule-surface binding energies obtained from experimental, traditional DFT, DFT with dispersion correction, and molecular mechanics values. The values and correlations reported here are primarily for molecular adsorption on carbon nanotubes. Available experimental and DFT computational binding energies were correlated with representative determinations of binding energies based on MM3 molecular mechanics [43].

Most of the reported values used were for an isolated molecule on a carbon nanotube surface [53–68]. However, some molecule-graphite surface interactions [69] and molecules in supramolecular systems [70–72] were also considered. Simple linear regressions were used to compare experimental, DFT, DFT with dispersion correction, and other molecular mechanics results with values computed with MM3 calculations. The MM3 calculations required creating a model such as a specific CNT or flat graphene surface to match the experimental or computed surface. The binding energies were determined as positive desorption values by simply taking the steric energy differences as shown in Eq. (5.2) to find ΔE. The isolated molecule and carbon surface steric energies, E_m and E_s, were calculated along with the neutral molecule adjacent to and attracted to the surface due to vdW forces, E_{ms} [43].

Actual reported experimental values (E^*, $n=15$) correlated well with our MM3 calculated values, ΔE, giving the equation $E^*=1.037\ \Delta E$ with $R^2=0.93$. In contrast, 29 reported DFT calculations of binding energy gave $\Delta E(\text{DFT})=1.043\ \Delta E$ with $R^2=0.28$. Given the reasonable comparison of our molecular mechanics values with actual experimental values, this poor correlation indicates that these reported DFT calculations are giving incorrect gas-surface interaction energies. These DFT calculations do not account well for vdW interactions that are primarily responsible for neutral molecule to carbon surface attractions. However, a plot of ΔE from 28 DFT calculations that each included a dispersion correction gave

ΔE(DFT-corrected) = 1.027 ΔE with $R^2 = 0.87$. In other words, these traditional electronic structure calculations using density functional theory (DFT) showed a poor correlation of ΔE unless a dispersion correction was included [43].

These correlations indicate that the vdW interactions must be explicitly considered in some way and further that the MM3 calculations are an effective means to estimate molecule-carbon surface binding energies. The surface geometry can be represented in the model, and the standard MM3 parameters for the vdW attractions of the surface carbon atoms to the molecular atoms near that surface effectively represent the molecule-surface attraction.

4.2 Interactions on layered graphene and graphene

We now turn to the data referred to in the experimental section. For this comparison of experimental organic molecule-graphitic surface binding energies E^* versus molecular mechanics, MM2 parameters were used as compiled from published values and summarized elsewhere [42]. As discussed previously, a total of 118 different E^* binding energies for organic molecules with a variety of structures and functionality were gathered from the literature. These experimental values used GSC on graphitic surfaces to find experimental binding energies [42].

For calculations, the model graphene surface was made of 702 benzene rings in a flat plane with no hydrogen atoms on the outermost rings. The outermost rings are used to hold the structure in place but the adsorbed molecule only interacts with the interior rings because of the size of the layer. The graphite model consisted of three of these layers arranged in the form of Bernal graphite with the atoms of the first and third layers directly above each other and the middle layer offset by half a benzene ring. Molecules that had their geometry optimized in isolation were placed on the surface in such a way to optimize the interaction by placing the maximum number of atoms close to the surface and by favoring more polarizable atoms as possible. The calculation was initiated with the molecule slightly too close to the carbon atoms of the surface so that the molecule was pushed away from the surface by vdW repulsion to the ideal location.

Using this approach, the E_{ms} values were found for 118 different molecules on the graphite and graphene model surfaces. The MM steric energy E_m for 118 isolated molecules and the steric energy E_s for the model surface were found. Using Eq. (5.2), 118 computed values of the molecule-surface binding energy on three layers, $\Delta E(3)$, and on single-layer graphene, $\Delta E(1)$, were determined. More than three layers were not needed, because the vdW forces have about 90% or more contribution due to the first graphene layer and only about 1% contributed from the third layer [42].

This first approach gave a linear regression of the results of E^*(graphite) versus $\Delta E(3)$ for the 118 molecules of E^*(graphite) = 0.9321 $\Delta E(3)$ with $R^2 = 0.8906$. The origin was set at (0,0) so the values scale down appropriately to a y-intercept of zero.

Table 5.1 shows the second approach used with a division of the 118 molecules into 11 different functional groups. For each of these subsets of similar molecules, plots of E^*(graphite) versus $\Delta E(3)$ were used to determine the slope and R^2. These subsets don't give the correlations that one would desire so further consideration was given to adjust what must be somewhat high estimates of $\Delta E(3)$ based on MM2 since the subset slopes, for the most part, were below a 1.0 value. Note that the R^2 values ranged from about 0.873 to 0.976 and the slopes

TABLE 5.1 The 118 molecules associated with experimental molecule-graphite binding energy values could be divided into 11 functionally similar groups

Functional group	Slope	R^2	n
Aldehyde	0.9057	0.9308	9
Alkane	0.8291	0.9971	8
Alkene/alkyne	0.8751	0.9756	7
Alkyl alcohol	0.8564	0.9043	13
Alkyl amine	0.8288	0.96	7
Ketone	0.8362	0.9421	10
Cycloalkane	0.8581	0.8726	4
Aromatic amine	1.0169	0.9394	22
Benzene	0.9633	0.8971	24
Chloroaromatic	1.0312	0.9257	8
Thiophene	0.9936	0.9728	6

Here are the experimental E* *versus computed* △E *linear correlations within each group [42].*

varied from 0.873 to 1.031. The slope below 1.0 indicates that the $\Delta E(3)$ values along the x-axis are too large relative to the experimental E^* values along the y-axis.

The third approach was to divide the set of 118 molecules into two sets of 60 rigid molecules and 58 flexible molecules. The rigid structures were characterized by sp^2 carbons in ring structures, and the flexible structures were characterized by sp^3 carbons in chain structures. Based on this division, the correlations were

$$E^*(\text{graphite}) = 0.9918\,\Delta E(3) \tag{5.3}$$

with $R^2 = 0.9130$ and

$$E^*(\text{graphite}) = 0.8500\,\Delta E(3) \tag{5.4}$$

with $R^2 = 0.9621$, for rigid and flexible, respectively. While this results in an improved fitting, the division into two sets is not perfect since some ring structures have side chains and there are various functional groups present as well. However, it did seem clear that the default MM2 carbon atom vdW forces parameters for the surface carbon atoms are better suited to directly determine the interaction of the molecular sp^2 atoms than the sp^3 atoms [42].

The fourth approach used was to do a formal separation of all nonhydrogen atoms into two groups where f_{Csp3} was the fraction of all sp^3 carbon atoms in each molecule and f_{other} was the fraction of all other nonhydrogen atoms in the same molecule. So each molecule has a value of f_{Csp3} and f_{other} where the sum of these two fractions was exactly one. After linear regression analysis of the 118 molecules to match calculated and experimental values, it was found that the best equation was

$$E^*(\text{graphite}) = 1.0000\,(0.8180 f_{\text{Csp3}} + 1.0221 f_{\text{other}})\,\Delta E(3) \tag{5.5}$$

with $R^2 = 0.9647$. The standard MM2 parameters were maintained while this simple modification reflects an adjustment in the vdW interaction that is about 20% overestimated for molecular sp^3 carbon atoms and very slightly (about 2%) underestimated for other atoms. All hydrogen atom interactions were left unmodified in any way. This result based on Eq. (5.2) provided an excellent way (Fig. 5.3) to correlate and to estimate organic molecule interactions with graphite surfaces [42].

Calculations of one-layer graphene and three-layer graphite models for the 118 molecules examined previously gave the correlation of

$$\Delta E(1) = 0.9349\,\Delta E(3) \tag{5.6}$$

with $R^2 = 0.9998$. So combining Eqs. (5.5) and (5.6) gave as the prediction for E^* for a graphene surface of

$$E^*(\text{graphene}) = (0.9349)\left(0.8180 f_{Csp3} + 1.0221 f_{other}\right)\Delta E(3). \tag{5.7}$$

Or more directly

$$E^*(\text{graphene}) = \left(0.8180 f_{Csp3} + 1.0221 f_{other}\right)\Delta E(1) \tag{5.8}$$

where the calculation of $\Delta E(3)$ used a three-layer graphite model and $\Delta E(1)$ used only a single layer to calculate values for any molecule-graphene interactions [42].

Our results suggest that the interaction of a molecule with graphene is about 94% of the value it would be on multilayer graphite. However, three layers of graphite are adequate to represent the graphite since the vdW interaction is quickly attenuated (see above). Using this approach allows one to make good estimates of both molecule-graphite and molecule-graphene binding energies.

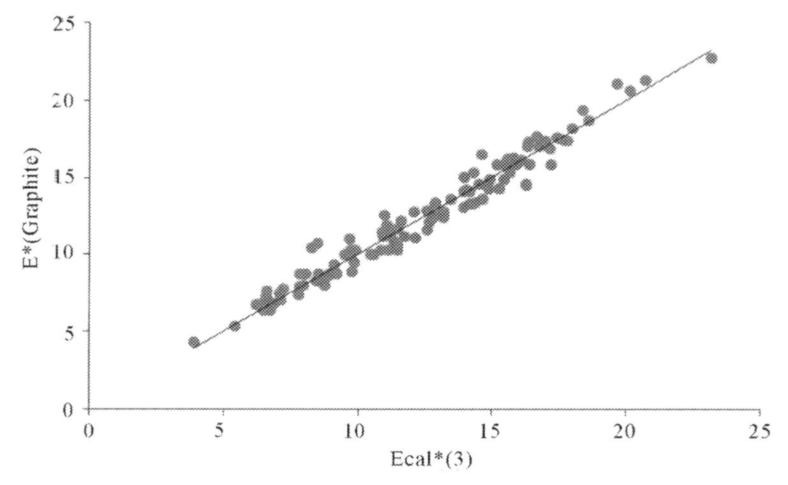

FIG. 5.3 For 118 organic adsorbate molecules on graphite: experimental binding energies E^* versus modified three-layer calculated binding energies Ecal*(3) using MM2 parameters. E^*(graphite) = 1.0000 Ecal*(3) with $R^2 = 0.965$ [42]. Here Ecal*(3) represents the modification of ΔE in Eq. (5.5), where Ecal*(3) = $(0.8180\, f_{Csp3} + 1.0221\, f_{other})\,\Delta E(3)$.

4.3 Explosive molecules on bilayer hydroxylated pore

We now turn to an example of a modified surface where experimental data are lacking so molecular mechanics was used to estimate the surface interactions of explosive molecules on the adsorption site consisting of a hydroxylated flat pore in a graphene bilayer. In prior work, first, a graphene bilayer based on Bernal stacking of 127 benzene rings per layer was used (Fig. 5.4) with MM2 force field parameters [41]. Initially, trinitrotoluene (TNT) adsorbate molecule was used as a typical representative explosive molecule. TNT has functional similarities with many other explosive molecules [36,37]. Benzene was also used as an adsorbate to compare to TNT. The binding energy for TNT was 17.9 kcal/mol. The binding energy for benzene was 9.4 kcal/mol.

Second, a bilayer pore was created with 18 perimeter carbons (Fig. 5.4, hydrogen atoms in the pore not shown). Each perimeter carbon had an H atom in the sp^2 plane of the upper graphene layer. Within this pore, the TNT binding energy increased to 24.5 kcal/mol.

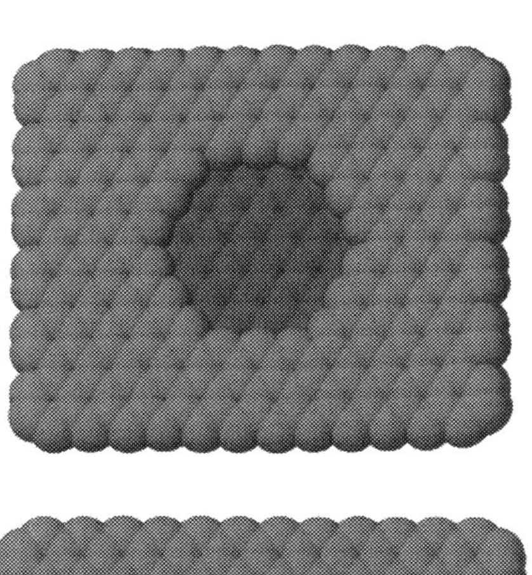
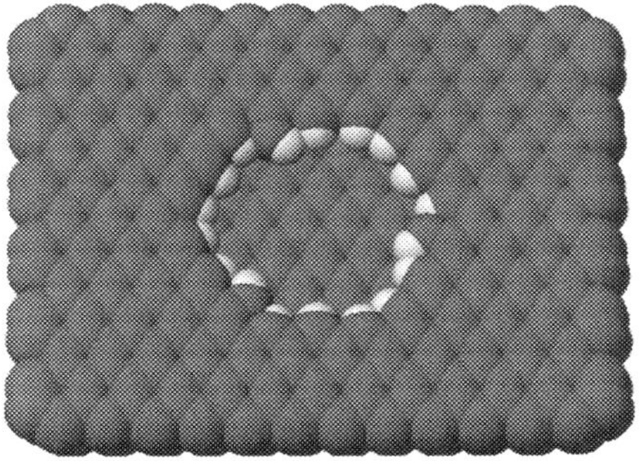

FIG. 5.4 Surface models made of graphene bilayers with two layers of 127 rings each: flat pristine graphene, hexagonal pore with 18 perimeter carbon atoms (bonded interior hydrogen atoms not shown), and three OH groups forming hydroxylated pore [41].

Third, the bilayer pore was modified to include three hydroxyl groups. This hydroxylated model pore (Fig. 5.4) gave a TNT-surface binding energy of 42.3 kcal/mol. So the models discussed above gave ΔE(TNT) of 17.9, 24.4, and 42.3. The hydrogen bonding contribution to ΔE increased from 0 on the flat bilayer to 20.9 kcal/mol in the hydroxylated pore. The overall contribution due to vdW forces had only a small increase in comparison. Since the hydroxylated pore showed an increase in molecule-surface interaction with TNT, it was predicted that other explosive molecules with similar functionality would also have enhanced adsorption. The nitro functional group is common in explosive molecules so including hydroxyls in the surface pore could increase the interaction for many explosive molecules via hydrogen bonding [41].

Given the common presence of nitro groups, it was expected that the pore that had enhanced the TNT binding energy would also do so for many explosive molecules in general. 22 explosive molecules were considered [36,37], namely, TNT, DADNE, CL-20, DNT, HMX, HNS, NTO, PETN, RDX, TATB, TNB, TNR, BDNFA, BTTN, EGDN, K10A, K10B, MTN, NIMMO, NG, TETRYL, and TNAZ. The abbreviations and names are given in Table 5.2. These 22 molecules are manufactured for various military and civilian uses [36,37,40] as shown in Table 5.3.

TABLE 5.2 Abbreviations and names for the 22 explosive molecules modeled [41]

	Abbreviation	Molecule name
1	TNT	2-Methyl-1,3,5-trinitrobenzene
2	DADNE	2,2-Dinitroethene-1,1-diamine
3	CL-20	2,4,6,8,10,12-Hexanitrohexaazaisowurtzitane
4	DNT	1-Methyl-2,4-dinitrobenzene
5	HMX	Octahydro-1,3,5,7-tetranitro-1,3,5,7-tetrazocine
6	HNS	1,3,5-Trinitro-2-[2-(2,4,6-trinitrophenyl)ethenyl]benzene
7	NTO	5-Nitro-1,2-dihydro-1,2,4-triazol-3-one
8	PETN	[3-Nitrooxy-2,2-bis(nitrooxymethyl)propyl] nitrate
9	RDX	1,3,5-Trinitro-1,3,5-triazinane
10	TATB	2,4,6-Trinitrobenzene-1,3,5-triamine
11	TNB	1,3,5-Trinitrobenzene
12	TNR	2,4,6-Trinitrobenzene-1,3-diol
13	BDNFA	Bis-dinitropropylformal
14	BTTN	Butane-1,2,4-triol trinitrate
15	EGDN	Ethylene glycol dinitrate
16	K10A	2,6-Dinitroethylbenzene
17	K10B	2,4,6-Trinitroethylbenzene

TABLE 5.2 Abbreviations and names for the 22 explosive molecules modeled [41]—Cont'd

	Abbreviation	Molecule name
18	TMETN	[2-Methyl-3-nitrooxy-2-(nitrooxymethyl)propyl]nitrate
19	NIMMO	3-Nitratomethyl-3-methyl oxetane
20	NG	1,2,3-Trinitroxypropane
21	Tetryl	2,4,6-Trinitrophenylmethylnitramine
22	TNAZ	1,3,3-Trinitroazetidine

TABLE 5.3 Common uses of the 22 explosive compounds modeled [41]

	Molecule abbreviation	Common use of molecule
1	TNT	Component of dynamite
2	DADNE	Detonation and gun propellant
3	CL-20	Rocket propellant
4	DNT	Precursor of TNT, plasticizer for explosives
5	HMX	Polymer-bonded explosives for nuclear detonation
6	HNS	Mortar grenades
7	NTO	Pressed thermoplastic explosives
8	PETN	Plastic explosive
9	RDX	Plastic explosive mix
10	TATB	Nuclear warhead charge/explosive charges
11	TNB	Commercial mining explosive
12	TNR	Lead salts/ignition agent
13	BDNFA	Energetic plasticizer
14	BTTN	Propellant
15	EGDN	Precursor
16	K10A	Polymer-bonded explosives
17	K10B	Polymer-bonded explosives
18	MTN	Plasticizer and propellant
19	NIMMO	Plasticizer
20	NG	High explosive
21	Tetryl	Base charge of blasting caps
22	TNAZ	Solid rocket and gun propellant

After geometry optimization, the explosive molecules of interest (see Table 5.2) were each placed in the hydroxylated pore site and ΔE found from Eq. (5.2) for each of the 22 explosive molecules. The geometry optimization used the conjugate gradient to locate the energy minimum. Van der Waals interactions between atoms separated by greater than 0.9 nm were excluded. Optimization continued until the energy change was less than 0.001 kcal/mol. A comparison of the bilayer and hydroxylated pore binding energy values is given in Table 5.4. Significant enhancement of the molecule-surface interaction energy was observed in going from the bilayer graphene to the hydroxylated pore. The stabilization or binding energy was found for each of the 22 molecules.

TABLE 5.4 Molecule-surface binding energies ΔE for the molecules on the flat graphene bilayer surface and in the hydroxylated pore; and the increase in ΔE due to hydroxylated pore adsorption [41]

Molecule	ΔE (flat bilayer) kcal/mol	ΔE (hydroxy pore) kcal/mol	ΔE (enhancement) kcal/mol
TNT	17.9	42.3	24.4
DADNE	14.4	22.5	8.1
CL-20	14.9	38.3	23.4
DNT	15.5	35.3	19.8
HMX	15.1	37.3	22.2
HNS	24.7	32.3	7.6
NTO	11.2	20.5	9.3
PETN	14.7	35.2	20.5
RDX	7.2	21.5	14.3
TATB	24.5	51.5	27.0
TNB	23.7	53.5	29.8
TNR	19.2	44.3	25.1
BDNFA	15.9	33.4	17.5
BTTN	14.2	27.0	12.8
EGDN	10.8	20.7	9.9
K10A	17.3	27.2	9.9
K10B	19.5	48.5	29.0
MTN	12.7	34.6	21.9
NIMMO	10.2	24.4	14.2
NG	11.8	37.3	25.5
Tetryl	15.5	32.7	17.2
TNAZ	17.2	23.6	6.4

5 Discussion

5.1 Comparisons of methods

The MM2 and MM3 force field methods provide no electronic details, but they are useful for correlating known E^* binding energy values and predicting unknown E^* values where the carbon surface has been modeled appropriately. Traditional DFT calculations do not account well for vdW interactions as demonstrated by a variation within values for the same system [73]. However, DFT calculations that include a dispersion correction factor can be effective determiners of molecule-surface binding energies [43].

5.2 Interactions on layered graphene and graphene

Graphite and graphene surface interactions, Eq. (5.7) and Eq. (5.8), respectively, should provide good estimates of molecule-graphite and molecule-graphene interaction energies. Thierfelder et al. reported on various methods of calculating the methane-graphene interaction energy [74]. As a reference point, they used an estimate of 0.12 to 0.14 eV for methane-graphite binding energy and then estimated that the methane-graphene binding energy should be in the range of 0.11–0.13 eV.

To make a comparison, based on our MM2 calculations and modifications, a calculation of $\Delta E(3)$ with Eq. (5.7) modification gave a prediction of 3.40 kcal/mol or 0.12 eV [42]. A calculation of $\Delta E(1)$ with Eq. (5.8) modification gave a prediction of 2.8 kcal/mol or 0.11 eV [42]. Methane was not a molecule in the data set of 118 molecules used to establish correlation. These results are in agreement with the Thierfelder estimate [74] and support the value of classical molecular mechanics to provide useful estimates of molecule-graphene binding energies.

5.3 Explosive molecules in bilayer hydroxylated pore

For the explosive molecule bilayer pore study, related experimental values such as vdW interaction energy contributions and hydrogen bonding contributions can be compared [41]. As shown in Table 5.4, all the molecules exhibited significant increases in ΔE from the flat bilayer graphene to the hydroxyl bilayer pore [41]. For the flat bilayer, the average ΔE of the 22 explosives was 15.8 kcal/mol. For the hydroxy pore, the average ΔE of the 22 explosives was 33.8 kcal/mol. The average of the enhancements per molecule was 18.0 kcal/mol [41].

To investigate the molecular mechanics hydrogen bonding energy contributions per hydrogen bond, a comparison with water experimental data was made. From Raman spectroscopy, the energy contribution of hydrogen bonding in water clusters was reported to be about 2.2–2.6 kcal/mol [75]. To find if the hydrogen bond energy contribution from MM2 was within this range, small water clusters were modeled. A pair of water molecules, three water molecules, four water molecules, and seven water molecules were modeled. Each cluster of molecules was optimized and the stabilization energy due to hydrogen bonding observed. The total hydrogen bonding energy for the cluster was divided by the number of bonds in the cluster to generate the energy per hydrogen bond. The interaction energy per hydrogen bond was found to be 2.5, 2.6, 2.3, and 2.6 kcal/mol for two, three, four, and seven water molecules, respectively [41]. These calculated values were within the range (2.2–2.6 kcal/mol) of the reported experimental values. MM2 calculations provide reasonable estimates of hydrogen bonding contributions.

Based on acceptable hydrogen bond distances, the number of hydrogen bonds between the molecule and surface pore was determined. The average hydrogen bond energies found was compared to the accepted range. The average organic molecular hydrogen bond was reported to be approximately 3.0–5.0 kcal/mol [76].

Examples of the hydrogen bond formation within the pore are shown for TNT and TATB (Fig. 5.5). The TNT formed 6 molecule-surface hydrogen bonds. The TATB in the hydroxylated pore had 6 internal hydrogen bonds and formed 6 new molecule-surface hydrogen bonds. Of course, only the new molecule-surface hydrogen bonds enhance adsorption. The hydrogen bonds formed per molecule ranged from 2 to 6, and the hydrogen bond strength per hydrogen bond varied from the two lowest of 3.0 and 3.2 and the two highest

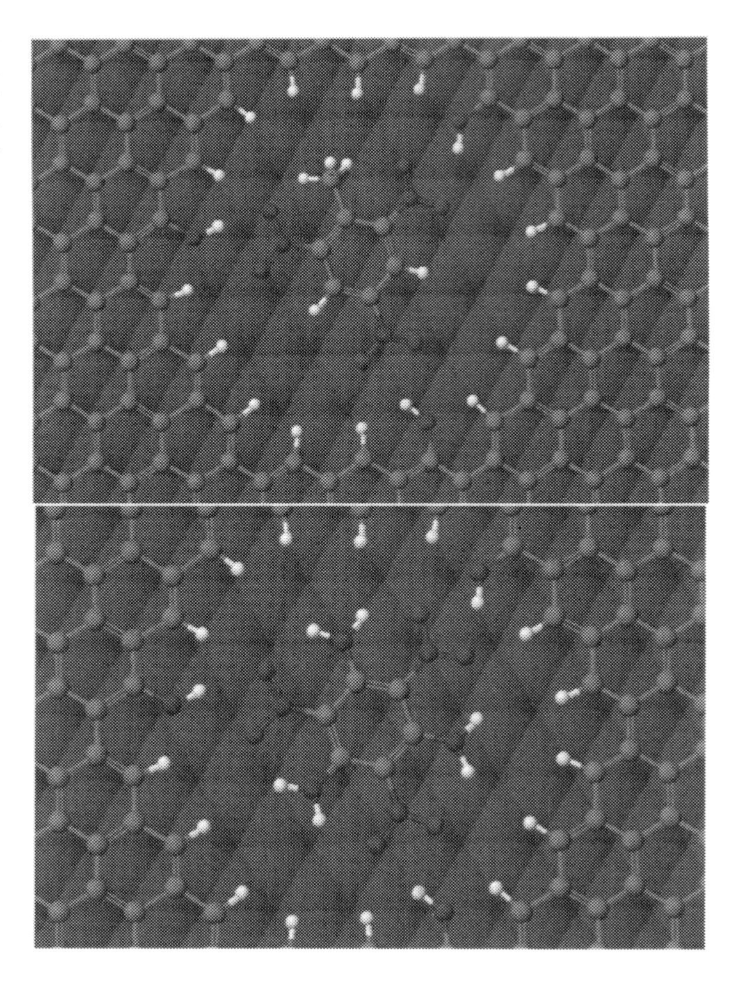

FIG. 5.5 TNT in hydroxylated pore (*top*) with 6 molecule-surface hydrogen bonds. TATB in hydroxylated pore (*bottom*) with 6 internal hydrogen bonds and 6 molecule-surface hydrogen bonds [41].

of 5.9 and 5.4 kcal/mol. The other 18 values varied from 3.5 to 5.1 kcal/mol. This range is in good agreement with the reported average organic molecular hydrogen bond in the 3.0–5.0 kcal/mol range [75]. A nanosensor based on this surface structure could greatly enhance explosive interaction relative to unmodified graphene.

6 Conclusion

Calculations based on force field parameters have clear limitations in terms of providing electronic details or predictions of certain properties that may be determined only from DFT or other electron structure calculations. However, as we have demonstrated here, molecular mechanics can provide useful information with regard to matching experimental results for gas-surface interactions where noncovalent interactions dominate, can provide some insights into surface structures, and can be used as a way to explore practical applications of modified graphene surfaces.

In one other example from past work, porous and nonporous structure determinations were made based on how well experimental binding energies for a range of molecules were compared to calculations based on surface models that varied in structure and porosity [16]. Interactions increase as the model changed from a flat multilayer graphene plate to a porous structure with multiple graphene-layer plates that can surround a molecule or a surrounding surface such as in the interior of a CNT or multiwalled CNT. For example, in this prior work, a double parallel plate (DPP) model with surfaces on four sides of a molecule gave a better fit than a parallel plate (PP) model with surfaces on two sides of a molecule or a flat plate (P) model with only one surface [16]. Each flat plate consisted of 127 benzene rings per layer and three layers per plate.

Consider the correlations of E^*(experimental) versus ΔE (calculated) for 10 organic molecules (alkanes and halogenated alkanes). For E^* versus ΔE, the R^2 value was 0.784 for the flat surface model with a relative standard deviation (ΔE compared to E^*) of 16.3% [16]. For the same experimental data using the best DPP model to calculate ΔE, the R^2 was 0.978 and had a relative standard deviation (ΔE compared to E^*) of 4.4%. For the P model, the slope of the linear correlation of E^* versus ΔE was 2.223; for the best PP model, the slope was 1.256; and for the best DPP model, the slope was 1.025. In other words, not only did E^* and ΔE correlate well but they agreed in the actual binding energy values. Additionally, the best DPP model using MM2 parameters gave a pore diameter of 0.55 nm, which matched the experimentally reported uniform pore width of 0.55 nm for the carbon powder used, Carbosieve SIII (Supelco) with a BET surface area of 995 m^2/g [16].

Molecular mechanics may also deal with structures such as multiple CNTs or many graphene layers that may have thousands of atoms. We have demonstrated that interactions among CNTs and between molecules and CNTs are well represented by MM3 parameters [43]. The method is computationally convenient and effective to determine interactions between carbon surfaces or molecules and carbon surfaces but ideally is based on first determining that some existing experimental values can be correctly correlated. Because interesting nanostructures may involve many thousands of atoms, computationally quicker methods are useful if focused on noncovalent interactions. Structures made of CNTs and linker molecules provided an example that has been effectively modeled and examined [43]. Force field calculations do not provide electronic details and are based on empirical values used to optimize parameters for some property of interest such as three-dimensional molecular structure. However, these classical molecular mechanics calculations can provide useful estimates of molecule interactions with carbon surfaces such as pristine graphene, modified graphene, carbon nanotubes, etc. For molecule-surface interaction calculations, a comparison with experimental values is essential to know if the values are appropriate. It has been observed that past DFT calculations of the molecule-surface interactions can be incorrect if the dispersion forces that dominate neutral organic molecule-graphene surfaces are not accounted for correctly [43].

In one example based on past work, enhancement of molecule-surface interactions was produced by modeling a model graphene bilayer hydroxylated pore. In comparison with an unmodified graphene bilayer, there was a significant increase in the surface binding energy. Most of the enhanced molecule-surface interaction was due to a significant increase in hydrogen bonding [41].

For 2-methyl-1,3,5-trinitrobenzene (TNT), the surface interaction increased from 17.9 kcal/mol for the flat graphene bilayer to 42.3 kcal/mol for the hexagonal hydroxylated pore. All of the 22 explosive molecules considered had increased adsorption in the hydroxylated pore due to hydrogen bonding with the average increase in binding energy of 17.0 kcal/mol (0.737 eV or 71.1 kJ/mol) [41]. For the 22 explosive molecules examined, the average interaction energy of the ΔE(pore) was more than double that of the ΔE(flat bilayer) [41].

Hydroxylated modified graphene or a graphene-like surface could be used as part of a nanosensor to detect trace amounts of explosive molecules in the air. Adsorbed molecules on graphene can change measurable properties such as electrical resistance. An increased attraction would provide higher sensitivity and aid in detecting trace amounts of explosives or other molecules of interest relative to background molecules.

The computational examples given here suggest a continued but specific role for molecular mechanics calculations related to estimates of molecule-surface adsorption interactions. The molecular mechanics calculations have the positive features of being fast, allowing thousands of atoms to be included, providing reasonable estimates of molecule-graphene and modified graphene interaction energies, and providing ΔE estimates and correlations of E^* for other carbon surfaces if appropriately modeled. Force field calculations have a specific role to play in modeling and developing future practical devices and applications of graphene and modified graphene surfaces.

Acknowledgment

We gratefully acknowledge the past support provided by the Grote Chemistry Fund and the Wheeler Odor Research Fund at the University of Tennessee at Chattanooga.

References

[1] W. Choi, J.-W. Lee (Eds.), Graphene Synthesis and Application, CRC Press, Boca Raton, 2012.
[2] V. Singh, D. Joung, L. Zhaiad, S. Das, S.I. Khondaker, S. Seala, Graphene based materials: past, present and future, Prog. Mater. Sci. 56 (8) (2011) 1178–1271, https://doi.org/10.1016/j.pmatsci.2011.03.003.

[3] M.-f. Li, Y.-g. Liu, G.-m. Zeng, N. Liu, S.-b. Liu, Graphene and graphene-based nanocomposites used for antibiotics removal in water treatment: a review, Chemosphere 226 (2019) 360–380. https://doi.org/10.1016/j.chemosphere.2019.03.117.

[4] M. Bonfanti, R. Martinazzo, Comment on theoretical study of the dynamics of atomic hydrogen adsorbed on graphene multilayers, Phys. Rev. B 97 (11) (2018) 117401. https://doi.org/10.1103/PhysRevB.97.117401.

[5] M.J. Lü, J. Li, X.Y. Yang, C.A. Zhang, J. Yang, H. Hu, X.B. Wang, Applications of graphene-based materials in environmental protection and detection, Chin. Sci. Bull. 58 (22) (2013) 2698–2710.

[6] A. Verma, A. Parashar, M. Packirisamy, Role of chemical adatoms in fracture mechanics of graphene nanolayer, Mater. Today Proc. 11 (2019) 920–924.

[7] M.S.M. Shukri, M.N.S. Saimin, M.K. Yaakob, M.Z.A. Yahya, M.F.M. Taib, Structural and electronic properties of CO and NO gas molecules on Pd-doped vacancy graphene: a first principles study, Appl. Surf. Sci. 494 (2019) 817–828. https://doi.org/10.1016/j.apsusc.2019.07.238.

[8] M. Musielak, A. Gagor, B. Zawisza, E. Talik, R. Sitko, Graphene oxide/carbon nanotube membranes for highly efficient removal of metal ions from water, ACS Appl. Mater. Interfaces 11 (31) (2019) 28582–28590, https://doi.org/10.1021/acsami.9b11214.

[9] S. Ahmadi, M.S.Y. Mardoukhi, M. Salehi, S. Sajjadi, A.H. Keihan, Molecular dynamics simulation of lactate dehydrogenase adsorption onto pristine and carboxylic-functionalized graphene, Mol. Simul. 45 (16) (2019) 1305–1311. https://doi.org/10.1080/08927022.2019.1632447.

[10] M. Gosika, S. Sen, A. Kundagrami, P.K. Maiti, Understanding the thermodynamics of the binding of PAMAM dendrimers to graphene: a combined analytical and simulation study, Langmuir 35 (28) (2019) 9219–9232. https://doi.org/10.1021/acs.langmuir.9b01247.

[11] X. Yin, B. Li, S. Liu, Z. Gu, B. Zhou, Z. Yang, Effect of the surface curvature on amyloid-β peptide adsorption for graphene, RSC Adv. 9 (18) (2019) 10094–10099, https://doi.org/10.1039/C8RA10015B.

[12] G. Ersan, O.G. Apul, T. Karanfil, Predictive models for adsorption of organic compounds by Graphene nanosheets: comparison with carbon nanotubes, Sci. Total Environ. 654 (2019) 28–34. https://doi:10.1016/j.scitotenv.2018.11.029.

[13] B. Szczesniak, J. Choma, M. Jaroniec, Gas adsorption properties of hybrid graphene-MOF materials, J. Colloid Interface Sci. 514 (2018) 801–813. https://doi:10.1016/j.jcis.2017.11.049.

[14] D.-E. Jiang, V.R. Cooper, S. Dai, D. Sheng, Porous graphene as the ultimate membrane for gas separation, Nano Lett. 9 (12) (2009) 4019–4024. https://doi.org/10.1021/nl9021946.

[15] T.R. Rybolt, C.E. Wells, Molecule-surface binding energies from molecular mechanics: nucleobases on graphene, in: H.E. Chan (Ed.), Graphene and Graphite Materials, Nova Publishers, New York, 2009, pp. 95–112.

[16] T.R. Rybolt, M.C. Trentle, M.J. Rice, H.E. Thomas, Graphene layer pore models for molecule-surface binding energies, in: P. Watkins (Ed.), Molecular Mechanics and Modeling, Nova Publishers, New York, 2015, pp. 73–107.

[17] T.R. Rybolt, H.E. Thomas, Henry's law behavior in gas-solid chromatography: a virial approach, in: E. Pefferkorn (Ed.), Interfacial Phenomena in Chromatography, in: M.J. Schick, A.T. Hubbard (Eds.), Surfactant Science Series, vol. 80, Marcel Dekker, New York, 1999, pp. 1–40.

[18] J. Zhang, L. Song, Z. Zhang, N. Chen, L. Qu, Environmentally responsive graphene systems, Small 10 (11) (2014) 2151–2164. https://doi.org/10.1002/smll.201303080.

[19] D. Yi, L. Senesac, T. Thundat, Speciation of energetic materials on a microcantilever using surface reduction, Scanning 30 (2008) 208–212. https://doi.org/10.1002/sca.20096.

[20] I. Rahim, M. Shah, M. Iqbal, A. Khan, Synthesis, structural, optical, morphological and multi sensing properties of graphene based thin film devices, Mater. Res. Exp. 5 (2018) 096403/1–096403/15. https://doi.org/10.1088/2053-1591/aac98b.

[21] F. Yavari, N. Koratkar, Graphene-based chemical sensors, J. Phys. Chem. Lett. 3 (13) (2012) 1746–1753. https://doi.org/10.1021/jz300358t.

[22] A. Wisitsoraat, A. Tuantranont, Graphene-based chemical and biosensors, in: A. Tuantranont (Ed.), Applications of Nanomaterials in Sensors and Diagnostics, Springer Science, 2013, pp. 103–141. https://doi:10.1007/5346_2012_47.

[23] M.G. Chung, D.H. Kim, H.M. Lee, T. Kim, J.H. Choi, D.K. Seo, J.-B. Yoo, S.-H. Hong, T.J. Kang, Y.H. Kim, Highly sensitive NO_2 gas sensor based on ozone treated graphene, Sensors Actuators B Chem. 166-167 (2012) 172–176. https://doi.org/10.1016/j.snb.2012.02.036.

[24] S. Park, M. Park, S. Kim, S.-G. Yi, M. Kim, J. Son, J. Cha, J. Hong, K.-H. Yoo, NO_2 gas sensor based on hydrogenated graphene, Appl. Phys. Lett. 111 (2017) 213102/1–213102/5. https://doi.org/10.1063/1.4999263.
[25] H. Ye, E.C. Nallon, V.P. Schnee, C. Shi, K. Jiang, J. Xu, S. Feng, H. Wang, Q. Li, Enhance the discrimination precision of graphene gas sensors with a hidden Markov model, Anal. Chem. 90 (2018) 13790–13795. https://doi.org/10.1021/acs.analchem.8b04386.
[26] E.C. Nallon, V.P. Schnee, C. Bright, M.P. Polcha, Q. Li, Chemical discrimination with an unmodified graphene chemical sensor, ACS Sens. 1 (2016) 26–31. https://doi.org/10.1021/acssensors.5b00029.
[27] J. Lee, C.-J. Lee, J. Kang, H. Park, J. Kim, M. Choi, H. Park, Multifunctional graphene sensor for detection of environment signals using a decoupling technique, Solid State Electron. 151 (2019) 40–46. https://doi.org/10.1016/j.sse.2018.10.014.
[28] C. Melios, V. Panchal, K. Edmonds, A. Lartsev, R. Yakimova, O. Kazakova, Detection of ultralow concentration NO_2 in complex environment using epitaxial graphene sensors, ACS Sens. 3 (2018) 1666–1674. https://doi.org/10.1021/acssensors.8b00364.
[29] A. Hannon, Y. Lu, J. Li, M. Meyyappan, A sensor array for the detection and discrimination of methane and other environmental pollutant gases, Sensors 16 (8) (2016) 1163/1–1163/11. https://doi.org/10.3390/s16081163.
[30] C. Rattanabut, W. Wongwiriyapan, W. Muangrat, W. Bunjongpru, M. Phonyiem, Y.J. Song, Graphene and poly(methyl methacrylate) composite laminates on flexible substrates for volatile organic compound detection, Jpn. J. Appl. Phys. 57 (4S) (2018) 04FP10/1–04FP10/5. https://doi.org/10.7567/JJAP.57.04FP09.
[31] O. Leenaerts, B. Partoens, F.M. Peeters, Adsorption of H_2O, NH_3, CO, NO_2, and NO on graphene: a first-principles study, arXiv:0710.1757v1 [cond-mat.mtrl-sci]. Preprint Archive, Condens. Matter (2007) 1–6. https://doi.org/10.1103/PhysRevB.77.125416.
[32] Y. Peng, J. Li, Ammonia adsorption on graphene and graphene oxide: a first-principles study, Front. Environ. Sci. Eng. 7 (2013) 403–411. https://doi.org/10.1007/s11783-013-0491-6.
[33] M. Rouhani, DFT study on adsorbing and detecting possibility of cyanogen chloride by pristine, B, Al, Ga, Si and Ge doped graphene, J. Mol. Struct. 1181 (2019) 518–535. https://doi.org/10.1016/j.molstruc.2019.01.006.
[34] Z. Zheng, H. Wang, Different elements doped graphene sensor for CO_2 greenhouse gases detection: the DFT study, Chem. Phys. Lett. 721 (2019) 33–37, https://doi.org/10.1016/j.cplett.2019.02.024.
[35] J. Zhang, E.P. Fahrenthold, Graphene-based sensing of gas-phase explosives, ACS Appl. Nano Mater. 2 (2019) 1445–1456. https://doi.org/10.1021/acsanm.8b02330.
[36] J. Akhavan, The Chemistry of Explosives, third ed., RSC Publishing, Cambridge, UK, 2011.
[37] T.M. Klapotke, Chemistry of High-Energy Materials, De Gruyter Publishers, Berlin, Germany, 2011.
[38] D.S. Moore, Instrumentation for trace detection of high explosives, Rev. Sci. Instrum. 75 (2004) 2499–2512. https://doi.org/10.1063/1.1771493.
[39] D.S. Moore, Recent advances in trace explosives detection instrumentation, Sens. Imaging 8 (2007) 9–38. https://doi.org/10.1007/s11220-007-0029-8.
[40] J. Yinon, Field detection and monitoring of explosives, Trends Anal. Chem. 21 (2002) 292–301. https://doi.org/10.1016/S0165-9936(02)00408-9.
[41] R.S. Holt, T.R. Rybolt, Modeling enhanced adsorption of explosive molecules on a hydroxylated graphene pore, Graphene 8 (2019) 1–18. https://doi.org/10.4236/graphene.2019.81001.
[42] J.H. Son, T.R. Rybolt, Force field based MM2 molecule-surface binding energies for graphite and graphene, Graphene 2 (2013) 18–34. https://doi.org/10.4236/graphene.2013.21004.
[43] C.W. Frye, T.R. Rybolt, Nanohashtag structures based on carbon nanotubes and molecular linkers, Surf. Sci. 669 (2018) 34–44. https://doi.org/10.1016/j.susc.2017.11.005.
[44] N.L. Allinger, Conformational analysis. 130. MM2. A hydrocarbon force field utilizing V1 and V2 torsional terms, J. Am. Chem. Soc. 99 (25) (1977) 8127–8134, https://doi.org/10.1021/ja00467a001.
[45] J. Lii, N.L. Allinger, Molecular mechanics. The MM3 force field for hydrocarbons. 3. The van der Waals' potentials and crystal data for aliphatic and aromatic hydrocarbons, J. Am. Chem. Soc. 111 (23) (1989) 8576–8582, https://doi.org/10.1021/ja00205a003.
[46] F. Jensen, Introduction to Computational Chemistry, John Wiley & Sons, Chichester, 1999.
[47] T.R. Rybolt, C.E. Wells, C.R. Sisson, C.B. Black, K.A. Ziegler, Evaluation of molecular mechanics calculated binding energies for isolated and monolayer organic molecules on graphite, J. Colloid Interface Sci. 314 (2) (2007) 434–445, https://doi.org/10.1016/j.jcis.2007.05.083.
[48] T.R. Rybolt, K.A. Ziegler, H.E. Thomas, J.L. Boyd, M.E. Ridgeway, Adsorption energies for a nanoporous carbon from gas-solid chromatography and molecular mechanics, J. Colloid Interface Sci. 296 (1) (2006) 41–50, https://doi.org/10.1016/j.jcis.2005.08.057.

[49] T.R. Rybolt, R.A. Hansel, Determining molecule-carbon surface adsorption energies using molecular mechanics and graphene nanostructures, J. Colloid Interface Sci. 300 (2) (2006) 805–808, https://doi.org/10.1016/j.jcis.2006.04.057.
[50] T.R. Rybolt, C.E. Wells, H.E. Thomas, C.M. Goodwin, J.L. Blakely, J.D. Turner, Binding energies for alkane molecules on a carbon surface from gas-solid chromatography and molecular mechanics, J. Colloid Interface Sci. 325 (1) (2008) 282–286, https://doi.org/10.1016/j.jcis.2008.06.043.
[51] T.R. Rybolt, K.T. Bivona, H.E. Thomas, C.M. O'Dell, Comparison of gas-solid chromatography and MM2 force field molecular binding energies for green-house gases on a carbonaceous surface, J. Colloid Interface Sci. 338 (1) (2009) 287–292, https://doi.org/10.1016/j.jcis.2009.06.001.
[52] E.G. Pogorelov, A.I. Zhbanov, Y.-C. Chang, S. Yang, Universal curves for the van der Waals interaction between single-walled carbon nanotubes, Langmuir 28 (2012) 1276–1282. https://doi.org/10.1021/la203776x.
[53] T. Pankewitz, W. Klopper, Ab initio modeling of methanol interaction with single-walled carbon nanotubes, J. Phys. Chem. C 111 (2007) 18917–18926. https://doi.org/10.1021/jp076538.
[54] A. Bauza, A. Frontera, T.J. Mooibroek, 1,1,2,2-Tetracyanocyclopropane (TCCP) as supramolecular synthon, Phys. Chem. Chem. Phys. 18 (2016) 1693–1698. https://doi.org/10.1039/C5CP06350G.
[55] J. Goering, U. Burghaus, Adsorption kinetics of thiophene on single-walled carbon nanotubes (CNTs), Chem. Phys. Lett. 447 (2007) 121–126. https://doi.org/10.1016/j.cplett.2007.09.015.
[56] I.M. Jauris, S.B. Fagan, M.A. Adebayo, F.M. Machado, Adsorption of acridine orange and methylene blue synthetic dyes and anthracene on single wall carbon nanotubes: a first principle approach, Comput. Theor. Chem. 1076 (2016) 42–50. https://doi.org/10.1016/j.comptc.2015.11.021.
[57] V. Machado de Menezes, E. Michelon, J. Rossato, I. Zanella, S.B. Fagan, Carbon nanostructures interacting with vitamins A, B3 and C: ab initio simulations, J. Biomed. Nanotechnol. 8 (2012) 345–349. https://doi.org/10.1166/jbn.2012.1434.
[58] S.G. Stepanian, V.A. Karachevtsev, A.Y. Glamazda, U. Dettlaff-Weglikowska, L. Adamowicz, Combined Raman scattering and ab initio investigation of the interaction between pyrene and carbon SWNT, Mol. Phys. 101 (2003) 2609–2614. https://doi.org/10.1080/0026897031000154284.
[59] H.A. Witek, B. Trzaskowski, E. Malolepsza, K. Morokuma, L. Adamowicz, Computational study of molecular properties of aggregates of C60 and (16,0) zigzag nanotube, Chem. Phys. Lett. 446 (2007) 87–91. https://doi.org/10.1016/j.cplett.2007.08.051.
[60] A. de Juan, A. Lopez-Moreno, J. Calbo, E. Orti, E.M. Perez, Determination of association constants towards carbon nanotubes, Chem. Sci. 6 (2015) 7008–7014. https://doi.org/10.1039/C5SC02916C.
[61] M.D. Ganji, A. Afsari, Interaction of alkanethiols with single-walled carbon nanotubes: first-principles calculations, Physica E (Amsterdam, Neth.) 41 (2009) 1696–1700. https://doi.org/10.1016/j.physe.2009.06.002.
[62] H. Liu, Y. Bu, Y. Mi, Y. Wang, Interaction site preference between carbon nanotube and nifedipine: a combined density functional theory and classical molecular dynamics study, J. Mol. Struct. THEOCHEM 901 (2009) 163–168. https://doi.org/10.1016/j.theochem.2009.01.021.
[63] H.-J. Lee, G. Kim, Y.-K. Kwon, Molecular adsorption study of nicotine and caffeine on the single-walled carbon nanotube from first principles, Chem. Phys. Lett. 580 (2013) 57–61. https://doi.org/10.1016/j.cplett.2013.06.033.
[64] V.A. Karachevtsev, E.S. Zarudnev, S.G. Stepanian, A.Y. Glamazda, M.V. Karachevtsev, L. Adamowicz, Raman spectroscopy and theoretical characterization of nanohybrids of porphyrins with carbon nanotubes, J. Phys. Chem. C 114 (2010) 16215–16222. https://doi.org/10.1021/jp104093q.
[65] M.D. Ganji, M. Mohseni, A. Bakhshandeh, Simple benzene derivatives adsorption on defective single-walled carbon nanotubes: a first-principles van der Waals density functional study, J. Mol. Model. 19 (2013) 1059–1067. https://doi.org/10.1007/s00894-012-1652-4.
[66] B. Yilmaz, J. Bjorgaard, C.L. Colbert, J.S. Siegel, M.E. Kose, Effective solubilization of single-walled carbon nanotubes in THF using PEGylated corannulene dispersant, ACS Appl. Mater. Interfaces 5 (2013) 3500–3503. https://doi.org/10.1021/am400442z.
[67] J. Goclon, M. Kozlowska, P. Rodziewicz, Noncovalent functionalization of single-walled carbon nanotubes by aromatic diisocyanate molecules: a computational study, Chem. Phys. Lett. 598 (2014) 10–16. https://doi.org/10.1016/j.cplett.2014.02.042.
[68] W. Orellana, Single- and double-wall carbon nanotubes fully covered with tetraphenylporphyrins: stability and optoelectronic properties from ab initio calculations, Chem. Phys. Lett. 634 (2015) 47–52. https://doi.org/10.1016/j.cplett.2015.05.055.

[69] R. Zacharia, H. Ulbricht, T. Hertel, Interlayer cohesive energy of graphite from thermal desorption of polyaromatic hydrocarbons, Phys. Rev. B Condens. Matter 69 (2004) 155406. https://doi.org/10.1103/PhysRevB.69.155406.
[70] A. Ambrosetti, D. Alfe, R.A. DiStasio, A. Tkatchenko, Hard numbers for large molecules: toward exact energetics for supramolecular systems, J. Phys. Chem. Lett. 5 (2014) 849–855. https://doi.org/10.1021/jz402663k.
[71] A. Tkatchenko, D. Alfe, K.S. Kim, First-principles modeling of non-covalent interactions in supramolecular systems: the role of many-body effects, J. Chem. Theory Comput. 8 (2012) 4317–4322. https://doi.org/10.1021/ct300711r.
[72] A. Sygula, F.R. Fronczek, R. Sygula, P.W. Rabideau, M.M. Olmstead, A double concave hydrocarbon buckycatcher, J. Am. Chem. Soc. 129 (2007) 3842–3843. https://doi.org/10.1021/ja070616p.
[73] J. Klimeš, A. Michaelides, Perspective advances and challenges in treating van der Waals dispersion forces in density functional theory, J. Chem. Phys. 137 (2012) 120901. https://doi.org/10.1063/1.4754130.
[74] C. Thierfelder, M. Witte, S. Blankenburg, E. Rauls, W.G. Schmidt, Methane adsorption on graphene from first principles including dispersion interaction, Surf. Sci. 605 (7–8) (2011) 746–749, https://doi.org/10.1016/j.susc.2011.01.012.
[75] S.S. Xantheas, Cooperativity and hydrogen bonding network in water clusters, Chem. Phys. 258 (2000) 225–231.
[76] G.A. Jeffery, An Introduction to Hydrogen Bonding, Oxford University Press, Oxford, England, 1979. https://doi.org/10.1021/ja9756331.

CHAPTER

6

Structural and electronic properties of covalently functionalized graphene

Tharanga R. Nanayakkara[a,b], *U. Kushan Wijewardena*[a,b], *Asanga B. Arampath*[a], *Kelvin Suggs*[a], *Natarajan Ravi*[c], and *Xiao-Qian Wang*[a]

[a]Department of Physics and Center for Functional Nanoscale Materials, Clark Atlanta University, Atlanta, GA, United States [b]Department of Physics and Astronomy, Georgia State University, Atlanta, GA, United States [c]Department of Physics, Spelman College, Atlanta, GA, United States

1 Introduction

Graphene, a single layer of graphite, is emerging as an extremely versatile material with remarkable properties and promising potential applications [1–5]. The application of graphene transistors, integrated circuits, and biosensors requires improved nanoscale control of the structural and electronic properties of graphene. As graphene hosts a plethora of phenomena associated with the Dirac-like linear dispersion, it remains an interesting topic to investigate how the unusual features impact the physics of adatoms on graphene. Specifically, since conductivity cannot be turned on and off effectively, pristine graphene cannot be used as a transistor in logic applications, where high on/off ratios are required. Field-switching capabilities depend on the presence of a bandgap in the electronic structure.

The key to graphene transistors is the control over the electronic properties as well as developing scale-up industrial-level techniques. While mechanical exfoliation leads to very high sample quality, the yield of monolayer graphene is meager [6,7]. The chemical vapor deposition is another famous technique in industry and research fields for producing large-scale graphene for relatively small cost [8,9]. Within the realm of single-layer graphene, electron confinement that occurs when rolling up graphene into single-walled nanotubes or cutting its edges to form nanoribbons is an effective way to open a gap in the band structure [10–17]. In this case, the translational symmetry is partially broken with the hybridization of two Dirac points. The result of opening a gap is inversely proportional to the confinement length.

Theoretical and Computational Chemistry, Volume 21
https://doi.org/10.1016/B978-0-12-819514-7.00008-7

Alternatively, the production of epitaxial graphene by evaporation of surface layers of SiC requires a somewhat restrictive temperature and time conditions [18]. The reduction of graphene oxide derived by the solution of graphite has attracted considerable attention [19]. However, the reduction of graphene oxide to ideally clean graphene is somewhat problematic due to the high energy of the chemical binding of hydroxyl groups with graphene.

Many approaches have been developed for noncovalently and covalently functionalized graphene [12,13,20–31]. Developing chemical methods to tune the properties of the material has become one of the most challenging issues in exploring graphene-based technologies. Different chemical modification methods have been demonstrated not only to enhance graphene solubility but also to provide suitable properties for graphene-based nanoelectronics and nanophotonic devices. The adaptation of graphene electronic characteristics has been accomplished by the well-established chemical functionalization methods. The functional groups such as hydrogen, oxygen, hydroxyl, hydroxide, epoxide, or fluorine bind covalently to carbon atoms on graphene while transforming the trigonal sp^2 orbital to the tetragonal sp^3 orbital. Such transformations severely modify the electronic properties of pristine graphene [32–36].

Recent experimental studies have demonstrated an efficient technique to functionalize the pristine graphene using nitrophenyl group covalently [28–31]. During the past few years, extraordinary development has been achieved for carbon-based spintronics. A series of theoretical and experimental studies have been carried out to reveal the spin relaxation mechanisms and spin transport properties in carbon materials, mostly for graphene and carbon nanotubes [37–44].

The latest experimental results show that the nitrophenyl-functionalized graphene has unique magnetic properties, even at room temperature. These experimental results demonstrate that nitrophenyl-functionalized graphene can act as an organic molecular magnet with ferromagnetic and antiferromagnetic ordering that persists at temperatures above 400 K. The nitrophenyl functionalization orientation and degree of coverage directly affect the magnetic properties of the graphene surface. The aryl-radical functionalization of epitaxial graphene not only changes the electronic properties but leads to disordered magnetism in the graphene sheet, which consists of a mixture of ferromagnetism (ferrimagnetism), superparamagnetic, and antiferromagnetic regions [45]. At the functionalization process, the transformation of the carbon centers from sp^2 to sp^3 introduces a barrier to the electron by saturating the carbon atoms and opening a bandgap that allows the generation of insulating and semiconducting regions in graphene wafers [46].

Despite its importance, a systematic study of an in-plane chemical functionalization for graphene is still lacking. Most importantly, there remains a paucity of investigations on how the electronic structure depends on the adsorption configuration of functional groups, which is pivotal for the practical realization of the electronic structure for engineering. Another important issue concerns the electron-mediated exchange interactions and the associated nearly free-electron states (NFE). NFE delocalize on the surface outside the atomic centers. When they are close to the band edge, NFE become a determining factor of the bandgap width. Since NFE have unique space distribution, we expect their response to surface chemical functionalization to be different from other states.

Nitrogen doping has also been considered as another effective way to manipulate the properties of graphene and make its potential use for numerous applications. However, a novel experimental technique to bandgap manipulation in graphene utilizing a nitrogen-seeded

SiC surface is reported [47]. Unlike traditional doping techniques with post-seeding the graphene with nitrogen [48–51], the novel method shows that sub-monolayer concentration of nitrogen absorbed on SiC, before the graphene growth, causes a significant bandgap opening. Moreover, a 0.7-eV bandgap opening is reported for a sub-monolayer concentration of bonded nitrogen at the interface of graphene-SiC.

For bilayer graphene to be used for field-effect transistors, the graphene intercalation compounds (GIC) must decouple the adjacent graphene layers and decrease the interlayer interaction. Here, we demonstrate the evolution of the band structure of graphene treated with fluorinated olefins through covalent functionalization. The bonding of fluorine to the graphene surface results in the transformation of orbital hybridization from sp^2 to sp^3. We find that the modification of graphene's electronic properties by such a drastic change in hybridization can lead to the elimination of the bands near the Fermi level and the opening of a bandgap.

Furthermore, a systematic search performed using particle swarm optimization search along with density functional theory calculations leads to a low-energy conformation for the C_2O phase of graphene oxide, which is a semiconductor with a bandgap of 2.8 eV. The novel structure consists of a combination of 1,3 and 1,2-epoxide bridges along with carbonyl groups, hence the name epoxy-carbonyl combination. The oxidization of graphene also transforms sp^2-hybridized regions to sp^3. The simulated structural, vibrational, and electronic properties of the novel structure are in good agreement with the experimental observations. Our findings elucidate the promising electronic properties of graphene-based materials for future device applications.

2 Nitrophenyl-functionalized graphene

Nitrophenyl-functionalized graphene perturbs the π-conjugation of graphene, and the corresponding electronic properties change from metallic to semiconducting. Nitrophenyl functionalization not only exposes the route for bandgap tailoring but creates a magnetic material with antiferromagnetism and ferromagnetism. Our work asserts the unique opportunity of tailoring the bandgap of graphene with varying chemisorption composition.

We consider six different configurations for nitrophenyl-functionalized structures, as described in Fig. 6.1. The structural and electronic properties were examined using the first-principles density functional calculations, which are based on density functional theory as implemented in the DMol3 package [52]. Perdew-Burke-Ernzerhof (PBE) parameterization of the generalized gradient approximation (GGA) was used in the calculations. A supercell with a vacuum space of 26 Å normal to graphene plane was used. A kinetic energy change of 5.4×10^{-4} eV in the orbital basis and appropriate Monkhorst–Pack k-point grids of $4 \times 4 \times 1$ were sufficient to converge the integration of the charge density. The optimization of atomic positions proceeds until the change in energy is less than 1.0×10^{-3} eV per cell. While the GGA calculation systematically underestimates the bandgaps, we mainly focused on the mechanism of bandgap opening. The GGA calculation is expected to provide qualitatively correct information and remains the general choice for examinations of covalent functionalization [52].

To examine the effect of addend density on the electronic structures, we consider two structures by adding two nitrophenyl polymers onto a 5×5 rhombus cell. The cell constitutes

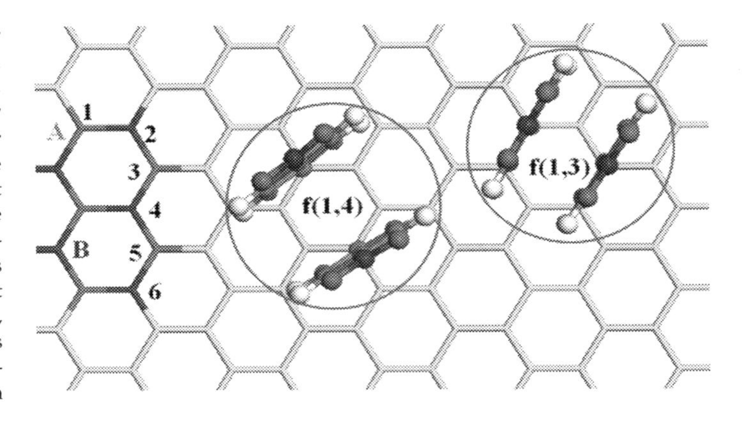

FIG. 6.1 Top view of the molecular structures of nitrophenyl-functionalized graphene. Carbon, nitrogen, oxygen, and hydrogen atoms are colored in *gray* (*green* (*dark gray* in the print version) for graphene), *light blue* (*light gray* in the print version), *red* (*dark gray* in the print version), and *white*, respectively. The letters A and B represent the two sublattices in the graphene sheet. Structures are labeled concerning the attachment of nitrophenyl groups. For example, f(1,4) means that first radical attaches to position 1, and then second radical attaches to position 4 in the same carbon ring in the graphene sheet.

50 carbon atoms for graphene, six carbon, two oxygen, one nitrogen, and four hydrogen atoms for each nitrophenyl molecule.

We consider the simplest Kondo perturbation, where the impurity is localized at a lattice site, and it only has an on-site Kondo interaction with the conduction electron spin. This type of conduction-electron-moderated spin coupling was first examined by Ruderman and Kittel [53], Kasuya [54], and Yosida [55], and thus, it is known as RKKY interaction. We consider a theorem for the RKKY interaction between these sites "impurities" in bipartite lattices with hopping between opposite AB sublattices at half-filling. The result is that the sign of the RKKY interaction depends only on whether the impurities are localized at opposite sublattices antiferromagnetically or on the same sublattices ferromagnetically. The sign is dictated by particle-hole symmetry and is thus valid on all length scales.

The conformations of nitrophenyl-functionalized graphene are shown in Fig. 6.1. The adduct increases the bond length between graphene and atoms adsorbed to it. The corresponding bond length between the C atom on graphene and the C atom of the adduct nitrophenyl is around 1.59 Å for f(1,3) configuration, and the corresponding value for f(1,4) configuration is 1.58 Å. The shortest distance between functionalized carbon atoms on graphene in f(1,3) configuration is around 2.58 Å, while this value for the f(1,4) configuration is about 2.89 Å. The latter C—C distance is significantly longer than the C—C bond length of 1.42 Å of pristine graphene with sp^2 hybridization and suggests a bond breaking. The C—C bond lengths in graphene beyond the nearest neighbors are found to be slightly changed by the nitrophenyl functionalization.

The interaction from the graphene-addend in the covalent functionalization has a straight influence on the electronic properties of pure graphene. Early theoretical studies examined the addition of functional groups as free radicals to graphene [30,56]. These functional groups severely change the geometries and electronic structures of pristine graphene by initiating native sp^3 hybridization defects, which induce the sp^3-type "impurity" state close the Fermi level. In the divalent functionalization, two sp^3 states induced by two adjacent functional groups are moved away from the Fermi level due to the rehybridization into bonding and antibonding states [57]. Hence, the native bonding arrangement can notably affect the electronic structure of functionalized graphene.

The calculated band structures for nitrophenyl-functionalized graphene are shown in Fig. 6.2, along with the pristine graphene for comparison. Interestingly, there is a 0.62-eV bandgap opening at the K point for f(1,4) configuration indicated in Fig. 6.2 *right panel*. After the nitrophenyl functionalization on the different sublattices, the π and π^* linear dispersion of pure graphene in the vicinity of the Dirac point (K) highly protects, even though a gap occurs between the π and π^* states. These electronic properties of nitrophenyl-functionalized compounds are in severe distinction to the sp^3 rehybridization and loss of π electrons found upon the addition of monovalent chemical sites in other functionalized graphene structures. The lack of sp^3 type (non-planar) "impurity" states in the locality of the Dirac point is consistent with the reasoning that the C—C bond between the two bridgehead atoms 1 and 4 (see Fig. 6.1) is either demolished or significantly weakened, leading to partial restoration of the π electron system.

Conversely, our results on nitrophenyl functionalization are peculiar to those with noncovalent functionalization. In noncovalent functionalization, there is a small adjustment of band structures close to the Fermi level, and the resultant band structure constitutes dispersed and flat bands that can be characterized as arising from the contributions of both nitrophenyl functional groups and pristine graphene. By contrast, the nitrophenyl-functionalized graphene displays extreme levels of hybridizations. The bandgap opening at the Dirac point indicates significant perturbations created by the functionalization. While attaching the nitrophenyl to the C atoms on graphene, it preserves the sp^2 hybridization on graphene, but the sp^2 hybridization angle is modified. Consequently, the electronic structure of graphene is undoubtedly affected by nitrophenyl functionalization. An essential

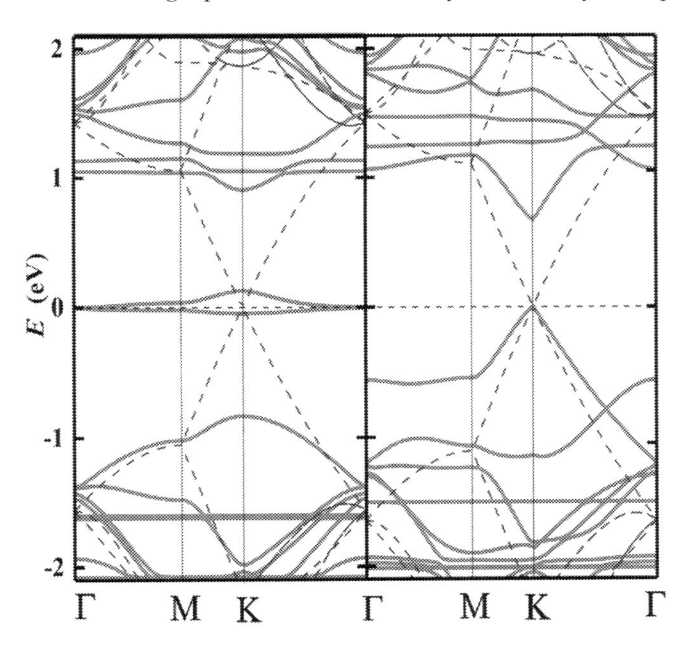

FIG. 6.2 Calculated band structures for nitrophenyl-functionalized graphene f(1,3) (*left panel*), f(1,4) (*right panel*), and pristine graphene (*dashed blue line; dark gray* in the print version) respectively. $\Gamma = (0,0)$, $K = (\pi/3a;\ 2\pi/=3a)$, and $M = (0;\ \pi/2a)$, where $a = 12.3$ Å for a 5×5 rhombus unit cell. The Fermi level is shifted to 0 eV (*dashed black line*).

consequence of the [2+1] cycloaddition-generated perturbation is that the adjustment in the electronic structure of graphene increases with incrementing nitrophenyl functionalization concentration. We examined the functionalization of graphene at a higher degree of coverage by including another nitrophenyl functional group in the unit cell. The bridgehead C—C bond destroying continues at higher concentrations of the functionalized groups according to results from geometry optimizations.

Careful examination of the band arrangements and dispersions near the Dirac point exposes that the gap opening is mainly due to the functionalization-induced modifications of the π conjugation. The interruption of the original π conjugation is displayed in the level hybridization, as detected in the band structure (Fig. 6.2). Notably, the lowest unoccupied molecular level (LUMO) and the highest occupied molecular orbital (HOMO) of nitrophenyl arrange with the π^* and π bands of graphene at about 1 and -1 eV, respectively. The band arrangement is such that the coupling between dispersed and flat bands leads to hybridization-persuaded level avoided-crossing, which directs to the splitting of π and π^* bands of graphene into two hybridized bands each.

Spin densities of the corresponding hybridized bands at the band center are shown in Fig. 6.3. For those states, the spin density distributions show spin-charge confinements on carbon atoms in graphene, which is bonded with the nitrophenyl addends for hybridized valence bands and conduction in f(1,4) conformation, which is to be contrasted to the f(1,3) conformation. Scrutiny of the spin density distributions in f(1,4) structure indicates the existence of π and π^* bonds in the hybridized states that contribute to the gap formation as well.

In this section, we have examined the electronic characteristics of nitrophenyl-functionalized graphene. The nitrophenyl addition to the pure graphene conserves the sp^2 hybridization network of the carbons on graphene. However, nitrophenyl functionalization significantly modifies the π conjugation of graphene near the Fermi level. The modification in π conjugation

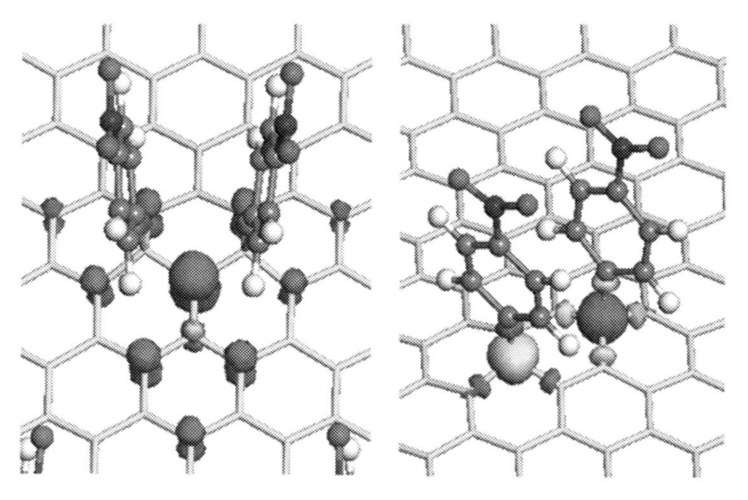

FIG. 6.3 The isosurface plot of the spin density of nitrophenyl-functionalized graphene. Ball-and-stick presentation of optimized structures of nitrophenyl-functionalized graphene with configuration f(1,3) and f(1,4) in the *left and right panels*, respectively. The isovalue is 0.025 a.u.

of graphene leads to the opening of a bandgap, which depends on the degree of coverage of nitrophenyl groups. This is a difference with the free-radical functionalization in graphene, where the sp^3-type band is induced close to the Fermi level. Such dependence of the electronic properties on the concentration of functionalization of graphene suggests a novel and controllable technique for the "bandgap engineering" of graphene. Our results on the nitrophenyl functionalization of graphene may serve as useful strategies for enabling the flexibility and optimization of graphene-based nanodevices for future applications.

3 Nitrogen-seeded twisted bilayer graphene

In this section, we examine a promising mechanism for bandgap engineering in twisted bilayer graphene using nitrogen atoms. Our results reveal that nitrogen bonded at the graphene-graphene interface leads to a ~0.75 eV bandgap opening in graphene. Besides, the bilayer graphene on top of 4H-SiC conserves the linear dispersion around the Dirac point. The 4H-SiC is one of the polytypes of SiC, which consists of four alternating layers along the [0001] direction. We find that there is a ~0.7 eV bandgap opening at the *K* point for nitrogen-seeded 4H-SiC with twisted bilayer graphene, which validates the recent experimental results [47].

Our first-principles calculations were based on GGA with the exchange-correlation of PBE parameterization and the many-body perturbation theory utilizing the *GW* approximation. We employed the dispersion correction using the Tkatchenko-Scheffler (TS) scheme, which exploits the relationship between polarizability and volume. The TS dispersion correction takes into account the relative variation in dispersion coefficients of various atomic bonding by weighing values extracted from the high-quality ab initio database with atomic volumes derived from partitioning of the self-consistent electronic density.

To model the nitrogen-seeded bilayer graphene, we used twisted bilayer graphene intercalated with two nitrogen atoms. A supercell with a vacuum space of 17.5 Å normal to the bilayer graphene plane was used. Nitrogen concentration of the model corresponds to 0.4 ML ($\sim 4.52 \times 10^{14}$ cm^{-2}) of the area covered by SiC. The optimization of atomic positions continued until the change in energy was less than 2×10^{-3} eV, and the forces were less than 5×10^{-2} eV/Å. Ultra-soft pseudopotentials and a kinetic energy cutoff of 500 eV were employed with Monkhorst–Pack meshes of $6 \times 6 \times 1$.

The GGA of the DFT is known to fail in describing electron–hole (*e–h*) and electron–electron (*e–e*) interactions. These are responsible for the formation of excitons and the quasiparticle excitations. In this regard, the GW-Bethe-Salpeter equation (*GW*-BSE) approach represents one of the state-of-the-art theories beyond DFT. The screened Coulomb interaction was calculated within the random-phase approximation (RPA) and the plasmon-pole approximation [58]. While RPA can be equivalent to the results of the DFT level, *GW*-RPA includes *e–e* interactions, and *GW*-BSE goes beyond RPA by including *e–e* and *e–h* interactions.

Shown in Fig. 6.4 is the prototype system for the nitrogen-seeded graphene bilayer. Two nitrogen atoms are connecting to the twisted bilayer graphene. Nitrogen has three bonds: two bonds are connected to one layer and one to the other of graphene. The corrugations of the top layer and the bottom layer are 0.46 Å due to the nitrogen intercalation. The calculated maximum and minimum layer distances are 3.7 and 2.7 Å, respectively. The bond

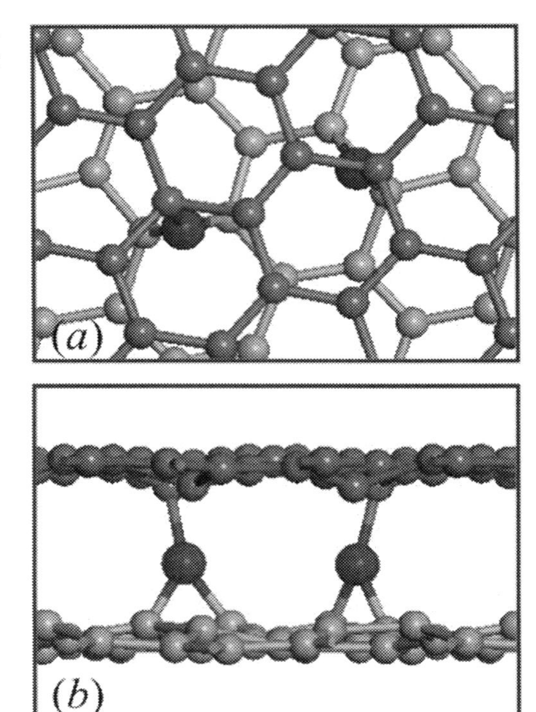

FIG. 6.4 Top (A) and side (B) views of the optimized conformation of nitrogen-seeded bilayer graphene with a twist angle of $\theta = 21.8°$, respectively.

lengths between carbon and nitrogen to the top and bottom layers are 1.5 and 1.4 Å, respectively. The bond lengths to the bottom layer are shorter due to two bonds connectivity to nitrogen.

The conformation was constructed from twisted bilayer graphene with an orientation angle of $\theta = 21.8°$. The twisted bilayer graphene is the smallest commensurate twisted structure of graphene. The twisted stacking substantially weakens the interlayer coupling. The corresponding band structure is characterized by degenerate linear dispersions reminiscent of monolayer graphene. The nitrogen atoms being intercalated between the layers mimic the nitrogen seeding effect observed experimentally. For maintaining the charge neutrality, a pair of nitrogen atoms was inserted simultaneously. The introduction of nitrogen atoms generates a sp^3 bond for the C-atoms. For each nitrogen atom, three bonds are introduced, linking the bilayers.

Depicted in Fig. 6.5 are the calculated band structures (A) and (B) for twisted bilayer graphene in dashed lines, along with nitrogen-seeded graphene in solid lines for comparison. It is observable that, after nitrogen seeding, the π and π^* linear dispersion of twisted bilayer graphene in the proximity of the Dirac point (K) is substantially modified, which leads to a bandgap opening of ~0.75 eV.

Shown in Fig. 6.6 are charge densities of the corresponding hybridized bands at the band center (the Γ point). The graphene layer highlighted with dark green has two nitrogen atoms

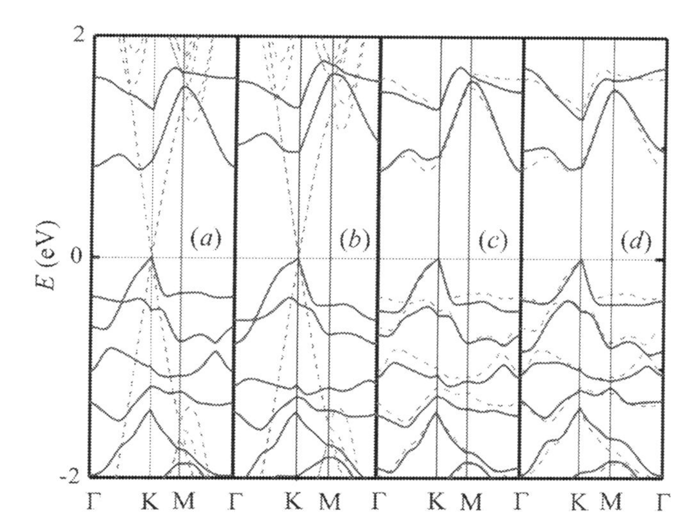

FIG. 6.5 Calculated band structure for the nitrogen-seeded bilayer graphene. *Solid lines* represent the results (A–D) with GGA, GW, GGA with 3% compression, and GGA with 3% tensile strain, respectively. *Dashed lines* refer to the results of a twisted bilayer graphene with an orientation angle of $\theta = 21.8°$ (A) and (B), and the results of the unconstrained conformation (C) and (D), respectively. $\Gamma = (0,0)\ \pi/a$, $K = (-\pi/3a, 2\pi/3a)$, and $M = (0, \pi/2a)$, where $a = 6.650$ Å for the unconstrained conformation. The Fermi level, highlighted by the *red line* (*dark gray* in the print version), is shifted to 0 eV.

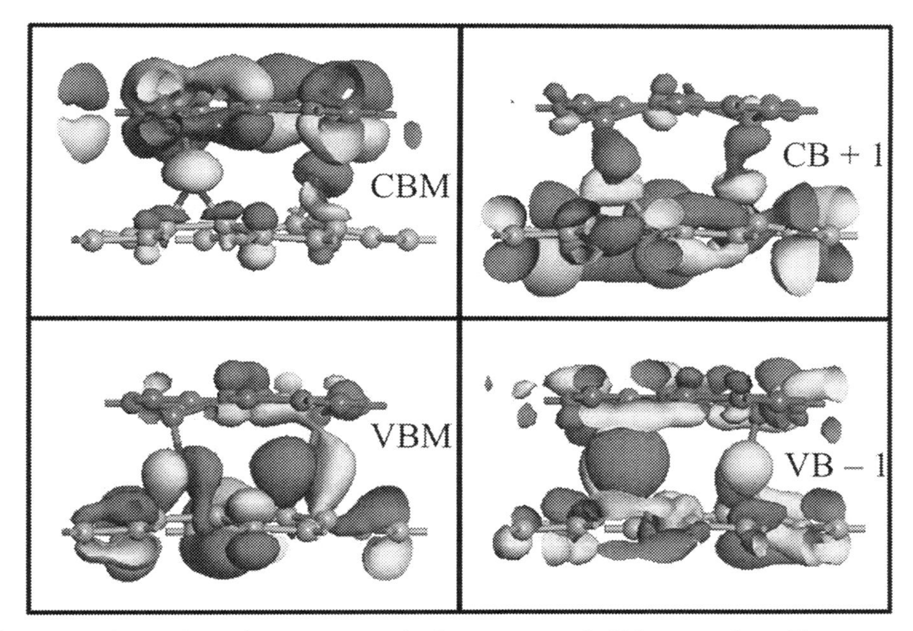

FIG. 6.6 Extracted isosurfaces of near-gap states for the nitrogen-seeded bilayer graphene. The components of the wave function are colored with red (*dark gray* in the print version) and cyan (*light gray* in the print version), respectively. Isosurface value is 0.03 a.u.

connected by single bonds. The other layer highlighted with orange has the nitrogen atoms connected by two single bonds. As seen from Fig. 6.6, the "orange layer" and "green layer" have profound charge density for the valence band maximum (VBM) and conduction band minimum (CBM), respectively. This is to be contrasted with the conjugated π and π^* patterns on graphene, where the symmetry of the charge density is preserved. The charge density distributions reveal the existence of σ and σ^* bonds in the hybridized states that contribute to the gap formation.

Summarized in Fig. 6.7 is the calculated in-planes photoabsorption spectra using *GW*-BSE, along with those using RPA and *GW*-RPA for nitrogen-seeded epitaxial graphene. The optical absorption spectra can be divided into two regions. The low-energy region goes up to 5 eV and is attributed to the transitions among the π and π^* bands. In the region beyond 10 eV, the spectra originate from the σ and σ^* transitions. Calculated photoabsorption spectra in Fig. 6.7 are in reasonable agreement with the predictions of bandgap opening 0.75 eV in the calculated band structure in Fig. 6.5. Specifically, a common spectroscopic attribute of the systems is the existence of a prominent π-π^* RPA peak at ~0.63 eV. *GW*-BSE, which compares with the results of *GW*-RPA, are at ~0.70 and ~0.82 eV, respectively.

The *right panel* of Fig. 6.8 shows the nitrogen-seeded 4H-SiC with bilayer graphene. The two nitrogen atoms are connected to C-face in 4H-SiC. The *middle panel* in Fig. 6.8 shows the calculated band structure for bilayer graphene on top of the 4H-SiC and nitrogen-seeded 4H-SiC with bilayer graphene. As seen from Fig. 6.8, the bilayer graphene on top of 4H-SiC preserves the linear dispersion around the Dirac point. However, in nitrogen-seeded 4H-SiC with bilayer graphene, there is a bandgap opening at the *K* point, which validates the recent experimental results [47].

In nitrogen-seeded twisted bilayer, graphene nitrogen preserves the sp^2 hybridization network of the carbons on graphene. Here, the π conjugation of graphene near the Fermi level is much disturbed by nitrogen insertion, which causes a bandgap opening that depends upon the addend concentration.

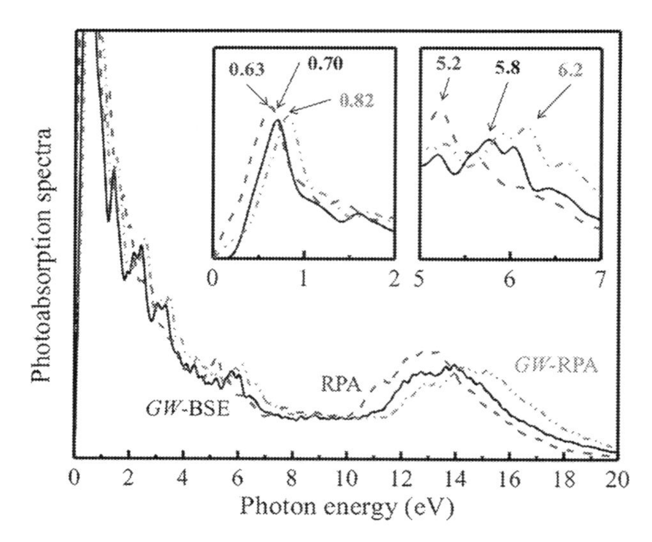

FIG. 6.7 Calculated photoabsorption spectra using *GW*-BSE (*solid black lines*), RPA (*blue dashed lines; dark gray* in the print version), and *GW*-RPA (*red dashed-double dotted lines; light gray* in the print version) for nitrogen-seeded epitaxial graphene. *Insets*: closeups in the regions of 0–2 and 5–7 eV, respectively.

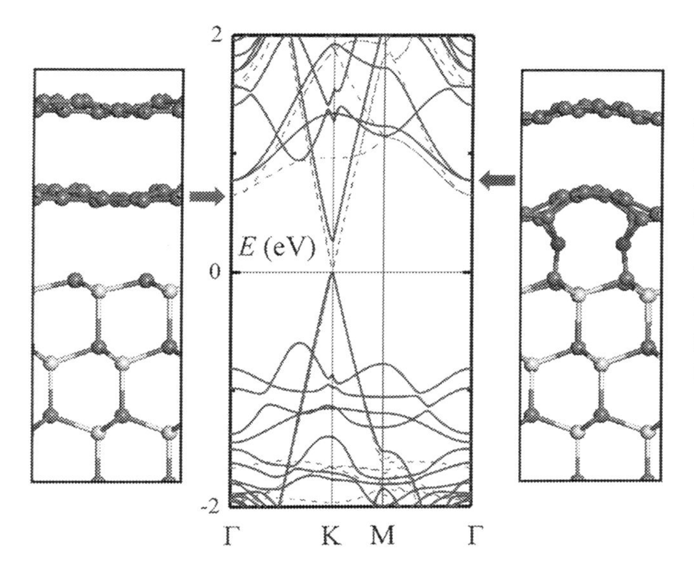

FIG. 6.8 The calculated band structure of the bilayer graphene on top of the 4H-SiC is represented by the *green dashed lines* (*dark gray* in the print version) and nitrogen-seeded 4H-SiC with bilayer graphene is represented by *solid blue lines* (*dark gray* in the print version). *Left and right panels* represent the side view of bilayer graphene on top of the 4H-SiC and nitrogen-seeded 4H-SiC with bilayer graphene, respectively. $\Gamma = (0,0)\ \pi/a$, $K = (-\pi/3a, 2\pi/3a)$, and $M = (0, \pi/2a)$, where $a = 6.160$ Å. The Fermi level, highlighted by the *red line* (*dark gray* in the print version), is shifted to 0 eV.

4 [2+2] Cycloaddition of graphene

Recent experimental works have demonstrated that cycloaddition of fluorinated olefins represents a practical approach to reduce the off currents of mixed nanotube materials for transistor applications [59–61]. We furthermore study the electronic structure characteristics of the corresponding [2+2] cycloaddition using dispersion-corrected density functional calculations. The bandgap opening in chemically functionalized graphene is associated with the sp^2 to sp^3 rehybridization. We predict/calculate that the experimentally observed suppression of semi-metallic conductivity can be attributed to asymmetry-aligned cycloaddition scheme that transforms semi-metallic graphene to semiconducting. Despite the exciting experimental findings, the mechanism of the elusive conversion remains unclear. Specifically, it is suggested that the change from semi-metallic to semiconducting behavior could be induced either through scattering centers associated with the covalent functionalization or through band structure modifications [62–64].

The cycloaddition of olefins involves an unusual thermally allowed [2+2] cycloaddition. Specifically, the highest occupied molecular orbital (HOMO) of the olefin, π, intersects with the conduction band of the graphene. In contrast, the lowest unoccupied molecular level (LUMO) of the olefin, π*, interacts with the valence band of graphene. Our findings demonstrate the nature of the [2+2] cycloaddition of perfluoro-(5-methyl-3,6-dioxane-1-ene) (PMDE) with graphene. The structural and electronic properties and optimization of geometry were calculated within the framework of the DFT as implemented in the DMol3 package. Dispersion- and gradient-corrected PBE was used [52]. The local density approximation

(LDA) approach is suitable for weakly interacting π systems, while dispersion-corrected GGA provides a more accurate description. The geometry optimization convergence criterion was satisfied when the total energy change was less than 3 $x10^{-5}$ eV. The GGA in exchange-correlation parameterization was used for band structure calculation. For the band structure computation, the selected path was *Γ–M–K–Γ*.

Although the GGA approach systematically underestimates the bandgaps, we were primarily interested in the mechanism of bandgap opening. For that purpose, the GGA approach was expected to provide qualitatively correct information. Periodic-boundary conditions were employed with a supercell in the planar plane large enough to eliminate the interaction between neighboring replicas. A double numerical basis was sufficient for the grid integration of the charge density to converge. All structures were relaxed with forces less than 0.01 eV/Å.

We illustrate in Fig. 6.9 the two optimized structures of PMDE, which we label pmde1 and pmde2. Fig. 6.9A shows PMDE adduct attached to the third and fourth carbon. Likewise, Fig. 6.9B shows the PMDE adduct attached to the second and third carbon. The adduct distorts the graphene surface in both configurations by lifting the two carbon atoms attached to it above the other atoms in the graphene lattice.

The adduct has increased in the bond lengths connecting the atoms on graphene to 1.50 Å. The length of the C—C bonds within the four-membered ring is ~1.59 Å for both PMDE orientations, including the deposition of metal or molecules on graphene and intercalation of metal clusters or molecules between the graphene layers. Fig. 6.10 depicts the calculated band structure for pmde1 and pmde2. The electronic properties are tunable from metallic to semiconducting with hydrogen intercalation. The preparation of high-quality graphene as host materials is pivotal for further device applications. As such, an in-depth understanding of the electronic properties of graphene derivatives is critical to the successful integration of graphene into future nanoelectronic devices [65–69].

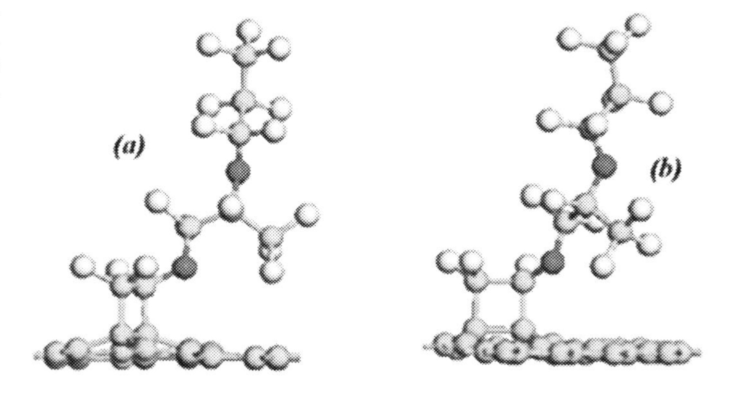

FIG. 6.9 Perspective view of optimized structures of two layers of pmde1 and pmde2 connecting with graphene is shown in the *left and right panels*, respectively.

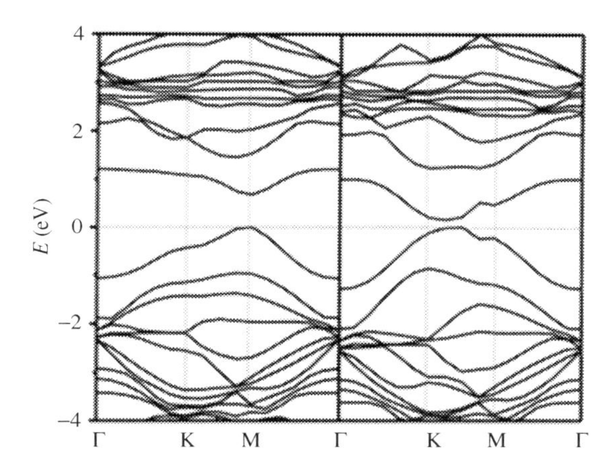

FIG. 6.10 Calculated band structure for pmde1 and pmde2, respectively. $\Gamma = (0,0)\ \pi/a$, $K = (-\pi/3a, 2\pi/3a)$, and $M = (0, \pi/2a)$, where $a = 4.887$ Å. The Fermi level, highlighted by the *dashed red line* (*light gray* in the print version), is shifted to 0 eV.

5 Graphene oxide (GO)

Graphene oxide (GO) forms by the oxidation of pristine graphene, which fundamentally alters the material characteristics such as electrical, mechanical, and thermal properties. It is advantageous to have an insight on such altered features of GO that enables the production and improvement of GO-based devices. Usage of graphene and reduced graphene oxides in electrochemical energy storage [70], production of micro supercapacitors [71], using in temperature sensor applications [72] can be given as a few examples. The solubility of GO in water and other polar solvent provides an avenue for application in the field of biomedicine such as drug delivery [73] and biosensing [74].

Although there are numerous theoretical studies on GO configurations [75–79], an in-depth explanation of the molecular structure of GO is demanding some more explorations. The complex composition variations and the inhomogeneous oxidation are the challenges that hinder the ability to identify a well-defined structure of GO. Previously proposed models for oxidation suggest arrangements that contain 1,2-epoxides [80], which lead to theoretical studies of structures such as a boat [81], twist-boat [82], and epoxy-pair [83] conformations. Furthermore, there are other conformations which involve hydroxyl groups, 1,3-epoxides, as well as some combinations of hydroxyl and carbonyl functional groups [80]. Among these different oxidation functional groups, the 1,2-epoxide arrangement is found to be more energetically favorable over the 1,3- and 1,4-epoxides due to the lesser disruption to the sp^2 graphene layer comparatively.

The calculations were made employing CALYPSO [84,85], with a search space of eight carbon and four oxygen atoms in a unit cell. Even though hydroxyl groups are found in GO, we are only focusing on the C_2O phase of GO. Our initial structure was arranged in three layers containing two oxygen atoms in both top and bottom layers and eight carbon atoms in the middle layer. Geometry and cell optimizations were achieved by performing the first-principles calculations using CASTEP with PBE parameterization of the GGA [86,87]. Ultra-soft pseudopotentials were used. DFT semi-empirical dispersion correction (DC) was employed [88]. We found that the PBE calculations with dispersion corrections were efficient. However, we confirmed our results by comparing them with PBE0 estimates. A cutoff of 700 eV for the kinetic energy in the plane-wave basis and Monkhorst–Pack *k*-points ($6 \times 6 \times 1$) were used to achieve the energy convergence in PBE calculations. The atomic positions were optimized until the energy difference got smaller than 1×10^{-4} eV per atom.

The DFT method with LDA is known to underestimate the bandgap. Therefore, PBE0 hybrid functionals were used even though the process is more computationally expensive. Norm-conserving pseudopotentials were employed. A kinetic energy cutoff of 400 eV and Monkhorst–Pack *k*-points ($6 \times 6 \times 1$) were used. DMol3 [52] with GGA and PBE parameterization with the double numerical precision (DNP) basis set were used for the vibrational modes calculations.

The results that we report here are obtained via a crystal-structure analysis by particle swarm optimization (CALYPSO) accompanying with density functional theory (DFT) calculations [84,85]. A conformation search performed systematically by generating around 3000 different atomic structures leads to a few low-energy arrangements. Closer examination of extracted structures revealed another low-energy conformation, the epoxy-carbonyl combination (ECC).

Fig. 6.11 shows the perspective views of the epoxy-carbonyl combination along with epoxy-pair, boat, and twist-boat, which we use for the comparison later. About 50% of carbon atoms in the graphene are oxidized in the ECC. Graphene is functionalized to form three epoxide bonds aggregated together. The surface roughness of the ECC is ~3.3 Å. The surface corrugation can be further increased if the ECC oriented in a manner that the carbonyl group is pointing toward both sides out of the plane. Such an alternating ECC has corrugations of ~5.2 Å, which is in good agreement with the experimentally observed roughness of ~6 Å [19]. The increased roughness can be associated with the lattice deformation caused by oxidation in the pristine graphene.

The fully oxidized twist-boat structure has epoxides running along the armchair direction, and boat configuration consists of epoxides running along the zigzag path. In these, epoxides are uniformly distributed on both the top and bottom sides of graphene. The corresponding surface corrugation is ~2.8 Å in these two conformations. The epoxy-pair conformation has a roughness of ~2.1 Å, which contains epoxide pairs running along the zigzag direction for the C_2O phase that amounts to 50% oxidation.

There are three epoxy bridges and a carbonyl group in the unit cell of ECC. Two of the epoxies are 1,2-epoxy, and one is 1,3-epoxy. The oxidation of graphene might break the bonds between carbon atoms that would cause a shift in atoms that lead to a 1,3-epoxide and a carbonyl group. These epoxy bridges cause an expansion of ~20% in the zigzag direction, while the oxidation happens along the armchair direction (Fig. 6.11). When considering the four prototype oxidation groups depicted in Fig. 6.11, the 1,2-epoxy was found to be the lowest in energy that might be related to the order of disturbance to the π conjugation.

Furthermore, the net charge neutrality between A and B sublattices is preserved when there is 1,2- or 1,4-epoxides. Therefore, 1,2-epoxy is found to be the predominant pattern for single oxygen oxidation. It is noteworthy that all the stable configurations studied so far maintain net charge neutrality. Therefore, looking at the charges induced by the functionalized oxygen may reveal vital information. The electronegativity of oxygen is higher than that of carbon, which would induce a slight negative charge on oxygen while making the

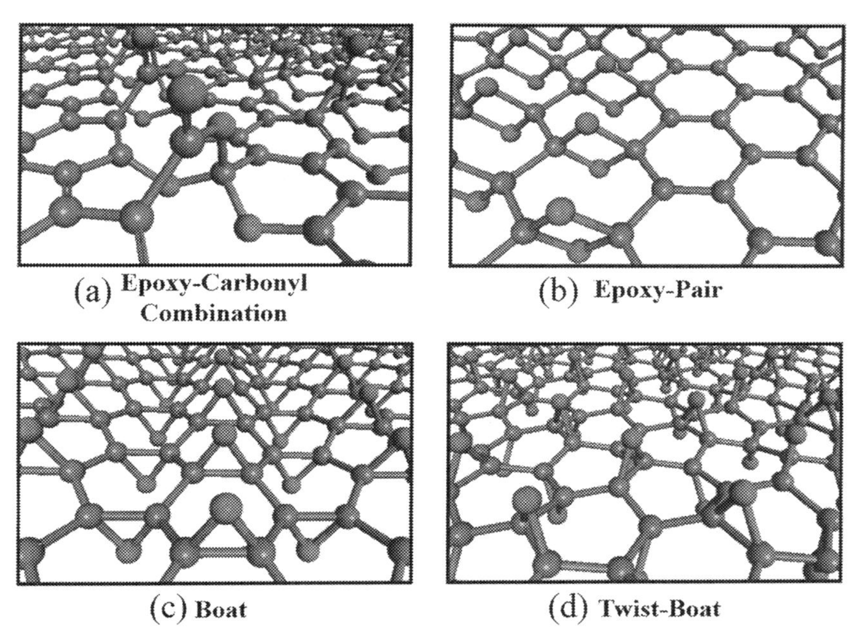

FIG. 6.11 Perspective views of chosen GO motifs of the C_2O phase. (A) Epoxy-carbonyl combination, which represents oxidization along the armchair direction. (B) GO configuration that corresponds to 50% oxidization, which contains pairs of epoxides along the zigzag direction. (C) Boat conformation that completely oxidized with only epoxide bridges along the zigzag direction on both top and bottom sides of the plane. (D) Twist-boat conformation that has epoxides along the armchair direction. *Reprinted (adapted) with permission from U.K. Wijewardena, S.E. Brown, X.Q. Wang, Epoxy-carbonyl conformation of graphene oxides, J. Phys. Chem. C 120 (2016) 22739–22743. Copyright (2016) American Chemical Society.*

adjacent carbon slightly positive. It is essential to understand how additional oxygen atoms would attach while keeping the structure still a low-energy conformation.

Epoxy-pair conformation includes two oxygen atoms per unit cell, A and B sublattices experience an equivalent deficiency of electrons due to the two 1,2-epoxy bridges, which preserves the neutrality of net charges. The low steric hindrance of the epoxy-pair is a favorable quality. Hence, epoxy-pair is the most energetically favorable structure when there are two oxidation groups [83]. It is worth noting that epoxy-pair supports the hypothesis of the tendency that oxygen has to aggregate when forming functionalization patterns. The aggregation can be associated with the strong, attractive interactions that oxygen atoms have among themselves [81,89–92].

On the other hand, epoxy-carbonyl combination has four oxygen atoms functionalized in a unit cell, which are aggregated closely, forming epoxide groups. The two 1,2-epoxides maintain the net charge neutrality between A and B sublattices as similar to the previous case. However, 1,3-epoxides would induce an electron deficiency on the sublattice A. At the same time, the carbonyl group stimulates a lack of electrons on sublattice B. The combination of all the functional groups would yield the net charge between the sublattice A and B to be neutral.

The solubility of GO is influenced by the localized charges formed by the polarization of epoxy bridges and carbonyl groups.

Charge densities follow the zigzag direction by arranging the network of π and π* in pristine graphene (see Fig. 6.12). The sp^2-conjugated chain is interrupted by functionalized oxygen in ECC. Charge densities at the center of the band (*Γ* point) for valence band maximum (VBM) and conduction band minimum (CBM) are shown in Fig. 6.13. The disruption to the

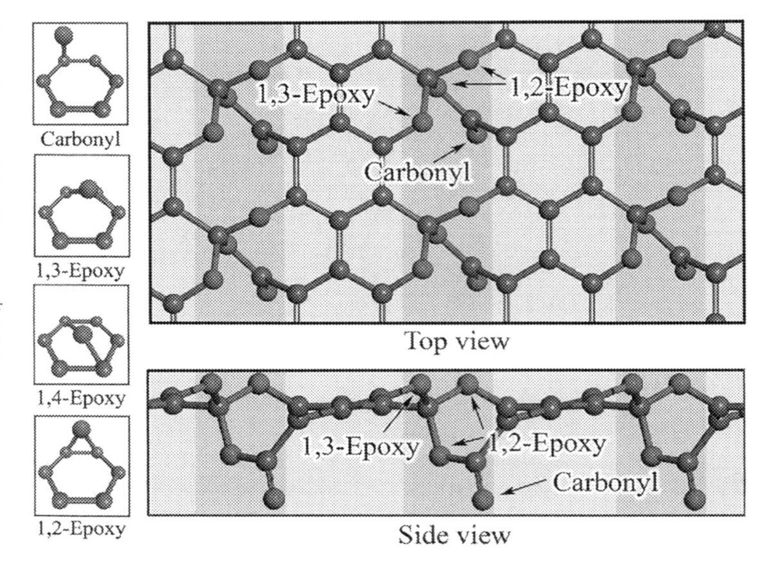

FIG. 6.12 *Left panel*: possible models of oxide functional groups on carbon. Two panels in the *right*: top and side views of the epoxy-carbonyl combination are shown in the top and bottom figures, respectively. *Pink color* (*light gray* in the print version) shows the area functionalized by oxygen. *Reprinted (adapted) with permission from U.K. Wijewardena, S.E. Brown, X.Q. Wang, Epoxy-carbonyl conformation of graphene oxides, J. Phys. Chem. C 120 (2016) 22739–22743. Copyright (2016) American Chemical Society.*

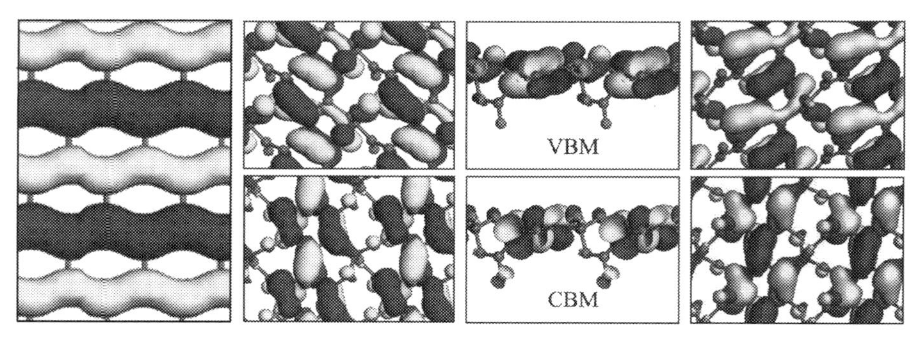

FIG. 6.13 *Left panel*: π and π* network in pristine graphene. The *second, third and fourth panels* from the *left* show the top, side, and bottom view of the charge density, respectively, on the epoxy-carbonyl combination. *Three panels in the top* represent VBM, and the *bottom three panels* are showing CBM (isovalue of 0.03 a.u.). *Reprinted (adapted) with permission from U.K. Wijewardena, S.E. Brown, X.Q. Wang, Epoxy-carbonyl conformation of graphene oxides, J. Phys. Chem. C 120 (2016) 22739–22743. Copyright (2016) American Chemical Society.*

TABLE 6.1 Binding energy per oxygen (E_B), bandgap (E_g), and associated unit cell ($a \times b$) for the boat, twist-boat, epoxy-pair, and epoxy-carbonyl combination

Structure	E_B^{PEE} (eV)	E_B^{PBE0} (eV)	E_g^{PEE}(eV)	E_g^{PBE0} (eV)	$a \times b$ (Å^2)
Boat	−23.69	−27.28	3.20	5.59	4.44 × 5.18
Twist-boat	−23.87	−27.46	4.37	7.16	5.10 × 4.40
Epoxy-pair	−24.02	−27.59	0.00	0.60	2.55 × 9.94
Epoxy-carbonyl	−24.37	−27.94	1.67	2.82	5.89 × 4.22

Reprinted (adapted) with permission from U.K. Wijewardena, S.E. Brown, X.Q. Wang, Epoxy-carbonyl conformation of graphene oxides, J. Phys. Chem. C 120 (2016) 22739–22743. Copyright (2016) American Chemical Society.

sp^2 chain produces an energy gap. The extracted binding energies per oxygen atoms are given in Table 6.1. The calculations were performed by using the PBE and PBE0 hybrid functionals.

Among the four structures studied here, the epoxy-carbonyl combination is most energetically favorable. The binding energy of ECC is 0.35, 0.5, and 0.68 eV/O lower than that of epoxy-pair, twist-boat, and boat configurations, respectively. The calculated values for the binding energies always depend on the exchange-correlation functional used. Nonetheless, the energy order of GO motifs is consistent with previous studies [81–83].

Moreover, the disruption to the π–π^* conjugated network appears to be correlated with the energy order of the conformations. The oxidation converts the sp^2 system in pristine graphene to sp^3 completely for boat and twist-boat conformations. However, the epoxy-pair structure has a 50% sp^2 network. The ECC has ~60% sp^2 and ~40% sp^3 regions, as shown in colored areas in Fig. 6.13.

As shown in Table 6.1, the unit cell lengths along armchair and zigzag are 5.89 and 4.22 Å, respectively. The cell lengths observed experimentally are 5.46 (±0.16) and 4.06 (±0.13) Å [93,94]. When comparing the cell dimensions among the structures we studied here, ECC is the only one that shows shrinkage along the armchair direction. Twist-boat and boat conformations predicted expansion in both directions. Epoxy-pair only expanded to the armchair direction. Epoxy-carbonyl combination, on the other hand, shows shrinkage along with the armchair and expansion along the zigzag, which agrees with experimental observations.

The band structures calculated using PBE and PBE0 functionals for the epoxy-carbonyl combination, twist-boat, boat, and epoxy-pair are depicted in Fig. 6.14. The PBE calculation predicted that the epoxy-pair conformation is semi-metallic, while the PBE0 calculation shows a band opening of 0.6 eV. The difference can be associated with the long exchange employed in the PBE0 hybrid functional. The smaller bandgap of the ECC compared to the boat and twist-boat can be due to the coexistence of sp^3- and sp^2-hybridized regions.

The bandgap calculated with PBE0 for the ECC is 2.82 eV, which is in perfect agreement with the experimental observation of 2.88 eV [95–98]. The experimental values were obtained from photoluminescence peaks for GO. The flat band along the Γ to Z direction on the band structure of ECC (Fig. 6.14) can be associated with the disruption to the sp^2 chain that reduces the electron mobility. It is worth noting that low carrier mobility is another experimental observation of GO that our findings are in agreement [99–101].

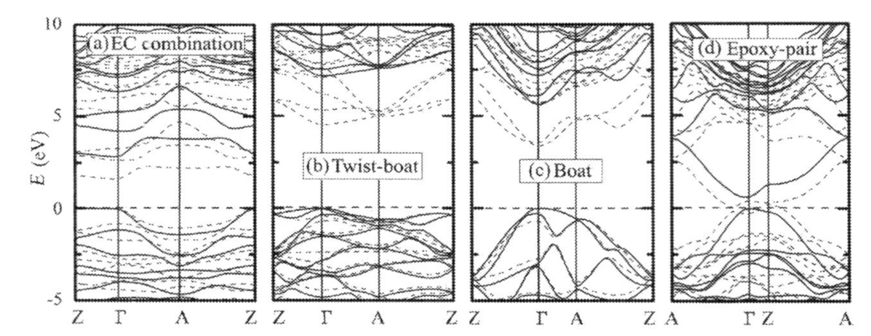

FIG. 6.14 Calculated band structures of GO in C2O phase: (A) epoxy-carbonyl combination, (B) twist-boat, (C) boat, and (D) epoxy-pair conformations. *Dashed and solid lines* represent the calculated band structures using PBE and PBE0 functionals, respectively. $\Gamma = (0, 0)$; $Z = (\pi/2a, \pi/2b)$; $A = (\pi/2a, 0)$. a and b are cell dimensions in zigzag and armchair directions, respectively. *Reprinted (adapted) with permission from U.K. Wijewardena, S.E. Brown, X.Q. Wang, Epoxy-carbonyl conformation of graphene oxides, J. Phys. Chem. C 120 (2016) 22739–22743. Copyright (2016) American Chemical Society.*

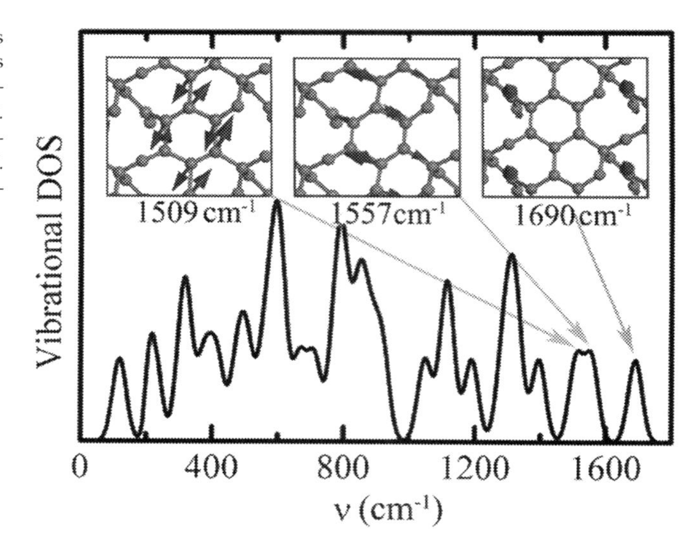

FIG. 6.15 Vibrational density of states and three of the highest vibrational motions for the epoxy-carbonyl combination. *Reprinted (adapted) with permission from U.K. Wijewardena, S.E. Brown, X.Q. Wang, Epoxy-carbonyl conformation of graphene oxides, J. Phys. Chem. C 120 (2016) 22739–22743. Copyright (2016) American Chemical Society.*

The vibrational density of states and the vibrational motions of the three highest frequencies are shown in Fig. 6.15. The corresponding frequencies at the band center are summarized in Table 6.2.

Experimental studies have shown that there are peaks around 1230–1320, 1500–1600, and 1600–1750 cm^{-1} in the Fourier transform IR spectrum of GO, which are assigned to vibrational modes of epoxy groups (C–O–C), sp^2-hybridized carbon (C=C), and the carbonyl (–C=O) functional, respectively [102–104]. We observed an infrared (IR) active peak around 1690 cm^{-1} with *Au* symmetry, which corresponds to the stretching of the carbonyl group.

TABLE 6.2 Vibrational frequencies for the epoxy-carbonyl combination of GO

Odd	IR active (cm^{-1})	Even	Raman active (cm^{-1})
A_u	375	A_g	602
	775		1120
	920		1394
	1690		1509
B_{1u}	226	B_{1g}	414
	559		507
	792		711
	1190		1048
B_{2u}	608		1311
	856	B_{2g}	320
	1327		887
B_{3u}	122		1110
	312		1557
	585	B_{3g}	187
	842		477
			665
			800
			1287

Reprinted (adapted) with permission from U.K. Wijewardena, S.E. Brown, X.Q. Wang, Epoxy-carbonyl conformation of graphene oxides, J. Phys. Chem. C 120 (2016) 22739–22743. Copyright (2016) American Chemical Society.

Peaks around 1557 and 1509 cm^{-1} relate to in-plane Raman active motions of sp^2 carbon with Ag and B_2g symmetries, respectively. Vibrational modes around 1327 cm^{-1} can be associated with IR active mode with a B_2u symmetry. The peaks near 1190 cm^{-1} with B_1u symmetry correspond to the motion of epoxides. We found that our calculated vibrational modes are in good agreement with the experimental observations [102–104].

Furthermore, the epoxy-carbonyl combination provides valuable insights into the unzipping mechanism of carbon nanotubes (Fig. 6.16). Oxidization of graphene is one of the methods used in unzipping carbon nanotubes [105]. The epoxides running along the zigzag direction in the epoxy-pair structure enable breaking the bonds along that direction. It makes it challenging to unzip a nanotube along the armchair direction. On the other hand, the epoxy-carbonyl combination, which consists of epoxides along the armchair direction, indicates the possibility of unzipping a nanotube along the armchair direction.

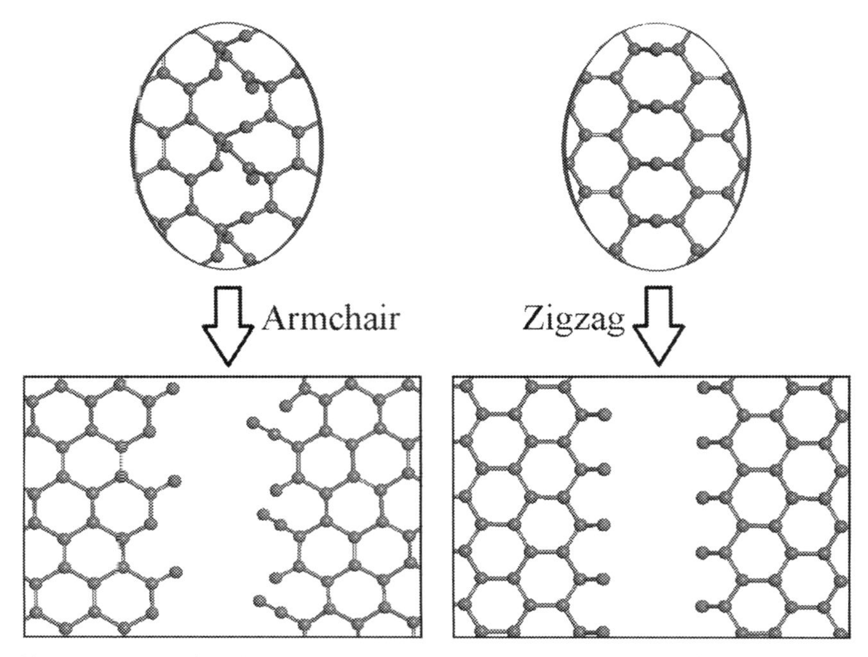

FIG. 6.16 Unzipping nanotubes along the armchair and zigzag directions in the *left and right panels*, respectively. *Reprinted (adapted) with permission from U.K. Wijewardena, S.E. Brown, X.Q. Wang, Epoxy-carbonyl conformation of graphene oxides, J. Phys. Chem. C 120 (2016) 22739–22743. Copyright (2016) American Chemical Society.*

6 Summary

In summary, many research approaches to graphene bandgap engineering have emerged in the past two decades. Experimentally and theoretically, various synthesis techniques and characterization methods have been explored to obtain and characterize covalently functionalized graphene. The nitrophenyl functionalization on graphene offers an effective way to tailor the properties of graphene while opening bandgap and introducing magnetic properties, thus making nitrophenyl-functionalized graphene a promising material for use in various applications. Besides, we discussed another approach to produce a semiconducting graphene that uses a nitrogen-seeded technique at the graphene-SiC interface. However, the experimental methods for the production of large-scale graphene-based materials with a controllable bandgap are still lacking; thus, current experimental methods need to be improved.

Further, we have studied the electronic properties of PMDE-functionalized graphene. We have shown that the [2+2] cycloaddition disturbs the π conjugation of graphene near the Fermi level, which leads to the opening of a bandgap, which depends on the position that PMDE attaches. We demonstrated the bandgap could be tuned from metallic to semiconducting. This describes a controllable method for the "band engineering" of graphene. Our findings on the nature of a PMDE-functionalization-induced bandgap produce useful insight for improvements of graphene-based nanodevices.

Here, we show that the epoxy-carbonyl conformation is a promising atomic structure that accounts for the structural, electronic, and vibrational properties of GO. The results we presented here are in good agreement with experimental findings. We have demonstrated the correlation between the disruption to the π–π^* conjugated network and the energy order of the conformations. The oxidation converts the sp^2 system in pristine graphene to sp^3, and we have shown how the percentage of oxidation, i.e., the regions with sp^2 and sp^3, influences the electrical properties of GO. The results we present here provide valuable information about how different oxide functional groups and their position in graphene can impact the energy bands, which ultimately results in interesting electronic properties.

Acknowledgment

This work was supported in part by the National Science Foundation Grant DMR-2122147 and Army Research Office Grants W911NF1810481 and W911NF1910502.

References

[1] M.J. Allen, V.C. Tung, R.B. Kaner, Honeycomb carbon: a review of graphene, Chem. Rev. 110 (1) (2010) 132–145.
[2] A.H. Castro Neto, F. Guinea, N.M.R. Peres, K.S. Novoselov, A.K. Geim, The electronic properties of graphene, Rev. Mod. Phys. 81 (1) (2009) 109–162.
[3] A.K. Geim, K.S. Novoselov, The rise of graphene, Nat. Mater. 6 (3) (2007) 183–191.
[4] C.N. Rao, A.K. Sood, K.S. Subrahmanyam, A. Govindaraj, Graphene: the new two-dimensional nanomaterial, Angew. Chem. Int. Ed. Engl. 48 (42) (2009) 7752–7777.
[5] Z. Chen, X.-Q. Wang, Stacking-dependent optical spectra and many-electron effects in bilayer graphene, Phys. Rev. B 83 (8) (2011) 081405.
[6] K.S. Novoselov, D. Jiang, F. Schedin, T.J. Booth, V.V. Khotkevich, S.V. Morozov, A.K. Geim, Two-dimensional atomic crystals, Proc. Natl. Acad. Sci. U. S. A. 102 (30) (2005) 10451–10453.
[7] M. Yi, Z.G. Shen, A review on mechanical exfoliation for the scalable production of graphene, J. Mater. Chem. A 3 (22) (2015) 11700–11715.
[8] X.S. Li, W.W. Cai, J.H. An, S. Kim, J. Nah, D.X. Yang, R. Piner, A. Velamakanni, I. Jung, E. Tutuc, S.K. Banerjee, L. Colombo, R.S. Ruoff, Large-area synthesis of high-quality and uniform graphene films on copper foils, Science 324 (5932) (2009) 1312–1314.
[9] T.R. Nanayakkara, U.K. Wijewardena, S.M. Withanage, A. Kriisa, R.L. Samaraweera, R.G. Mani, Strain relaxation in different shapes of single crystal graphene grown by chemical vapor deposition on copper, Carbon 168 (2020) 684–690.
[10] L. Jiao, L. Zhang, X. Wang, G. Diankov, H. Dai, Narrow graphene nanoribbons from carbon nanotubes, Nature 458 (7240) (2009) 877–880.
[11] X. Li, X. Wang, L. Zhang, S. Lee, H. Dai, Chemically derived, ultrasmooth graphene nanoribbon semiconductors, Science 319 (5867) (2008) 1229–1232.
[12] A. Nduwimana, X.Q. Wang, Energy gaps in supramolecular functionalized graphene nanoribbons, ACS Nano 3 (7) (2009) 1995–1999.
[13] O.O. Ogunro, C.I. Nicolas, E.A. Mintz, X.Q. Wang, Band gap opening in the cycloaddition functionalization of carbon nanotubes, ACS Macro Lett. 1 (4) (2012) 524–528.
[14] W. Yi, A. Malkovskiy, Y. Xu, X.-Q. Wang, A.P. Sokolov, M. Lebron-Colon, M.A. Meador, Y. Pang, Polymer conformation-assisted wrapping of single-walled carbon nanotube: the impact of cis-vinylene linkage, Polymer 51 (2) (2010) 475–481.
[15] O.O. Ogunro, X.-Q. Wang, Quantum electronic stability in selective enrichment of carbon nanotubes, Nano Lett. 9 (3) (2009) 1034–1038.

[16] H. Li, J.M. Melnyczuk, L.I. Lewis, S. Palchoudhury, J. Wu, P. Nagappan, I.I. Harruna, X.-Q. Wang, Selectively self-assembling graphene nanoribbons with shaped iron oxide nanoparticles, RSC Adv. 4 (62) (2014) 33127–33133.
[17] I. Sapkota, M.A. Roundtree, J.H. Hall, X.-Q. Wang, Tunable band gap in gold intercalated graphene, Phys. Chem. Chem. Phys. 14 (46) (2012) 15991–15994.
[18] G.R. Yazdi, T. Iakimov, R. Yakimova, Epitaxial graphene on SiC: a review of growth and characterization, Crystals 6 (5) (2016).
[19] S. Stankovich, D.A. Dikin, R.D. Piner, K.A. Kohlhaas, A. Kleinhammes, Y. Jia, Y. Wu, S.T. Nguyen, R.S. Ruoff, Synthesis of graphene-based nanosheets via chemical reduction of exfoliated graphite oxide, Carbon 45 (7) (2007) 1558–1565.
[20] Z.F. Chen, S. Nagase, A. Hirsch, R.C. Haddon, W. Thiel, P.V. Schleyer, Side-wall opening of single-walled carbon nanotubes (SWCNTs) by chemical modification: a critical theoretical study, Angew. Chem. Int. Ed. 43 (12) (2004) 1552–1554.
[21] J. Choi, K.J. Kim, B. Kim, H. Lee, S. Kim, Covalent functionalization of epitaxial graphene by azidotrimethylsilane, J. Phys. Chem. C 113 (22) (2009) 9433–9435.
[22] D.C. Elias, R.R. Nair, T.M.G. Mohiuddin, S.V. Morozov, P. Blake, M.P. Halsall, A.C. Ferrari, D.W. Boukhvalov, M.I. Katsnelson, A.K. Geim, K.S. Novoselov, Control of graphene's properties by reversible hydrogenation: evidence for graphane, Science 323 (5914) (2009) 610–613.
[23] H.K. He, C. Gao, General approach to individually dispersed, highly soluble, and conductive graphene nanosheets functionalized by nitrene chemistry, Chem. Mater. 22 (17) (2010) 5054–5064.
[24] L.H. Liu, M.M. Lerner, M.D. Yan, Derivitization of pristine graphene with well-defined chemical functionalities, Nano Lett. 10 (9) (2010) 3754–3756.
[25] L.H. Liu, M. Yan, Simple method for the covalent immobilization of graphene, Nano Lett. 9 (9) (2009) 3375–3378.
[26] K.S. Mali, J. Greenwood, J. Adisoejoso, R. Phillipson, S. De Feyter, Nanostructuring graphene for controlled and reproducible functionalization, Nanoscale 7 (5) (2015) 1566–1585.
[27] M. Quintana, K. Spyrou, M. Grzelczak, W.R. Browne, P. Rudolf, M. Prato, Functionalization of graphene via 1,3-dipolar cycloaddition, ACS Nano 4 (6) (2010) 3527–3533.
[28] X. Wang, Y. Ouyang, X. Li, H. Wang, J. Guo, H. Dai, Room-temperature all-semiconducting sub-10-nm graphene nanoribbon field-effect transistors, Phys. Rev. Lett. 100 (20) (2008) 206803.
[29] J.J. Zhao, Z.F. Chen, Z. Zhou, H. Park, P.V. Schleyer, J.P. Lu, Engineering the electronic structure of single-walled carbon nanotubes by chemical functionalization, ChemPhysChem 6 (4) (2005) 598–601.
[30] X.Q. Tian, J. Gu, J.B. Xu, Configuration-dependent electronic and magnetic properties of graphene monolayers and nanoribbons functionalized with aryl groups, J. Chem. Phys. 140 (4) (2014).
[31] R.N. Gunasinghe, D.G. Reuven, K. Suggs, X.-Q. Wang, Filled and empty orbital interactions in a planar covalent organic framework on graphene, J. Phys. Chem. Lett. 3 (20) (2012) 3048–3052.
[32] D.K. Samarakoon, Z. Chen, C. Nicolas, X.Q. Wang, Structural and electronic properties of fluorographene, Small 7 (7) (2011) 965–969.
[33] U.K. Wijewardena, S.E. Brown, X.Q. Wang, Epoxy-carbonyl conformation of graphene oxides, J. Phys. Chem. C 120 (39) (2016) 22739–22743.
[34] M. Fang, K.G. Wang, H.B. Lu, Y.L. Yang, S. Nutt, Covalent polymer functionalization of graphene nanosheets and mechanical properties of composites, J. Mater. Chem. 19 (38) (2009) 7098–7105.
[35] D.K. Samarakoon, X.-Q. Wang, Chair and twist-boat membranes in hydrogenated graphene, ACS Nano 3 (12) (2009) 4017–4022.
[36] D.K. Samarakoon, X.-Q. Wang, Tunable band gap in hydrogenated bilayer graphene, ACS Nano 4 (7) (2010) 4126–4130.
[37] M. Popinciuc, C. Jozsa, P.J. Zomer, N. Tombros, A. Veligura, H.T. Jonkman, B.J. van Wees, Electronic spin transport in graphene field-effect transistors, Phys. Rev. B 80 (21) (2009).
[38] B. Dlubak, M.B. Martin, C. Deranlot, B. Servet, S. Xavier, R. Mattana, M. Sprinkle, C. Berger, W.A. De Heer, F. Petroff, A. Anane, P. Seneor, A. Fert, Highly efficient spin transport in epitaxial graphene on SiC, Nat. Phys. 8 (7) (2012) 557–561.
[39] W. Han, K.M. McCreary, K. Pi, W.H. Wang, Y. Li, H. Wen, J.R. Chen, R.K. Kawakami, Spin transport and relaxation in graphene, J. Magn. Magn. Mater. 324 (4) (2012) 369–381.
[40] A.H. Castro Neto, F. Guinea, Impurity-induced spin-orbit coupling in graphene, Phys. Rev. Lett. 103 (2) (2009).

[41] I. Choudhuri, P. Bhauriyal, B. Pathak, Recent advances in graphene-like 2D materials for spintronics applications, Chem. Mater. 31 (20) (2019) 8260–8285.
[42] M.V. Kamalakar, A. Dankert, J. Bergsten, T. Ive, S.P. Dash, Spintronics with graphene-hexagonal boron nitride van der Waals heterostructures, Appl. Phys. Lett. 105 (21) (2014).
[43] W. Han, R.K. Kawakami, M. Gmitra, J. Fabian, Graphene spintronics, Nat. Nanotechnol. 9 (10) (2014) 794–807.
[44] L.A. London, L.A. Bolton, D.K. Samarakoon, B.S. Sannigrahi, X.Q. Wang, I.M. Khan, Effect of polymer stereoregularity on polystyrene/single-walled carbon nanotube interactions, RSC Adv. 5 (73) (2015) 59186–59193.
[45] J.M. Hong, E. Bekyarova, W.A. de Heer, R.C. Haddon, S. Khizroev, Chemically engineered graphene-based 2D organic molecular magnet, ACS Nano 7 (11) (2013) 10011–10022.
[46] E. Bekyarova, S. Sarkar, S. Niyogi, M.E. Itkis, R.C. Haddon, Advances in the chemical modification of epitaxial graphene, J. Phys. D. Appl. Phys. 45 (15) (2012).
[47] F. Wang, G. Liu, S. Rothwell, M. Nevius, A. Tejeda, A. Taleb-Ibrahimi, L.C. Feldman, P.I. Cohen, E.H. Conrad, Wide-gap semiconducting graphene from nitrogen-seeded SiC, Nano Lett. 13 (10) (2013) 4827–4832.
[48] E. Velez-Fort, C. Mathieu, E. Pallecchi, M. Pigneur, M.G. Silly, R. Belkhou, M. Marangolo, A. Shukla, F. Sirotti, A. Ouerghi, Epitaxial graphene on 4H-SiC(0001) grown under nitrogen flux: evidence of low nitrogen doping and high charge transfer, ACS Nano 6 (12) (2012) 10893–10900.
[49] D. Usachov, O. Vilkov, A. Gruneis, D. Haberer, A. Fedorov, V.K. Adamchuk, A.B. Preobrajenski, P. Dudin, A. Barinov, M. Oehzelt, C. Laubschat, D.V. Vyalikh, Nitrogen-doped graphene: efficient growth, structure, and electronic properties, Nano Lett. 11 (12) (2011) 5401–5407.
[50] F. Joucken, Y. Tison, J. Lagoute, J. Dumont, D. Cabosart, B. Zheng, V. Repain, C. Chacon, Y. Girard, A.R. Botello-Mendez, S. Rousset, R. Sporken, J.C. Charlier, L. Henrard, Localized state and charge transfer in nitrogen-doped graphene, Phys. Rev. B 85 (16) (2012).
[51] H.B. Wang, T. Maiyalagan, X. Wang, Review on recent progress in nitrogen-doped graphene: synthesis, characterization, and its potential applications, ACS Catal. 2 (5) (2012) 781–794.
[52] *DMol3*, 6.1, Accelrys Software Inc, San Diego CA, 2015.
[53] M.A. Ruderman, C. Kittel, Indirect exchange coupling of nuclear magnetic moments by conduction electrons, Phys. Rev. 96 (1) (1954) 99–102.
[54] T. Kasuya, A theory of metallic ferro- and antiferromagnetism on Zener's model, Prog. Theor. Phys. 16 (1) (1956) 45–57.
[55] K. Yosida, Magnetic properties of Cu-Mn alloys, Phys. Rev. 106 (5) (1957) 893–898.
[56] D.W. Boukhvalov, DFT modeling of the covalent functionalization of graphene: from ideal to realistic models, RSC Adv. 3 (20) (2013) 7150–7159.
[57] H. Park, J. Zhao, J.P. Lu, Effects of sidewall functionalization on conducting properties of single wall carbon nanotubes, Nano Lett. 6 (5) (2006) 916–919.
[58] A. Marini, C. Hogan, M. Grüning, D. Varsano, Yambo: an ab initio tool for excited state calculations, Comput. Phys. Commun. 180 (8) (2009) 1392–1403.
[59] C. Cioffi, S. Campidelli, F.G. Brunetti, M. Meneghetti, M. Prato, Functionalisation of carbon nanohorns, Chem. Commun. 20 (2006) 2129–2131.
[60] R. Singh, D. Pantarotto, L. Lacerda, G. Pastorin, C. Klumpp, M. Prato, A. Bianco, K. Kostarelos, Tissue biodistribution and blood clearance rates of intravenously administered carbon nanotube radiotracers, Proc. Natl. Acad. Sci. U. S. A. 103 (9) (2006) 3357–3362.
[61] K. Kostarelos, L. Lacerda, G. Pastorin, W. Wu, S. Wieckowski, J. Luangsivilay, S. Godefroy, D. Pantarotto, J.-P. Briand, S. Muller, M. Prato, A. Bianco, Cellular uptake of functionalized carbon nanotubes is independent of functional group and cell type, Nat. Nanotechnol. 2 (2) (2007) 108–113.
[62] M. Orlita, C. Faugeras, P. Plochocka, P. Neugebauer, G. Martinez, D.K. Maude, A.L. Barra, M. Sprinkle, C. Berger, W.A. de Heer, M. Potemski, Approaching the Dirac point in high-mobility multilayer epitaxial graphene, Phys. Rev. Lett. 101 (26) (2008) 267601.
[63] X. Zhang, A. Hsu, H. Wang, Y. Song, J. Kong, M.S. Dresselhaus, T. Palacios, Impact of chlorine functionalization on high-mobility chemical vapor deposition grown graphene, ACS Nano 7 (8) (2013) 7262–7270.
[64] K. Suggs, D. Reuven, X.-Q. Wang, Electronic properties of cycloaddition-functionalized graphene, J. Phys. Chem. C 115 (8) (2011) 3313–3317.
[65] Q. Shao, G. Liu, D. Teweldebrhan, A.A. Balandin, High-temperature quenching of electrical resistance in graphene interconnects, Appl. Phys. Lett. 92 (20) (2008) 202108.

[66] C. Si, G. Zhou, Y. Li, J. Wu, W. Duan, Interface engineering of epitaxial graphene on SiC(0001) via fluorine intercalation: a first principles study, Appl. Phys. Lett. 100 (10) (2012) 103105.
[67] S. Malik, A. Vijayaraghavan, R. Erni, K. Ariga, I. Khalakhan, J.P. Hill, High purity graphenes prepared by a chemical intercalation method, Nanoscale 2 (10) (2010) 2139–2143.
[68] M.D. Williams, D.K. Samarakoon, D.W. Hess, X.-Q. Wang, Tunable bands in biased multilayer epitaxial graphene, Nanoscale 4 (9) (2012) 2962–2967.
[69] U.K. Wijewardena, T. Nanayakkara, R. Samaraweera, S. Withanage, A. Kriisa, R.G. Mani, Effects of long-time current annealing to the hysteresis in CVD graphene on SiO_2, MRS Adv. 4 (61–62) (2019) 3319–3326.
[70] W. Hooch Antink, Y. Choi, K.-D. Seong, J.M. Kim, Y. Piao, Recent progress in porous graphene and reduced graphene oxide-based nanomaterials for electrochemical energy storage devices, Adv. Mater. Interfaces 5 (5) (2018) 1701212.
[71] C. Couly, M. Alhabeb, K.L. Van Aken, N. Kurra, L. Gomes, A.M. Navarro-Suárez, B. Anasori, H.N. Alshareef, Y. Gogotsi, Asymmetric flexible MXene-reduced graphene oxide micro-supercapacitor, Adv. Electron. Mater. 4 (1) (2018) 1700339
[72] P. Sehrawat, Abid, S.S. Islam, P. Mishra, Reduced graphene oxide based temperature sensor: extraordinary performance governed by lattice dynamics assisted carrier transport, Sensors Actuators B Chem. 258 (2018) 424–435.
[73] Q. Pan, Y. Lv, G.R. Williams, L. Tao, H. Yang, H. Li, L. Zhu, Lactobionic acid and carboxymethyl chitosan functionalized graphene oxide nanocomposites as targeted anticancer drug delivery systems, Carbohydr. Polym. 151 (2016) 812–820.
[74] C. Chung, Y.-K. Kim, D. Shin, S.-R. Ryoo, B.H. Hong, D.-H. Min, Biomedical applications of graphene and graphene oxide, Acc. Chem. Res. 46 (10) (2013) 2211–2224.
[75] J.-A. Yan, W.Y. Ruan, M.Y. Chou, Phonon dispersions and vibrational properties of monolayer, bilayer, and trilayer graphene: density-functional perturbation theory, Phys. Rev. B 77 (12) (2008) 125401.
[76] J.-L. Li, K.N. Kudin, M.J. McAllister, R.K. Prud'homme, I.A. Aksay, R. Car, Oxygen-driven unzipping of graphitic materials, Phys. Rev. Lett. 96 (17) (2006) 176101.
[77] K.N. Kudin, B. Ozbas, H.C. Schniepp, R.K. Prud'homme, I.A. Aksay, R. Car, Raman spectra of graphite oxide and functionalized graphene sheets, Nano Lett. 8 (1) (2008) 36–41.
[78] D.W. Boukhvalov, M.I. Katsnelson, Modeling of graphite oxide, J. Am. Chem. Soc. 130 (32) (2008) 10697–10701.
[79] Z. Li, W. Zhang, Y. Luo, J. Yang, J.G. Hou, How graphene is cut upon oxidation? J. Am. Chem. Soc. 131 (18) (2009) 6320–6321.
[80] S. Mao, H. Pu, J. Chen, Graphene oxide and its reduction: modeling and experimental progress, RSC Adv. 2 (7) (2012) 2643–2662.
[81] J.-A. Yan, M.Y. Chou, Oxidation functional groups on graphene: structural and electronic properties, Phys. Rev. B 82 (12) (2010) 125403.
[82] D.K. Samarakoon, X.Q. Wang, Twist-boat conformation in graphene oxides, Nanoscale 3 (1) (2011) 192–195.
[83] H.J. Xiang, S.-H. Wei, X.G. Gong, Structural motifs in oxidized graphene: a genetic algorithm study based on density functional theory, Phys. Rev. B 82 (3) (2010) 035416.
[84] Y. Wang, J. Lv, L. Zhu, Y. Ma, Crystal structure prediction via particle-swarm optimization, Phys. Rev. B 82 (9) (2010) 094116.
[85] Y. Wang, M. Miao, J. Lv, L. Zhu, K. Yin, H. Liu, Y. Ma, An effective structure prediction method for layered materials based on 2D particle swarm optimization algorithm, J. Chem. Phys. 137 (22) (2012) 224108.
[86] J. Clark Stewart, D. Segall Matthew, J. Pickard Chris, J. Hasnip Phil, I.J. Probert Matt, K. Refson, C. Payne Mike, First principles methods using CASTEP, in: Zeitschrift für Kristallographie—Crystalline Materials, vol. 220, Oldenbourg Wissenschaftsverlag, München, Germany, 2005, p. 567. https://doi.org/10.1524/zkri.220.5.567.65075.
[87] J.P. Perdew, K. Burke, M. Ernzerhof, Generalized gradient approximation made simple, Phys. Rev. Lett. 77 (18) (1996) 3865–3868.
[88] E.R. McNellis, J. Meyer, K. Reuter, Azobenzene at coinage metal surfaces: role of dispersive van der Waals interactions, Phys. Rev. B 80 (20) (2009) 205414.
[89] B. Huang, H. Xiang, Q. Xu, S.-H. Wei, Overcoming the phase inhomogeneity in chemically functionalized graphene: the case of graphene oxides, Phys. Rev. Lett. 110 (8) (2013) 085501.

[90] M.-T. Nguyen, R. Erni, D. Passerone, Two-dimensional nucleation and growth mechanism explaining graphene oxide structures, Phys. Rev. B 86 (11) (2012) 115406.
[91] N. Lu, D. Yin, Z. Li, J. Yang, Structure of graphene oxide: thermodynamics versus kinetics, J. Phys. Chem. C 115 (24) (2011) 11991–11995.
[92] A. Bagri, C. Mattevi, M. Acik, Y.J. Chabal, M. Chhowalla, V.B. Shenoy, Structural evolution during the reduction of chemically derived graphene oxide, Nat. Chem. 2 (7) (2010) 581–587.
[93] D. Pandey, R. Reifenberger, R. Piner, Scanning probe microscopy study of exfoliated oxidized graphene sheets, Surf. Sci. 602 (9) (2008) 1607–1613.
[94] C. Gómez-Navarro, J.C. Meyer, R.S. Sundaram, A. Chuvilin, S. Kurasch, M. Burghard, K. Kern, U. Kaiser, Atomic structure of reduced graphene oxide, Nano Lett. 10 (4) (2010) 1144–1148.
[95] M.P. McDonald, A. Eltom, F. Vietmeyer, J. Thapa, Y.V. Morozov, D.A. Sokolov, J.H. Hodak, K. Vinodgopal, P.V. Kamat, M. Kuno, Direct observation of spatially heterogeneous single-layer graphene oxide reduction kinetics, Nano Lett. 13 (12) (2013) 5777–5784.
[96] F. Liu, T. Tang, Q. Feng, M. Li, Y. Liu, N. Tang, W. Zhong, Y. Du, Tuning photoluminescence of reduced graphene oxide quantum dots from blue to purple, J. Appl. Phys. 115 (16) (2014) 164307.
[97] T. Sakthivel, V. Gunasekaran, S.J. Kim, Effect of oxygenated functional groups on the photoluminescence properties of graphene-oxide nanosheets, Mater. Sci. Semicond. Process. 19 (2014) 174–178.
[98] C.-T. Chien, S.-S. Li, W.-J. Lai, Y.-C. Yeh, H.-A. Chen, I.-S. Chen, L.-C. Chen, K.-H. Chen, T. Nemoto, S. Isoda, M. Chen, T. Fujita, G. Eda, H. Yamaguchi, M. Chhowalla, C.-W. Chen, Tunable photoluminescence from graphene oxide, Angew. Chem. Int. Ed. 51 (27) (2012) 6662–6666.
[99] W. Gao, L.B. Alemany, L. Ci, P.M. Ajayan, New insights into the structure and reduction of graphite oxide, Nat. Chem. 1 (5) (2009) 403–408.
[100] K.P. Loh, Q. Bao, G. Eda, M. Chhowalla, Graphene oxide as a chemically tunable platform for optical applications, Nat. Chem. 2 (12) (2010) 1015–1024.
[101] Y. Zhu, S. Murali, W. Cai, X. Li, J.W. Suk, J.R. Potts, R.S. Ruoff, Graphene and graphene oxide: synthesis, properties, and applications, Adv. Mater. 22 (35) (2010) 3906–3924.
[102] D.W. Lee, V.L. De Los Santos, J.W. Seo, L.L. Felix, D.A. Bustamante, J.M. Cole, C.H.W. Barnes, The structure of graphite oxide: investigation of its surface chemical groups, J. Phys. Chem. B 114 (17) (2010) 5723–5728.
[103] Sudesh, N. Kumar, S. Das, C. Bernhard, G.D. Varma, Effect of graphene oxide doping on superconducting properties of bulk MgB2, Supercond. Sci. Technol. 26 (9) (2013) 095008.
[104] D.S.R. Josephine, B. Sakthivel, K. Sethuraman, A. Dhakshinamoorthy, Synthesis, characterization and catalytic activity of CdS-graphene oxide nanocomposites, ChemistrySelect 1 (10) (2016) 2332–2340.
[105] D.V. Kosynkin, A.L. Higginbotham, A. Sinitskii, J.R. Lomeda, A. Dimiev, B.K. Price, J.M. Tour, Longitudinal unzipping of carbon nanotubes to form graphene nanoribbons, Nature 458 (7240) (2009) 872–876.

CHAPTER

7

Reversible and irreversible functionalization of graphene

Y. Bhargav Kumar[a,b], *Ravindra K. Rawal*[a], *Ashutosh Thakur*[a], *and G. Narahari Sastry*[a,b]

[a]CSIR-North East Institute of Science and Technology, Jorhat, India [b]Academy of Scientific and Innovative Research (AcSIR), Ghaziabad, India

1 Introduction

Ever since the discovery of buckminsterfullerene, the interest in the chemistry and physics of carbon clusters has witnessed a remarkable resurgence. Fullerenes, carbon nanotubes, and their derivatives have found applications virtually in all fields ranging from chemistry, biology, and materials to medicine. Graphene is the newer entry in this class and made a spectacular entry into materials science and triggered the interest in the two-dimensional (2D) materials. Graphene is an extremely important synthetically isolated allotrope of carbon, which possesses commendable electronic, optical, mechanical, and chemical properties [1,2]. It is a 2D sheet-like molecule, and the single layer contains sp^2-hybridized carbon atoms arranged in a honeycomb crystal lattice (Fig. 7.1). Single-layer graphene has been successfully isolated by micromechanical cleavage of graphite (the so-called Scotch tape method) in 2004 by Geim and Novoselov for which they earned the 2010 Noble Prize in Physics [3,4]. Each sp^2-hybridized carbon atom in the graphene sheet is bonded with adjacent carbon atoms through three σ bonds on the plane and one π bond perpendicular to the plane. The connection between carbon atoms in graphene is very strong. Because of the extremely thin 2D structure, when an external force is applied, the graphene surface can easily deform to offset the external force without mechanical failure. Due to these intrinsic features, single-layer graphene is the thinnest, strongest, and stiffest material in the world. Moreover, it is highly transparent and an excellent conductor of both heat and electricity [1,2,5].

Graphene is one of the materials of extensive investigation due to its outstanding properties like large specific surface area, high degree of conjugation, great mechanical strength, and

https://doi.org/10.1016/B978-0-12-819514-7.00005-1

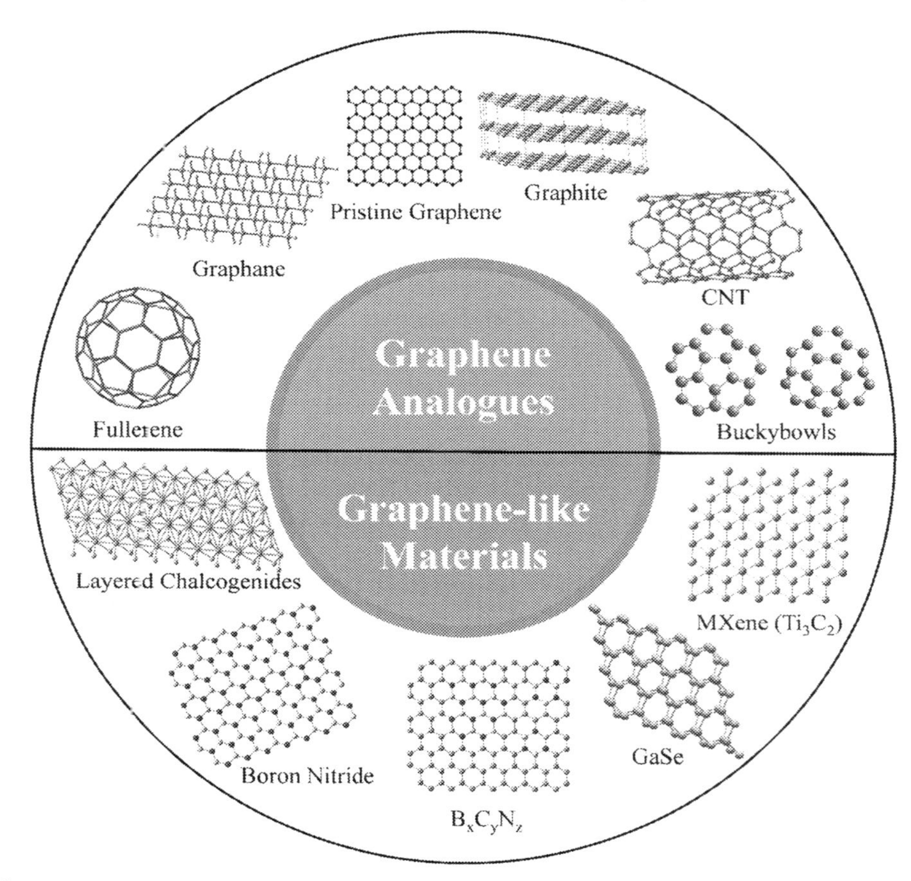

FIG. 7.1 Depiction of different graphene analogues and layered graphene-like materials.

exceptional thermal and electrical conductivity. Because of these extraordinary and superior properties, graphene has already demonstrated a great potential for a variety of applications such as use in electronic devices, capacitors, fuel cells, batteries, sensors, transparent conductive films, high-frequency circuits, and toxic material removal devices.

After the first successful isolation and characterization of graphene in 2004, a great deal of effort was devoted to developing new routes for the effective production of well-defined graphene sheets. The usually applied methods are the micromechanical cleavage or chemical exfoliation of graphite; chemical vapor deposition (CVD) growth; and chemical, electrochemical, thermal, or photocatalytic reduction of graphene oxide (GO) and fluorographene [4,6–8]. Despite the success in both synthesis of graphene and the exploitation of its various properties, graphene as a single-component material has several limitations. These limitations include zero bandgap, inertness to chemical reactions, the propensity to agglomerate, and difficulty in processing, which significantly limit the scope of graphene for various applications [1,2]. To overcome these limitations, functional modification of graphene and its derivatives such as GO has been extensively investigated [9–11]. The functionalization can be done based

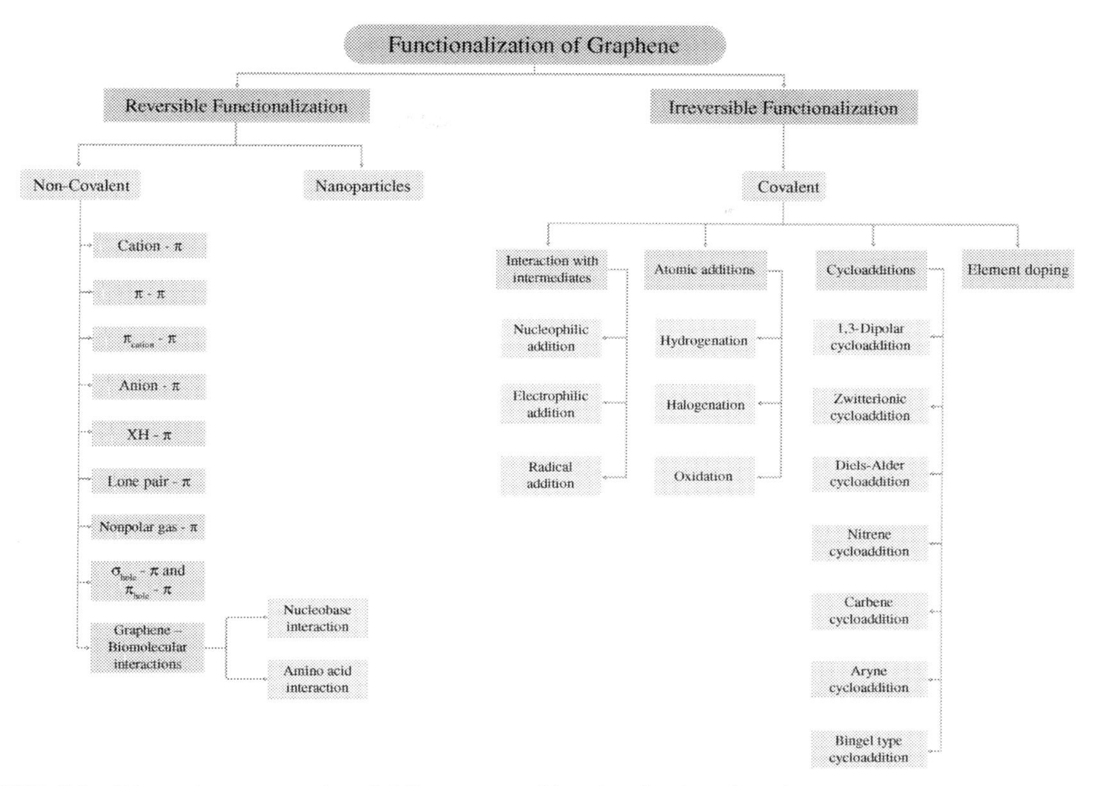

FIG. 7.2 Schematic representation of different types of functionalization of graphene.

on several interactions between graphene and GO sheets with atomic elements, ions, organic molecules, and inorganic compounds. Depending upon the strength of the interaction, the functionalization can be either reversible or irreversible as shown in Fig. 7.2 [12,13]. While graphene, GO, and graphane have found varying applications, functionalization of these classes of materials opens up new opportunities and is interesting in its own right.

Herein, we give a brief overview of several reversible and irreversible functionalization approaches of graphene and GO. To understand the necessity of functionalization of graphene and GO, we also provide an overview of pristine graphene, other graphene analogues of carbon, and 2D graphene-like substances in the next section (Table 7.1). The chronological order according to the discovery and evolution of various important allotropes of carbon is also depicted in Fig. 7.3.

2 Overview of graphene, its analogues, and graphene-like substances

2.1 Pristine graphene

Exfoliated graphene with minimum defects is called pristine graphene (Fig. 7.1). The quality of pristine graphene is important to utilize graphene and graphene-based functional

TABLE 7.1 Graphene, its analogues, and graphene-like materials are represented with their constituent atoms and structure

Material	Constituent atoms	Structure
Graphene	C	2D sheets
Graphene oxide	C, O, H	2D sheets
Graphite	C	3D stacked sheets
Graphane	C, H	2D puckered sheets
Carbon nanotubes	C	1D cylindrical tubes
Fullerenes	C	0D football/icosahedron
Buckybowls	C	3D bowl-like structures
Layered chalcogenides	Mo, W, S, Se	2D sheets
h-BN	B, N	2D sheets
Gallium selenide	Ga, Se	3D tetra-layered sheets
Borocarbonitrides	B, C, N	2D sheets
MXenes	Ti, V, Nb, C, N, etc.	2D sheets

materials for potential electronic, electrochemical energy, and other important applications. Therefore, the development of a simple, practical, and low-cost route to produce high-quality graphene at a larger amount is the topic of intense research. Apart from the scotch tape method, the usual chemical method for the synthesis of graphene is the reduction of GO. As GO is produced under highly oxidizing conditions, there remains a lot of defects in the exfoliated graphene obtained by this method. Presently, efforts are directed to obtain high-quality graphene by the use of graphite as starting material. In this approach, graphite is exfoliated and dispersed by ultrasonication in a variety of organic solvents such as o-dichlorobenzene, pyridine, chloroacetate, and perfluorinated aromatic molecules [6–8]. This method has been quite successful in obtaining monolayer graphene with minimum defects. Additionally, nanoribbons of pristine graphene can be produced by chemically unzipping carbon nanotubes [12–14]. For electronic applications, the bandgap of high-quality pristine graphene is required to be improved, and this can be effectively done by doping it with trivalent and pentavalent impurities [1].

2.2 Graphite

Graphite is the most common allotrope of carbon and is also a parent material of graphene (Fig. 7.1). Graphite is a 3D analogue of graphene and can be represented as the product of stacking several layers of pristine graphene bound by weak van der Waals interactions. The interlayer distance is reported to be 3.37 Å, and all the carbons are sp^2-hybridized, making a hexagonal crystal lattice [15]. Several strategies have been employed to isolate graphene from graphite by mechanical cleavage, chemical exfoliation, electrochemical exfoliation, etc.

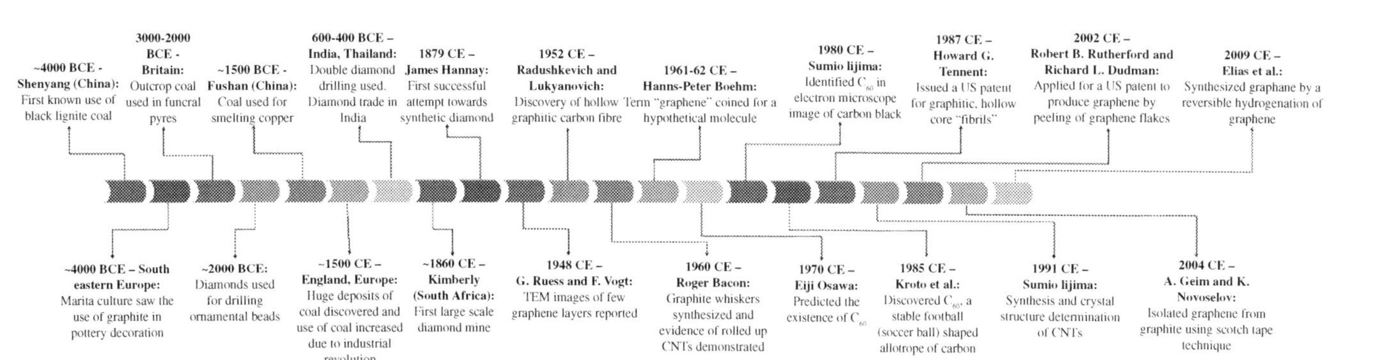

FIG. 7.3 Systematic representation of various carbonaceous materials (coal, graphite, diamond, fullerenes, carbon nanotubes, graphene, and graphane) in the chronological order of their discovery/synthesis/isolation.

2.3 Graphane

Graphane is another 2D analogue of graphene with interesting physical properties (Fig. 7.1). As the name suggests, graphane is a saturated analogue of graphene where the double bonds are saturated by the addition of hydrogen atoms. Computational studies by Sofo et al. suggested that the material is hexagonally arranged either in chair or in boat conformations [16]. The chair form is more stable than the boat form as confirmed by the theoretical studies of Elias et al. [17]. The experimental studies that were performed after Elias et al. enabled the synthesis of graphane for the first time by exposing pristine graphene sheets to hydrogen plasma [17]. The experimental results supported the computational results proving that graphane is indeed in the form of a puckered chair conformation with the hydrogen atoms above and below the sheet. Similar to graphene, graphane sheets can also be doped to improve the bandgap for a diverse scope of applications. For potential applications of graphane, quantum mechanical calculations were performed to understand its interactions with various molecules, metal ions, and onium ions [18]. For instance, Sastry and co-workers showed that the binding of H_2O molecules with graphane is stronger than that of graphene, revealing that graphane is more hydrophilic [19].

2.4 Carbon nanotubes

Carbon nanotubes (CNTs) are 1D hexagonal hollow cylindrical tubes of sp^2-hybridized carbon atoms, with diameters in the range of a few nanometers and lengths up to 500 mm (Fig. 7.1) [20]. They can be projected as the product of rolling a graphene sheet around any axis. The synthesis of CNTs was first reported by Iijima et al. in 1991 as a byproduct during the arc-discharge evaporation of carbon [21]. Due to high aspect ratios of CNTs, they can be fit together to obtain various morphologies, which make them useful for a wide variety of applications [22]. CNTs find their applications in the area of targeted drug delivery. However, quantum chemical calculations by Sastry and co-workers showed that curved CNTs exhibit weaker noncovalent interactions with various biomolecules than flat graphene [23,24]. The free electrons of sp^2-hybridized carbon atoms of CNTs participate in conjugation, which causes the material's high electrical conductivity. The CNTs are in general divided into single-walled CNTs (SWCNTs) and multi-walled CNTs (MWCNTs). The MWCNTs are obtained by concentrically nesting many SWCNTs one over the other. The SWCNTs are broadly classified into chiral and achiral SWCNTs based on the absence or presence of mirror plane. The achiral SWCNTs are further classified into armchair and zig-zag SWCNTs [25], which exhibit distinct properties such as metallic character by armchair and semiconductor-like by zig-zag SWCNTs. These structural differences occur due to the different possibilities to roll the plain graphene sheet to prepare SWCNTs. Based on the lattice parameters, these three different types of SWCNTs can be distinguished using (n, m) notation. Structurally similar atoms on a defect-free graphene sheet are the key to understanding and defining the SWCNTs into their respective types. Based on the orientation of choice, achiral SWCNTs roll up to form zig-zag SWCNTs represented by $(n,0)$, while armchair SWCNTs are represented by (n,n). The chiral SWCNTs are represented by (n, m), where n and m are the number of basis vector lengths form the origin point A_1 (any structurally similar atom) to the joining point A_2 (the tube is formed by precisely joining A_1 and A_2 by avoiding any chance to give way to defects).

2.5 Fullerenes

These are one of the earliest synthetically prepared allotropes of carbon (Fig. 7.1). These globular 0D molecules were first reported by Kroto et al. in 1985, for which they have been awarded a Noble Chemistry Prize in 1996 [26]. The first synthesized fullerene C_{60} was identified to have the shape of a football (or soccer ball) with a total of 12 pentagonal rings and 20 hexagonal rings [26]. The next bigger and stable homologues of C_{60} are C_{70}, C_{76}, C_{78}, C_{82}, C_{84}, C_{90}, C_{94}, and C_{96}, respectively, and the curvature in the structures is attributed to the presence of 5-membered rings in addition to the otherwise 6-membered hexagonal rings [25].

2.6 Buckybowls

Buckybowls are polyaromatic non-planar π-conjugated hydrocarbons, which can be considered as structural components of fullerenes and carbon nanotubes (Fig. 7.1). These molecules adopt a bowl-shaped conformation to minimize the ring strain of the fused aromatic structures. The two most widely studied buckybowls are C_{5v} symmetric corannulene ($C_{20}H_{10}$) and C_{3v} symmetric sumanene ($C_{21}H_{12}$) [27–29]. In the molecular structure of corannulene, a five-membered cyclopentane ring is fused with five benzene rings, while 3 five-membered rings are fused with three benzene rings in sumanene. Due to their unique structures and chemical as well as physical properties (such as bowl structure, bowl chirality, face selectivity, crystal packing, electron conductivity, optical properties, metal complexation, and gas absorption), corannulene and sumanene, as well as their derivatives, have received considerable attention both as model compounds for the buckminsterfullerene (C_{60}) and functional materials for advanced applications in materials science and engineering [30–42].

The development of practical synthetic methods for corannulene started in the 1990s. In contrast, most of the methods (those based on planar starting compounds) for sumanene were unsuccessful until the solution-phase method was developed by Sakurai et al. in 2003 [27,43]. Sastry and co-workers conducted a theoretical study and showed that sumanene's synthesis could be possible by choosing a proper starting material [44]. Sakurai et al. constructed a three-dimensional (3D) framework containing tetrahedral sp^3 carbons. Oxidative aromatization of this 3D framework resulted in the required sumanene structure. A large number of derivatives of buckybowls can also be synthesized by substitution in the aromatic ring, fusing additional five- and six-membered aromatic rings onto the rim structure, complexation with metal ions, and by interactions with host molecules such as C_{60}. Thus, with the functionalization of buckybowls, a new class of π-conjugated materials can be prepared, which have the potential as novel electrical materials, organometallic catalysts, hydrogen storage and lithium storage materials, supramolecules for chemical encapsulation and delivery of drugs, redox-active materials, etc.

2.7 Graphene-like substances

Layered Chalcogenides: The isolation of graphene has not only led to an instant surge of applications in the fields of electronics and optics but also paved the way for the study and synthesis of novel layered 2D materials of organic and inorganic origins [45–54]. Layered chalcogenides are inorganic analogues of graphene, which have layered structures and exhibit exemplary properties (Fig. 7.1). The most common examples of such materials are the sulfides and selenides of molybdenum and tungsten (MoS_2, $MoSe_2$, WS_2, and WSe_2). Similar to graphene, these layered materials also either exist as single-layer chalcogenides or aggregate to form few-layer chalcogenides [51]. MoS_2 occurs in nature as the mineral molybdenite but the synthesis of many-layer crystal of this material was reported by Frindt in 1966 [55]. Joensen et al. have reported the exfoliation of graphene-like MoS_2 after intercalation of lithium atoms in between the layers [56,57] and similar attempts were made to isolate WS_2 as well [58,59]. $MoSe_2$ and WSe_2 also have similar structures but are actually three-layered materials with the metal atom layer sandwiched between the layer of selenium atoms [60,61].

Boron nitride and Gallium selenide: Boron nitride (BN), an isoelectronic species of graphene, is also its structural analogue with each hexagonal unit consisting of 3B and 3N atoms in place of 6C atoms (Fig. 7.1). The first recorded synthesis of a monolayer BN film was by the pyrolysis of borazine on the surface of Ni [62]. Unlike graphene, BN is an insulator and hence is used in dielectric gates [63]. The same way carbon exhibits allotropism in the form of layered graphite and tetrahedral diamond, BN also exhibits polymorphism in the form of hexagonal BN (h-BN) and cubic BN (c-BN) [63]. Gallium selenide (GaSe) is another layered material similar to graphene, and its layers are stacked upon each other by weak van der Waals interactions, making the crystal more intriguing [64]. Unlike graphene, it crystallizes into four unique polymorphs, viz., β-(2H), ε-(2H), γ-(3R), and δ-(4H) GaSe [65].

Borocarbonitrides: Borocarbonitrides are graphene-like 2D materials, which possess the three atoms boron, carbon, and nitrogen usually in unequal proportions with a general formula $B_xC_yN_z$ (Fig. 7.1) [66]. These materials may contain the layered graphene sheets with BN rings replacing a few hexagonal carbon rings containing only C—B, C—C, C—N, B—N bonds, but no B—B and N—N bonds [67,68]. However, these materials are completely different from the B-doped and N-doped graphenes both structurally and chemically [67].

MXenes: These are the most recent class of 2D materials that can be classified as carbides, carbonitrides, or nitrides of early transition metals (Fig. 7.1). The first-ever isolation of MXene, namely, Ti_3C_2, was reported by Naguib et al. in 2011 by the exfoliation of Ti_3AlC_2 crystals in hydrofluoric acid [69]. The general formula of MXenes is given by $M_{n+1}X_n$, where M is an early transition metal atom and X is either carbon or nitrogen or sometimes both with $n = 1$–3. These materials are derived from the crystalline phases called MAX phases with a general formula $M_{n+1}AX_n$, where A is an element from group 13 or 14 like Al, Si, etc. [70]. The MXenes are chemically more stable compared to their respective MAX phases due to the high reactivity of A, which easily escapes from the phase and leads to the formation of MXenes [69]. A few well-known MXenes that are in use nowadays include Ti_3C_2, Ti_2C, Nb_2C, V_2C, $(Ti_{0.5}, Nb_{0.5})_2C$, $(V_{0.5},Cr_{0.5})_3C_2$, Ti_3CN, and Ta_4C_3 [71].

3 Reversible functionalization approaches

3.1 Functionalization by noncovalent interactions

Noncovalent interactions are weak, yet fundamentally important intermolecular forces responsible for the structure, stability, and function of molecular clusters as

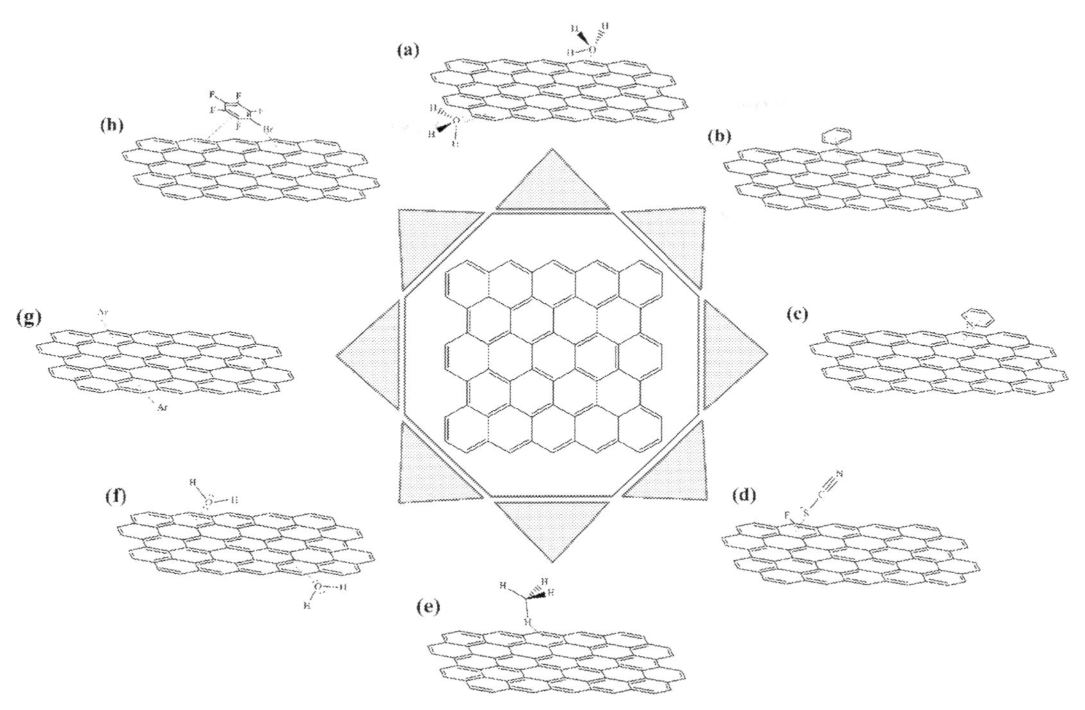

FIG. 7.4 Depiction of different types of noncovalent functionalization modes of graphene. (A) Cation-π interaction, (B) π-π interaction, (C) π_{cation}-π interaction, (D) anion-π interaction, (E) XH-π interaction {X = C, N, O}, (F) lone pair-π interaction, and (G) nonpolar gas-π interaction, and (H) σ_{hole}-π and π_{hole}-π interaction.

shown in Fig. 7.4 [72]. These interactions play a key role in enhancing the properties of graphene and GO, making them available for a range of applications in gas separation, biosensing, energy storage and conversion, electronics, CO_2 capture, and catalysis to name a few. Although identified over a century ago, these interactions were studied in detail only a few decades earlier [73].

3.1.1 *Cation-π interactions*

Cation-π interactions are one of the finely studied noncovalent interactions, which are often comparable and sometimes stronger than hydrogen bonding interactions (Fig. 7.4A) [74]. The interactions are typically due to the interaction of an atom or a molecule containing a positive charge with a π electron donor, which is electrically neutral. The cations bind to the negatively dense region of the π donor by a strong attractive force, which is driven by the electrostatic component in most of the cases except when the onium ions are considered, in which case the dispersion component dominates with a considerable contribution of the electrostatic component, and hence, these interactions are stronger than the interactions between two neutral molecules governed by a weaker dispersive interaction [1]. The binding strength of cation-π interactions is dependent both on the cation and on the π system. Size of the cation greatly influences the strength of the interaction. As we go down the group from Li to Cs in alkali metals or Be to Ba in alkaline earth metals, the size of their respective cation increases, thereby leading to an increase in the surface area of the interacting surfaces and hence reducing the strength of the interaction. The charge on the atom also plays a big role in the strength of the interactions as it shares a proportional relation to the strength of the interaction. The substitution on the cationic molecular ion also determines the strength in its own way. Hussain et al. have shown that the interaction energies decrease with an increase in the substitution of the onium ions, i.e., ammonium, phosphonium, hydronium, and sulfonium (NH_4^+, PH_4^+, OH_3^+, SH_3^+), by methyl groups when interacting with benzene and other heterocyclic aromatic systems [75]. The distance between cation and the π system has a direct relation to the interaction energy of the cation-π system. The geometry of the π-system is one more important factor upon which the strength of these interactions can be determined. The size and curvature effects of the π system govern the strength of the system. The planar π systems have a higher affinity to bind a cation than curved or puckered structures. The curved π systems tend to weakly interact with the cations, and this effect increases with the increase in the curvature. However, the transition metal cations interact in a way completely different from the alkali and alkaline earth metal cation interaction with the π systems. Metal ions like Na^+, Ag^+, and Cu^+ interact with the cation exactly in the centroid of the π system, while the cations like Au^+, Pt^+, Pd^+, and Hg^+ prefer an off-centered interaction due to the backward donation of electrons from the π system to the transition metal cations [1].

The delocalization of polarizable π electrons of graphene brings a significant change to the intrinsic characteristics of graphene, which is responsible for the performance of separation membrane for water treatment. The membrane stability and selectivity are enhanced by cross-linking effects due to cation–π interactions [76,77].

3.1.2 *π-π interactions*

The π-π interactions are one of the most fascinating and highly important noncovalent interactions in the carbonaceous materials like graphene, CNTs, fullerenes etc. due to the fact that the electrically neutral yet negatively rich electrostatic π regions of the interacting systems exhibit a finite attractive potential (Fig. 7.4B). The π-π interactions are weaker in comparison with cation-π interactions due to the absence of any ionic species taking part in the interaction [1,78]. These interactions were reported in great detail considering benzene dimer as a model system and it was identified that the interaction is dominated by a π-σ attraction (T-shaped orientation) where the interplanar angle is approaching 90° and the other mode of interaction, i.e., π-π, is repulsive and slowly slides into a relatively stable parallel displaced stacking orientation with energy as low as ~2 kcal/mol, comparable to the T-shaped orientation [1,79]. The occurrence of these interactions is also fairly dependent on the type of substituents attached to the aromatic moiety. When the carbon atoms of the benzene molecule are replaced by any heteroatoms like nitrogen or phosphorous, the shift in the interaction energy is obvious due to the change in the electron density around the center of the π cloud of the molecule. Instead, the addition of substituents to the carbons also alters the way in which the π systems interact. Adding electron-withdrawing groups to the π electron-containing system reduces the electron density above and below the plane of the π system, thereby making it possible for an unsubstituted π system to interact face to face with it. Conversely, if the substitution is done by electron-releasing groups, the electron density above and below the molecular plane is increased and leads to a greater repulsion or sliding between the interacting moieties. These interactions are dominated by the dispersion component, and the electrostatic term comes to play only after the substitution with other groups.

The behavior of π-π interactions in the presence of other noncovalent interactions was explained by Zhao and Zhu [77]. There is always a cooperative relationship between the π-π interactions and hydrogen bonding and likewise with cation-π interactions. Additionally, the cation has a higher affinity to bind to a π-π dimer than to a single π system [77]. Recent studies of π-π interactions with graphene by Arranz-Mascarós et al. have shown that tuning of the bandgap of graphene is possible by carefully interacting HIS (2-amino, 5-nitroso, 2,6-dihistidine-substituted pyrimidine derivative) residues over its surface as the bandgap is directly related to the amount of HIS interacting with graphene [72]. Yang et al. have suggested the importance of π-π interactions in the binding of perylene diimide to graphene sheets via self-assembly to form an aerogel composite at low temperatures [80]. This has shown a better visible-light photocatalytic activity of the self-assembled composite compared to the perylene diimide, which is attributed to an increase in carrier mobility and better separation of electron–hole pairs achieved due to π-π interactions between perylene diimide and graphene [80].

3.1.3 π_{cation}-π interactions

This is a special case of cation-π interaction where a neutral π system interacts with a cationic π electron-containing system (Fig. 7.4C). It is also sometimes referred to as π^+-π interactions, where π^+ represents the cationic π system. These interactions differ in energy from both cation-π and π-π interactions [1]. They typically have weaker interaction strengths than the conventional cation-π interactions due to the absence of a strong cationic metal ion to bind with the π system but have stronger interaction strengths compared to their π-π counterparts owing to the presence of a delocalized positive charge on its surface. In the case of classic carbocationic π systems or nitrogen-substituted cationic π systems interacting with the neutral π system, the preferred orientation is T-shaped π^+-π(T) complex due to the positively charged H^+ atom attached to the C/N atom [81]. The π^+-π(T) complexes have a higher contribution from the electrostatic component followed by the dispersion and induction energies, whereas the π^+-π(D) (or parallel displaced) complexes exhibit the dominance of the dispersion component and relatively weaker electrostatic and induction energies [81].

3.1.4 Anion-π interactions

Anion-π interactions are quite unusual to happen when we trust our chemical intuition, which suggests the obvious impossibility for an anionic species to attractively interact with a π system, which also has a huge negative charge density above its molecular plane (Fig. 7.4D). These interactions occur between anionic species and the electron-deficient neutral π systems. The interactions between the anions and π systems can happen in three mutually competitive modes, viz., (i) anion interacting with the electron-deficient hydrogen attached to the π system, (ii) anion covalently forming a σ-type interaction with carbon atoms making a Meisenheimer complex, and (iii) anion interacting with the electron-deficient region above the molecular plane of the π system [1,82]. The interaction strengths of these unnaturally strong interactions are comparable to typical cation-π systems. Although it is necessarily an ion-neutral molecule-type interaction, the total strength of the interaction is governed by the dispersion component [83] rather than the anticipated electrostatic term [1]. However, conflicting remarks were made by Frontera et al. regarding the component of energy having the maximum contribution, wherein they have suggested that the electrostatic and ion-induced polarization components are the major contributors for the interaction energy of anion-π complexes [82]. Naturally, there is also a contribution from the exchange-repulsion component due to the obvious repulsions between the two electron-dense systems approaching each other [83].

Theoretical studies performed by Maiyelvaganan et al. on the adsorption of ionic liquids on sheet-like (coronene derivative) and curved (CNTs) carbonaceous surfaces have revealed some interesting insights into the effect of curvature on these cation-π and anion-π interactions [84]. Ionic liquids with their abundance of both cationic and anionic residues lead to a competitive mode of interaction with the carbonaceous materials. While the cationic 1-methyl-3-methylimidazolium residue individually has a good affinity to bind with both the sheet and curved interfaces, the anions play a key role in the adsorption of these ionic liquids. Strong anions like Cl^- (which are repelled by the π electron cloud of the surface) interact strongly with the cation and decrease the cation interaction with graphene/CNTs, while weaker anions tend to adsorb readily to the surface along with their respective cation. Alternately, in the case of curved surfaces, the stronger anions are adsorbed strongly due to stronger electrostatic interactions between the anion and the curved surface [84]. Further, theoretical calculations performed by Saha and Bhattacharyya by interacting oxoanions (NO_3^-, PO_4^{3-}, and ClO_4^-) with coronene (undoped, B, Al, N, P, Si, BN, and 3BN doped) as a representative molecule for graphene have shown a clear distinction between the effect of doped and undoped coronene systems on the binding affinities of oxoanions [85]. Also, they have demonstrated the existence of physisorption due to anion-π interactions in the case of undoped, N, BN, and 3BN doped coronenes, while the other doped coronenes undergo chemisorption [85].

3.1.5 XH-π interactions

The XH-π-type interactions are a special type of hydrogen bonding interactions with an exception that the electron-deficient hydrogen interacts attractively to the electron-dense cloud over the π system (Fig. 7.4E). The atom attached to the hydrogen can be a strongly electronegative atoms like fluorine and oxygen or can be as weak as a carbon whose electronegativity is only a little greater than the hydrogen atom. Usually, these interactions occur between π systems and the hydrogen attached to boron, carbon, nitrogen, oxygen, and fluorine atoms. The most widely studied systems both experimentally and computationally exhibiting XH-π-type interactions are CH-π systems due to their prominence in a multitude of biological processes [86]. Unlike typical hydrogen bonds, these interactions are not dominated by the electrostatic term completely but also depend on the type of XH-π interaction. In the case of polar XH-π interactions, the electrostatic term dominates the interaction energy, while the nonpolar XH-π interactions are predominantly dispersive in nature [87]. Theoretical studies of CH-π systems have shown the increase in interaction energy with the shift of hybridization of carbon from sp^3 to sp. [86].

Li and Chen have conducted theoretical calculations employing dispersion-corrected density functional theory calculations and confirmed the existence of XH-π interactions between graphene and its partial (C_4H graphene) or fully hydrogenated (graphane) derivatives [88]. They further explained that such an interaction helped in increasing the bandgap in graphene/C_4H or graphene/graphene bilayer. Studies conducted by de Moraes et al. using density functional theory-based calculations have given insights into the comparative study of π-π interactions and XH-π interactions [89]. Aromatic molecules like benzene, phenol, catechol, and dopamine were subjected to interactions with themselves and then with graphene leading to both π-π and XH-π types of interactions. Interaction of the aforementioned aromatic molecules with graphene has clearly shown that the face-to-face interaction or π-π interaction is more favored than the XH-π interactions possibly due to the lesser π electron density over graphene compared to simple aromatic molecules [89].

3.1.6 Lone pair-π interactions

Lone pair-π interactions are another class of counterintuitive noncovalent interactions in addition to the anion-π interactions occurring in nature (Fig. 7.4F). One of the first reported studies of these interactions was by Egli and Gessner in 1995, where they observed a lone pair-π interaction between the lone pair of oxygen and a nucleotide within a Z-DNA, thereby stabilizing it, and they called it "$n \rightarrow \pi^*$ hyperconjugation" [90]. Although overlooked by the scientific community for several decades, these interactions turned out to be useful in stabilizing the biological macromolecules [91]. One more interesting finding was reported in 2004 by Stollar et al. where they have observed multiple cases of lone pair-π interactions between the lone pairs in oxygen of water and the aromatic residues of the amino acids [92]. Theoretical studies have shown that the lone pair-π interactions are possible between the neutral electron-rich lone pair-containing system and the electron-deficient aromatic system [93]. A lone pair-π interaction is considered strong where the angle between a lone pair and the centroid of the aromatic ring is in the range of 75–90°, while, in the remaining cases, the interaction is either a moderate lone pair-π interaction or a covalent σ-type interaction [91].

Güryel et al. in their computational study have observed that lone pair-π interactions are one of the most significant contributors toward the interaction between pristine graphene and functional groups (alkyl, hydroxyl, aldehyde, carboxyl, amino, and nitro) in addition to XH-π and π-π interactions [94]. They also have confirmed that the interaction of the nitro group with GO is almost comparable to that of a hydrogen bond [94].

3.1.7 Nonpolar gas-π interactions

Nonpolar gas-π interactions are the interactions between the nonpolar gases and π electron-rich surfaces of molecules (Fig. 7.4G). One of the earliest studies on such an interaction was performed by Hobza et al. in 1992 where they theoretically determined the structure and energetics of benzene-X (X = He, Ne, Ar, Kr, Xe) systems using second- and fourth-order Møller-Plesset calculations [95]. Although they reported the interactions to be mere van der Waals interactions, the recent surge of theoretical and experimental investigations have revealed a completely new type of noncovalent interaction. Hobza et al. have shown that the trend of stabilization energies of benzene-X complexes is in correlation with the experimental values available for these interactions [95]. Tarakeshwar et al. have conducted similar studies on argon with benzene and two fluoro-substituted benzene derivatives at the second-order Møller-Plesset level [96]. They have calculated the interaction energies, geometries, vibrational frequencies, and electron densities systematically whose values were well in agreement with the experimental numbers. They have shown that the addition of electron-withdrawing groups to benzene can lead to a decrease in the π electron density above and below the molecular plane of benzene, in turn leading to a significant decrease in the exchange repulsion [96]. Recent theoretical studies by Hwang et al. with the focus on greenhouse gases (CH_4 and CO_2) adsorption on graphene and fluorographene have depicted the possibility of graphene-based materials as very good gas-sensing materials [97]. While CO_2 was observed to bind more effectively to the graphene-based materials than CH_4, the binding affinity of both CH_4 and CO_2 with fluorographene has been higher than that of pristine graphene. The efficient adsorption of CO_2 ($(-14.27 \pm 0.61) \times 10^{-40}$ cm^2) is attributed to its higher quadrupole moment than CH_4 (0 cm^2) [97].

3.1.8 σ_{hole}-π and π_{hole}-π interactions

Recently, σ- and π-holes have come up as fascinating concepts in the noncovalent interactions area as shown in Fig. 7.4F [98,99]. Both σ- and π-holes represent the electronic density, which results in positive electrostatic potentials. The σ-hole found at the end of covalently bonded halogen atom (C—Br) surface develops as an area of positive molecular surface electrostatic potential (MSEP) [100]. This positive MSEP may interact with the electronegative atom (N or O atom as halogen bond) or with the rich π-electron region as a σ_{hole}-π interaction [101]. However, the π-hole develops when the strong electron-withdrawing substituents are attached to the aromatic π system and result in a positive MSEP area on the π ring. So this π-hole may interact with the π-electron-rich region as π_{hole}-π interaction. The MSEP of C_6F_5Br exhibits that it possesses the σ-hole at the end of its C—Br bond and π-hole on its aromatic π ring. In general, the π_{hole}-π interactions are stronger than the σ_{hole}-π interactions. The physisorption property of graphene was notably affected by σ-hole and π-hole characters of bromopentafluorobenzene (C_6F_5Br) and the π-electron-rich region of graphene due to the formation of π_{hole}-π and σ_{hole}-π interactions [102].

3.1.9 Graphene-biomolecular interactions

Graphene and its analogues widely interact with biomolecules containing π residues. DNA, the basic building blocks of life and proteins, are made of aromatic monomers called nucleic acids and amino acids, respectively (Fig. 7.5). While all the four nucleic acids contain π electrons ready for a surface-to-surface interaction, only four of the 20 naturally occurring amino acids are aromatic and take part in functionalization over graphene and its analogues. These interactions are to be studied with utmost importance as these graphene-based biosensors would be helpful as biomarkers for the early detection of many diseases [103]. Because of its extraordinary structural properties, graphene can be integrated with biomolecules like nucleosides, proteins, peptides, enzymes, etc. with applications extended to biosensing, nanotechnology, and medicine [104].

Binding of nucleobases and nucleosides: The binding of DNA nucleobases to graphene is a topic of absolute interest due to its double-helical structure and biocompatibility (Fig. 7.5) [104]. The four DNA nucleobases adenine (A), guanine (G), cytosine (C), and thymine (T) can readily interact with layered graphene sheet. Graphene can interact with DNA nucleobases in various modes of interaction like π-π interactions, hydrogen bonds, and electrostatic interactions [104]. However, the strength of these interactions is governed by a variety of factors, including water/fat solubility of the nucleobase, distance from the surface of graphene etc. Nucleobases were experimentally interacted with graphene in strict aqueous conditions. The order of decreasing relative interaction strengths between graphene and

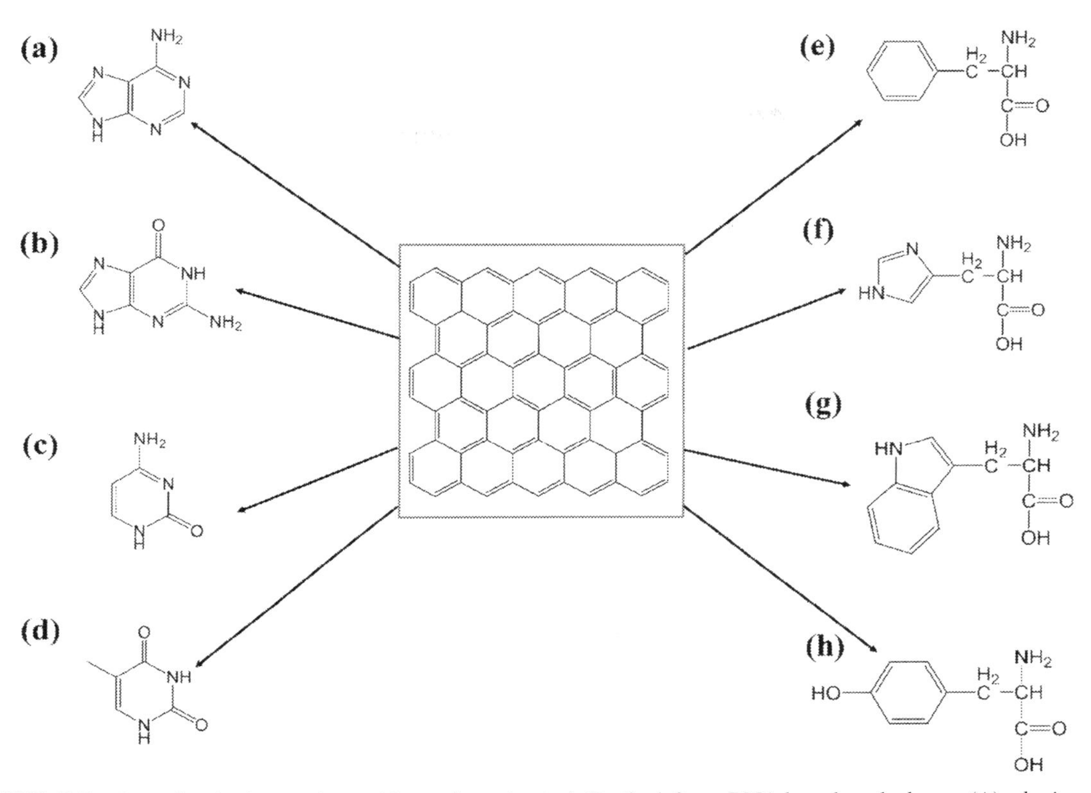

FIG. 7.5 Biomolecular interactions with graphene (center). On the left are DNA-based nucleobases: (A) adenine, (B) guanine, (C) cytosine, and (D) thymine, while on the right are aromatic amino acids, (E) phenyl alanine, (F) histidine, (G) tryptophan, and (H) tyrosine.

the four nucleobases is G > A > C ≥ T [105]. A theoretical-level comparison between binding ability of graphene and CNTs was reported by Sastry and co-workers. [19,23]. The curvature of nanostructure also plays a vital role in the interaction strengths with the nucleobases. Graphene, with its 2D structure clearly, has an added advantage toward binding to these nucleobase residues compared to the curved 1D CNTs.

Binding of peptides and amino acids: Similar to nucleobases, amino acids also interact with graphene and its analogues (Fig. 7.5). Due to the abundance of delocalized π electrons, graphene can interact with the naturally occurring aromatic amino acid residues. There are, however, only four such residues (PHE, TRP, TRY, and HIS) available to bind with graphene and its analogues. A comparative study was reported by Sastry and co-workers in 2015 where the binding affinity of the four aromatic amino acids with graphene is compared to its hydrogenated partner graphane. Graphane, due to the distorted planar structure caused by the addition of hydrogen atoms, interacts with the aromatic amino acids in the CH...π mode [19]. The order of interaction energies of the four aromatic amino acids with graphene is observed to be TRP > HIS ~ TYR > PHE. Zou et al. have reported the experimental interaction of graphene with two peptides C-terminus cysteine-modified cecropin P1(CP1C) and N-terminus cysteine-modified MSI-78 (MSI-78 (C1)) followed by theoretical validation by using molecular dynamic simulations [103]. They have hypothesized that the orientation of peptides on graphene depends on the competitive interaction of planar aromatic residue and hydrophilic amino acid residue with the surface of graphene [103].

3.2 Functionalization by nanoparticles

Due to its high specific surface area and excellent electronic, physical, and chemical properties, graphene has emerged as an ideal support material for metal or metal oxide nanoparticles (NPs, Fig. 7.6). Graphene decorated with NPs, such as Au, Ag, Pt, Pd, TiO_2, ZnO_2, Fe_3O_4, etc., exhibits unique properties due to the synergistic effect between graphene and inorganic NPs. Therefore, graphene-based nanocomposites have become promising for various applications such as chemical and biological sensors; photocatalyst; and optical, energy storage, and conversion devices [106–110]. The functionalization of graphene with NPs enhances the properties of graphene and stabilizes graphene sheets, thereby preventing them from aggregation by strong van der Waals interactions.

The preparation of graphene-based nanocomposites is carried out in two ways [107]: post immobilization of nanoparticles called ex situ hybridization and in situ bindings of nanoparticles called in situ crystallization. The former method involves the mixing of separate solutions of reduced graphene oxide (rGO) nanosheets and pre-synthesized NPs. Prior to mixing, both/either the NPs and/or rGO sheets can be surface-functionalized to enhance the stability and processability of the final products. However, this method usually suffers from a low-density and nonuniform coverage of NPs on graphene sheets. In the in situ crystallization

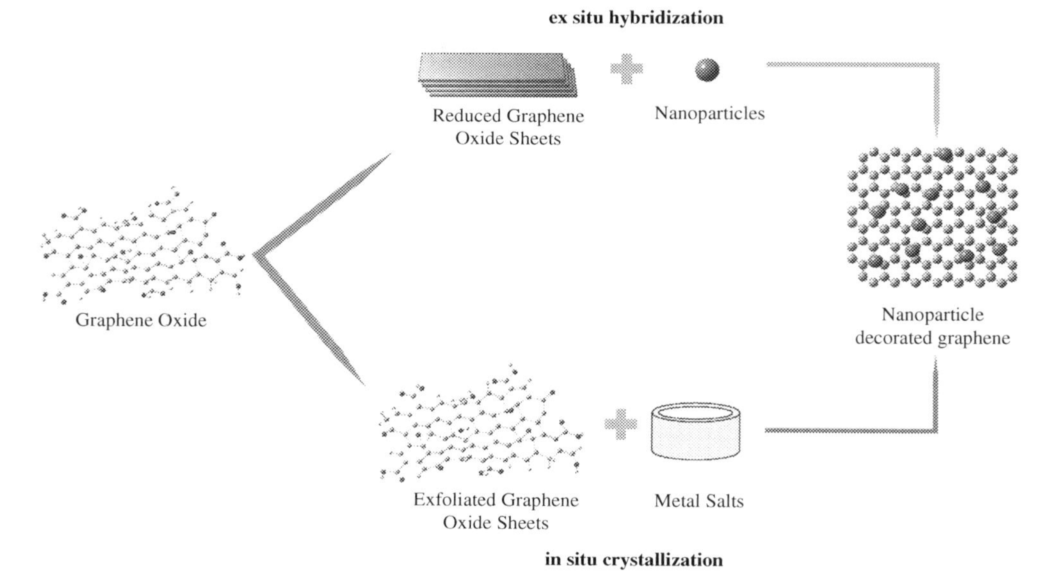

FIG. 7.6 Nanoparticle interaction with graphene oxide via ex situ hybridization and in situ crystallization.

method, the preparation of graphene-based nanocomposites is done by concomitant reduction of GO and the respective metal salt(s) using a reducing agent.

4 Irreversible functionalization approaches

Covalent functionalization is studied in detail due to the fact that such functionalization would enhance the electrical properties of graphene to a great extent [111]. The covalent functionalization over the graphene is not feasible at any carbon on the surface of the basal plane, but at the edges, defects and the places with little out-of-plane deformations due to the higher reactivity at those regions (Fig. 7.2) [111]. There are two main types of irreversible covalent functionalizations that can be performed on graphene, viz., (i) functionalization by covalent bonding and (ii) functionalization by element doping.

4.1 Functionalization by covalent bonding

Covalent functionalization of graphene is mostly irreversible. The most reported examples of such irreversible functionalization approaches on graphene are nucleophilic addition, electrophilic addition, free radical addition, atomic covalent functionalization (hydrogenation, halogenation, and oxidation) reaction, and cycloaddition reactions. Covalent functionalization of graphene leads to the conversion of any sp^2 carbon taking part in bonding to sp^3, possibly leading to the disruption of two-dimensional sheet-like properties [9]. They significantly change the structure and electronic properties of graphene. Covalent functionalization of redox-active molecules tends to have a higher stability than their noncovalent counterparts [10].

4.1.1 Nucleophilic addition

Nucleophilic addition reactions involve the addition of negatively charged nucleophiles to graphene (Fig. 7.7). Nucleophilic addition on graphene was achieved by Zhen Li et al. [112] by the introduction of negatively charged nitrogen ion obtained from a carbazole derivative (poly-9,9′-diheylfluorene carbazole or PDC) on reduced graphene oxide. The nitrogen is generated by treating PDC with sodium hydride as it is known to be a good reducing agent and hence can easily deprotonate a hydrogen from PDC generating the nucleophilic carbazole derivative. This ion now readily binds to rGO, making nucleophilic addition. A similar approach was followed by Zhan et al., where poly(*N*-vinylcarbazole) is treated with sodium hydride, thereby abstracting protons from carbon instead of nitrogen [113]. The carbanion thus created interacts with the surface of reduced graphene oxide leading to a nucleophilic addition.

4.1.2 Electrophilic addition

Electrophilic addition reactions involve the addition of positively charged electrophiles to graphene (Fig. 7.7). Yan et al. proposed that rGO undergoes electrophilic addition by the addition of n-BuLi followed by bromo-substituted triethylamine [114]. Adding n-BuLi to rGO may lead to the formation of two possible intermediate products. In the first case,

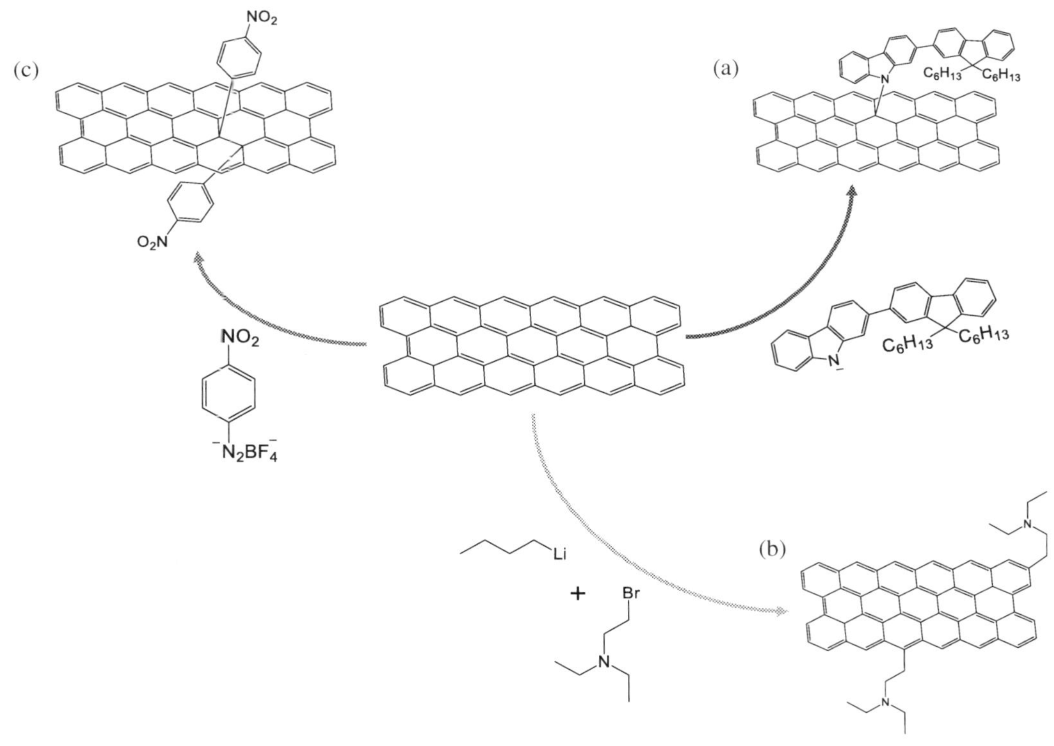

FIG. 7.7 Interaction of reaction intermediates with graphene (center): (A) nucleophilic addition, (B) electrophilic addition, and (C) radical addition reactions with graphene.

deprotonation occurs, which leads to substitution of lithium on graphene surface, while in the case of carbometalation, both lithium and the n-butyl group get substituted on graphene by replacing a C—C double bond. In either case, the substitution of lithium on graphene is the most important step because this helps in the attack of electrophile on rGO followed by the removal of lithium bromide.

4.1.3 Radical reactions

Free radicals are chemically reactive species, which readily interact with sp^2 carbon atoms of graphene having free π electrons (Fig. 7.7). One of the earlier reports of such work was proposed by Tour et al. by decorating graphene surfaces with nitrophenyl radicals [12,13]. They have generated the graphene by chemically unzipping the CNTs and nitrophenyl radicals by heating the diazonium salt. The nitrophenyl radicals attack the sp^2 centers of graphene and convert them into sp^3 centers by getting attached to them. Niyogi et al. [115] have shown that the attachment of nitrophenyl radicals on graphene creates a band gap, which is controllable, due to which these nitrophenyl-decorated graphenes could potentially be used as semiconducting materials.

4.1.4 *Atom covalent functionalization*

The addition of single atoms to the surface of graphene results in many amazing changes in the properties of it. The most common examples of these single-atom introductions are hydrogen, halogen, and oxygen atom introductions (Fig. 7.8). These introductions typically involve the conversion of sp^2 hybrid carbon atoms of graphene to their sp^3 counterparts, thereby allowing free movement which is otherwise restricted due to the C=C double bond of the sp^2 carbon. Also, the carbon atoms prone to these introductions tend to move either above or below the layer of graphene due to two main reasons, i.e., (i) the change in C=C bond length of 1.42 Å to higher values of C—C and C—F bond lengths and (ii) the change in C—C—C bond angle from 120° to 109° 28′ [116–119]. Even though the above changes take place due to the single atom introductions, there is no change in the overall 2D nature of the functionalized graphene sheet.

4.1.4.1 HYDROGENATION

The introduction of hydrogen atoms onto the surface of graphene to form either graphane or hydrogenated graphene was first predicted by Sofo et al. based on their theoretical calculations (Fig. 7.8) [16]. Experimentally, this feat was achieved by exposing pristine graphene to cold hydrogen plasma to form graphane and hydrogenated graphene by two separate groups in the subsequent years [17,120]. It has been proven that although graphene has preserved its hexagonal symmetry unaltered, there has been a significant change in the electronic structure

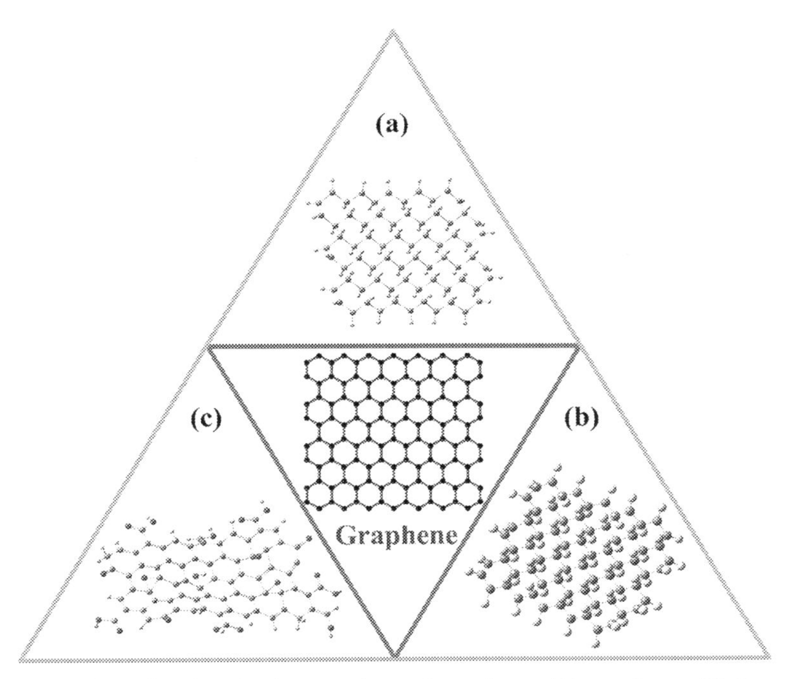

FIG. 7.8 Atomic covalent functionalization reactions of graphene: (A) graphane, (B) fluorographene, and (C) graphene oxide.

locally. The C=C bond changes to a C—C bond, leading to an increase of the bond length from 1.42 Å to 1.52–1.56 Å. Due to these changes, the flat graphene when converted to distorted graphane loses its free π-electrons and hence behaves as an insulator [121]. Instead, if the graphene is led to partial hydrogenation, the resulting hydrogenated graphene behaves as a semi-metal due to the presence of a few free π-electrons. The properties of graphene can hence be properly tuned by partial hydrogenation [122,123]. It was also observed that the introduction of hydrogen over graphene can be reversible in specific conditions. For instance, the dehydrogenation of hydrogenated graphene or graphane can be performed by annealing these materials at 450°C in a neutral Ar atmosphere [17]. Sun et al. have made an amazing contribution by incorporating ferromagnetism into the otherwise diamagnetic graphene [124]. They have partially hydrogenated graphene by ball-milling method using acetic acid as the hydrogenating agent. The resultant few layered partially hydrogenated graphene sheets have shown a magnetic moment of 0.274 Am^2/kg, indicating that they are ferromagnetic in nature. The magnetic behavior may be the result of both the partial hydrogenation and zig-zag edges of graphene [124]. A photocatalytic method of hydrogenation of graphene was suggested by Yuan et al. using tilted Pd nanocones arranged exactly below a single-layer graphene sheet [125]. These cones are excited, which in turn dissociates H_2 to form highly reactive hydrogen atoms. These reactive hydrogen atoms interact with single-layer graphene lying above the Pd nanocones, leading to a partial hydrogenation. Similar kind of light-induced reactions on graphene may lead to further new applications of graphene.

4.1.4.2 HALOGENATION

Similar to hydrogen atoms, the halogen atoms also take part in the functionalization of graphene by an irreversible covalent route (Fig. 7.8). Even though all the halogen atoms can be introduced onto the surface of graphene, the most studied are fluorine atom introductions. The addition of fluorine atoms onto the surface of graphene yields fluorographene/graphene fluoride. The synthesis of fluorographene is carried out by the reaction of graphene with XeF_2 at room temperature [126]. The fluorination of graphene can be performed in two different ways giving two different types of fluorinated graphene sheets. The fluorination of graphene grown by chemical vapor deposition on copper support leads to the fluorination on one side of the sheet in a stoichiometric ratio of C_4F_1. On the other hand, complete double-sided fluorination of graphene with a stoichiometric ratio of C_1F_1 is obtained when the fluorination is done on graphene on the silicon-on-insulator support [1]. There are some less well-known methods to prepare fluorographene from graphite. The pristine graphite fluoride can be extended to mechanical exfoliation, a method similar to the one used to isolate graphene from graphite to give fluorographene [127,128]. Another exfoliation method called the chemical exfoliation of pyrolytic graphite fluoride yields a variant of fluorographene in the stoichiometric ratio $C_{0.7}F_1$. This method is unique for yielding defective fluorographene, as high-temperature pyrolysis was involved in preparing the graphite fluoride material [126,129]. Unlike the hydrogen atom introduction reactions, fluorine atom introduction leads to the change not only in the structure but also in their electronic and optical properties. As evident from the aforementioned reasons, the hybridization of carbon atoms in graphene changes from sp^2 to sp^3 as it converts to fluorographene. In addition, this also leads to changes in the bond length of bonded sp^2 carbon atoms from 1.42 Å to C—C bond length of 1.57–1.59 Å and C—F bond length of 1.41–1.45 Å, respectively. Similar to the effect of hydrogen atom introduction, the fluorine atom introduction also opens the bandgap due to the loss of free π-electrons on the surface. It is to be noted that thermal defluorination of fluorographene can be performed at a high temperature (400–600°C), which leads to the decomposition of the layered structure into individual C—F products like CF_4, C_2F_4, C_2F_6 etc. On the contrary, the defluorination at a lower temperatures of about 100–200°C with hydrazine vapor or by the interaction of KI with fluorographene colloidal dispersion in sulfolane at 240°C forms graphene [126]. Similar to the introduction of fluorine atom on the graphene surface, other halogen atoms can also be introduced, thereby forming stoichiometric C_1X_1 (X = F, Cl, Br, I)-type halographenes and their stability is in the order $F > Cl > Br > I$, where fluorographene is observed to be more stable than graphene itself [116,118,119]. Khan et al. have reported the use of brominated graphene as a starting material in the functionalization of graphene to enhance the surface and electronic properties [10]. Olanrele et al. have reported the interaction of graphene with diatomic halogen molecules by the use of boron- and nitrogen-doped graphene [130]. Theoretical calculations were performed to understand the binding affinities of different diatomic halogen molecules (Cl_2, Br_2, I_2) with both pristine graphene and graphene doped with trivalent boron and pentavalent nitrogen impurities [130]. Doping has effectively enhanced the binding affinity of halogen atoms toward graphene. A comparison between the humidity-sensing ability of fluorographene and chlorographene was made both experimentally and theoretically by Hajian et al. [131]. It was revealed that the fluorine-functionalized graphene is more sensitive toward humidity sensing than the chlorine-functionalized graphene.

4.1.4.3 OXIDATION

Unlike the addition of hydrogen or halogens, the addition of oxygen leads to the formation of two covalent bonds, making this addition extremely important (Fig. 7.8) [132]. The most widespread use of such an addition is observed in the synthesis of graphite oxide by Hummer's method. Although the focus of this chapter is functionalization of graphene, the abovementioned method stands as a great example for a covalent addition of oxygen. Theoretical studies involving the interaction of atomic oxygen with the layered graphene surface have anticipated the formation of an epoxide joining two adjacent carbon atoms of graphene [133,134]. This type of an epoxidation is possible only when the oxygen interacts with a particular surface of graphene that is defect and wrinkle-free. On the contrary, when the oxygen atom interacts with defective regions of graphene, it may lead to the formation of other functional groups like carboxylic acids, esters, and carbonyls due to the possible presence of hydrogen atoms. The experimental procedure for making graphene epoxide involves careful exposure to oxygen plasmas and atomic oxygen beams obtained by cracking of oxygen molecules at high temperatures [135]. Kudus et al. have reported a simplified version of Hummer's method to oxidize graphene to form GO even without the use of strong oxidizing agent like $KMnO_4$ [136]. However, this method has led to the formation of a few defects on the surface of graphene oxide. Araújo et al. have shown that the surface properties of graphene flakes can be tuned by using different oxidants to interact with graphene flakes [137]. They have shown that the reaction with nitric acid (HNO_3) has led to carboxylic acid groups dominating the graphene surface, while interaction with oxidation agents like potassium permanganate ($KMnO_4$), m-chloroperbenzoic acid (m-CPBA), and ozone (O_3) yields either hydroxyl or epoxy functionalization on graphene.

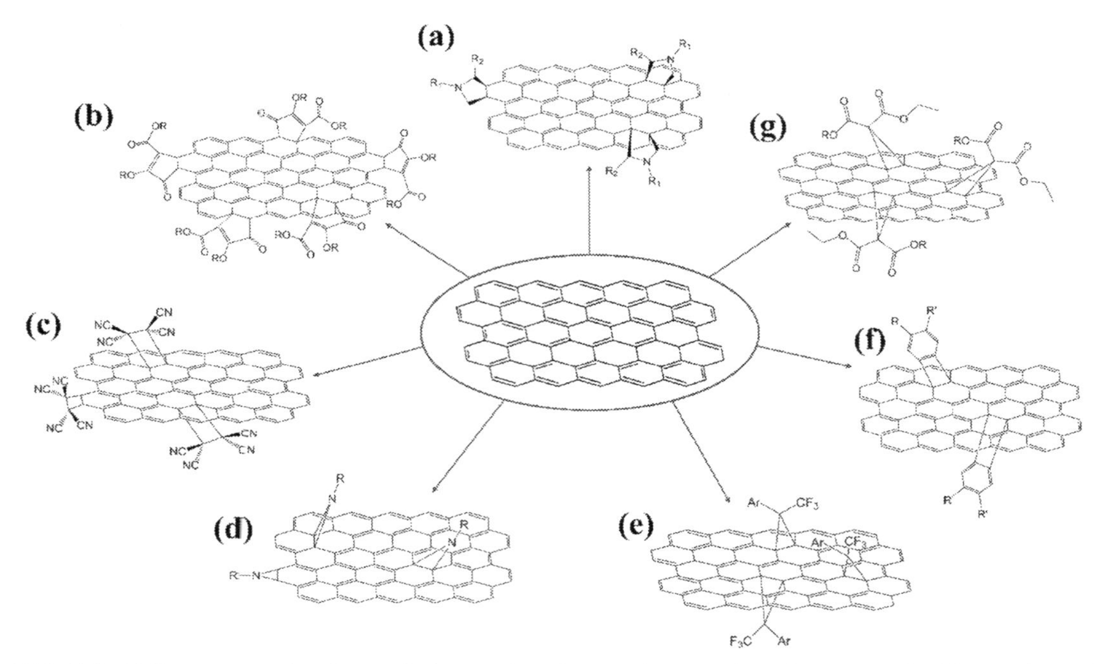

FIG. 7.9 Different types of surface and edge covalent functionalization of graphene via cycloadditions, (center) graphene, (A) 1,3-dipolar addition, (B) addition of zwitterionic substrates, (C) Diels–Alder-type cycloaddition, (D) nitrene cycloaddition, (E) carbene cycloaddition, (F) aryne cycloaddition, and (G) Bingel-type cycloaddition.

4.1.5 Cycloaddition reactions

Cycloaddition reactions are a type of pericyclic reactions that involve the interaction of two π electron-rich reactants to give a cyclic product by the formation of two new σ-bonds (Fig. 7.9) [138]. Graphene being inherently π electron-rich can be involved in interacting with other such species to give similar cyclic adducts.

4.1.5.1 1,3-DIPOLAR CYCLOADDITION OF AZOMETHINE YLIDE

1,3-Dipolar addition usually occurs between a species containing two charges at first and third positions (1,3-dipole) and an unsaturated carbon–carbon linkage (dipolarophile) (Fig. 7.9). The double bonds on the rings of graphene act as the dipolarophiles when it tends to interact with a 1,3-dipole of azomethine ylide. In this type of functionalization, the three-centered 1,3-dipole and two-centered dipolarophile yield a closed five-membered rings attached perpendicular to the surface of graphene [139]. In the present case, the reaction leads to the formation of pyrrolidine rings on the graphene surface. This type of interaction is of fundamental importance in tuning the properties of pristine graphene for applications in varied fields. This is attributed to the fact that the azomethine ylide can have two positions for being substituted or modified by alkyl and other organic functional groups, viz., one on the nitrogen atom of the pyrrolidine ring (*N*-alkyl/aryl-substituted pyrrolidine derivatives) and the other on any of the two carbon atoms adjacent to the nitrogen. Hence, the functionalization by 1,3-dipolar addition can happen in two different pathways leading to the two above-mentioned products. Substitution on the nitrogen atom of α-glycine followed by its addition to formaldehyde in the presence of thermal energy leads to the substitution on the nitrogen atom, which is almost perpendicular to the graphene surface, while the substitution on the aldehydic hydrogen with an *N*-methylated α-glycine gives the substitution on the adjacent carbon, which is closer to the surface. This functionalization can be applied on either pristine graphene or GO as both of them have a surplus of reactive sp^2 carbons.

4.1.5.2 CYCLOADDITION BY ZWITTERIONIC INTERMEDIATE

Cycloaddition reaction can also happen by the reaction of zwitterionic intermediates with graphene or graphene oxide. In fact, such a reaction of zwitterions with graphene was reported by Zhang et al. in 2012 (Fig. 7.9) [140]. The required zwitterion was produced upon the condensation of substituted acetylene dicarboxylate with p-dimethylamino pyridine. Depending on the choice of substituents on the carboxylate functionalities, the resultant functionalized graphene nanostructures would become dispersible either in water or in organic solvents.

4.1.5.3 DIELS–ALDER CYCLOADDITION

Diels–Alder cycloaddition or Diels–Alder reaction is one of the most well-known reactions in synthetic organic chemistry (Fig. 7.9). It is widely famous as one of the few reactions that can contribute to 100% atom economy. The cycloaddition involves a cyclization between diene and dienophile in the presence of either heat or light in a single step. Graphene, being a source of a plethora of doubly bonded carbon atoms, can be functionalized using this reaction. Due to the aforementioned reason, graphene can act both as a diene and a dienophile with respect to the interacting reagent. Several studies on the reactions of graphene nanosheets were performed and were identified to be due to the Diels–Alder cycloaddition by Haddon et al. in 2011 [141]. With the interacting partners like maleic anhydride and tetracyanoethylene, graphene behaves like a diene, while with reactants like 2,3-dimethoxy-1, 3-butadiene and 9-methyl anthracene, it acts as a dienophile. Haddon et al. have also reported that the resultant of the cycloaddition was a six-membered ring always perpendicular to the plane of graphene [141]. The reaction with tetracyanoethylene is reversible under carefully employed conditions retrieving the former electronic properties of graphene. A similar strategy was applied by Yuan et al. to chemically graft long chains of polyethylene glycol (PEG) on rGO [142]. PEG chains are attached to the cyclopentadienyl groups, which in turn would participate in Diels–Alder cycloaddition with rGO where the former acts as a diene and the latter as a dienophile. The reaction would result in the further cyclization of the cyclopentadienyl groups, thereby forming bridged (bicyclic) rings perpendicular to the molecular plane of rGO with long chains of PEG attached to them.

4.1.5.4 NITRENE CYCLOADDITION

Nitrenes, one of the most important reactive intermediates, are obtained by exposing the organic azides ($R\text{-}N_3$) to thermal or photochemical conditions, thereby removing nitrogen molecule (Fig. 7.9). They are neutral monovalent atoms of nitrogen. Nitrenes are highly reactive, and they interact with C=C bonds to form azirine rings by a cycloaddition mechanism. The organic portion of the now formed azides can be chosen and changed as per our need for applications. Many scientific reports were published by replacing the organic group by aromatic rings [143,144], polymeric chains [145], and simple aliphatic chains that may contain several functional groups like hydroxyl and carboxyl groups—which, if they contain at least one replaceable hydrogen, may be further functionalized [146,147].

4.1.5.5 CARBENE CYCLOADDITION

Similar to the nitrenes, carbenes are also well known as highly reactive intermediates (Fig. 7.9). They are electron-deficient species analogous to nitrenes, but can interact with not only the sp^2-hybridized C=C bonds on the surface of graphene but also on the singly bonded sp^3 centers of carbon on the edges and defects. A well-known functionalization method uses the dichlorocarbene produced by reacting chloroform sodium hydroxide on the nanoplatelets of graphene to form three-membered cyclic rings substituted on the surface of graphene [139]. Ismaili et al. in 2011 have reported a more complex method wherein the reactive carbene was generated upon the photochemical treatment of a 3-aryl-3 (trifluoromethyl)-diazirine derivative linked to gold nanoparticles by Au—S bond [148]. The diazirine compounds have a three-membered ring wherein the carbon as a sp^3 hybrid is attached to the two doubly bonded nitrogen atoms. Upon photochemical treatment, the two nitrogen atoms are removed as N_2 gas, leaving behind a reactive lone pair containing carbon-based intermediate, carbene, which interacts with the surface of doubly bonded sp^2-hybridized C=C groups to form three-membered cyclic rings perpendicular to the surface [148].

4.1.5.6 ARYNE CYCLOADDITION

Arynes are one of the most important types of reactive intermediates, which can be produced by removing the substituents/hydrogens from the two adjacent doubly bonded carbon atoms of an aromatic ring (Fig. 7.9). This reactive species can interact with the surface of graphene, which here serves as the diene by undergoing a cyclic [2+2] addition. The product of this reaction is the conversion of interacting sp^2 carbon atoms to their respective sp^3 hybrids and the subsequent formation of four-membered cycles on the surface of graphene. The functionalization of graphene by aryne intermediate to form the cyclic four-membered substitutions on the surface of graphene proceeds through an elimination-addition mechanism. The first step is the elimination of the substituents attached to the two adjacent sp^2 carbon atoms, making the way for the formation of a reactive triply bonded pair of sp-hybridized carbon atoms, which is the aryne. The next step is the addition of C=C of graphene sheet to the aryne via cyclic transition state resulting in the formation of a new four-membered ring-substituted on the surface of graphene. The most common example of such a cycloaddition reaction is benzyne cycloaddition over graphene [149]. The above-mentioned case was reported by Ma et al. [150] where the substituted benzyne gives three different derivatives labeled as H-hybrid, F-hybrid, and Me-hybrid corresponding to hydrogen, fluorine, and methyl substitutions, respectively.

4.1.5.7 BINGEL-TYPE CYCLOADDITION

A special type of [2+1] cycloaddition was discovered by Bingel [151] where a bromo-substituted active methylene group interacts with fullerene with diazabicyclo[5.4.0] undec-7-ene (DBU) as a catalyst to form a cyclopropane substitution on the surface of the fullerene (Fig. 7.9). The cycloaddition selectively occurs on the shorter C=C double bond existing between two hexagonal rings of the fullerene. A similar type of approach was extended to several other carbon nanostructures like CNTs, nanohorns etc. in the preceding years. Interestingly, as reported by Tagmatarchis et al. in 2010, this type of a cycloaddition was found to occur between graphene and a bromo-substituted active methylene group using the catalyst DBU under microwave conditions [152]. The microwave radiation does not only reduce the reaction time but also produces lesser byproducts. The graphene nanoplatelets were first dispersed in benzylamine by sonication, which is used as a starting material for the cycloaddition with diethyl malonate and monosubstituted tetrathiafulvalene diethyl malonate under the microwave-irradiated conditions.

4.2 Element doping

Element doping in graphene can be achieved by annealing heat treatment, arc discharge, and ion bombardment to incorporate different elements such as B, P, and N into graphene, which results in the substitution of defects and vacancy defects, while maintaining its intrinsic 2D structure (Fig. 7.10). Element doping changes the surface properties by adjusting its energy band structure that increases the performance of the material [153]. But the element doping in graphene is very complex to control quantitatively. The N-doped graphene (6.5% at N) was prepared by thermal annealing to treat NH_4NO_3 and graphene by Duan et al. [154], which was used for catalytic oxidative degradation of phenol. The reported catalytic activity of the N-doped graphene was 5.4 times greater than that of unmodified graphene. Another CVD technique was used first time to prepare substitutionally N-doped graphene by Wei et al. [155].

5 Comparison of functionalization of graphene with its analogues

The extent of functionalization of a small molecule or a biomolecular residue is not only dependent on its size, charge, number of π electrons but also on the curvature of the carbonaceous material. Curvature plays a vital role in the functionalizing with certain functionalities. Theoretical studies were conducted by Sastry and co-workers where they have clearly shown the inverse proportionality between the curvature and the binding energy of the physisorbed residue [23,24,156]. The comparison between the binding affinity of different carbon nanotubes and graphene toward the nucleobases adenine (A), guanine (G), cytosine (C), thymine (T), and uracil (U) was demonstrated. For the sake of a better comparison, only armchair-type CNTs of different sizes, i.e., (3,3), (4, 4, 5, 5), were considered, thereby consecutively increasing the area of contact and reducing the curvature. For a more refined understanding, both the pyrimidine and imidazole rings of the purines were let to interact separately with the nanostructures. The binding energies obtained at all the considered theoretical methods vividly portrayed greater binding for the nanostructures with minimal curvature. Even though CNTs are also similar to graphene in composition, they fundamentally vary in the physical and chemical properties due to their structural differences. Graphene has

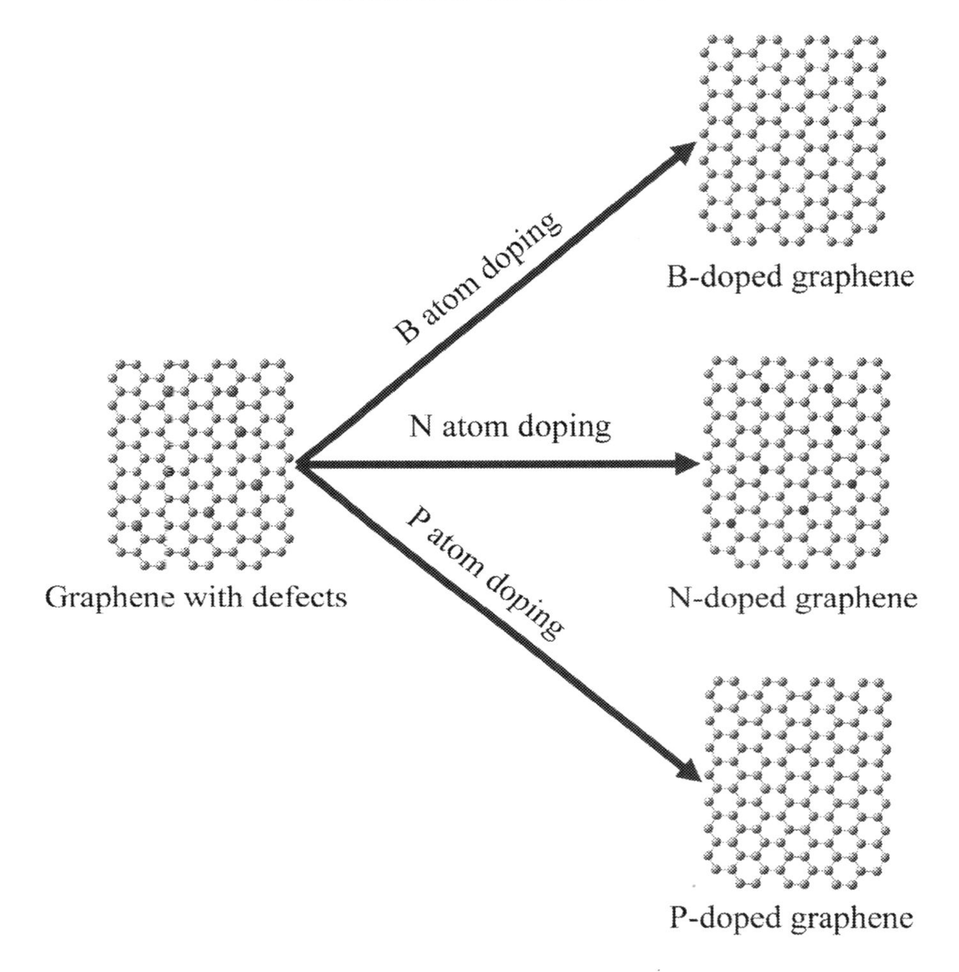

FIG. 7.10 Element doping in the defective sites of graphene by boron, nitrogen, and phosphorous.

abundant surface area of contact due to its sheet-like structure, while its 1D analogue even though infinitely long cannot provide a large surface to interact with other molecules. Further studies by the same group provided more insights into the interaction between alkali and alkaline earth metals on both graphene nanoribbons and CNTs as a comparative study [156]. Alkali and alkaline earth metals interact readily with the electron-rich graphene-like substances via cation-π mode. Several CNTs both zig-zag and armchair-type and graphene nanoribbons of different sizes were led to interact with a few alkali and alkaline earth metal ions. The curvature played an important role in effective binding of the ions to both CNTs and graphene. Similar to the pattern observed in the case of adsorption of nucleobases on CNTs and graphene, binding energies of metal ion interaction with CNTs and graphene have also shown an inverse proportional relation with the curvature of the material.

Comparative studies were conducted on the binding affinities of small molecules with planar carbonaceous substances and the concave and convex regions of fullerenes. C_{60} and C_{70} were cut into readily identifiable fragments sumanene, corannulene, and coronene [34,42]. It is already well known that coronene is used in the synthesis of graphene by following a bottom-up approach [157,158]. Upon small-molecule interaction with the concave, convex, and flat surfaces of the fragments, it was observed that the concave surface showed a greater affinity to bind to the small molecules than the convex and coronene. This has led to a number of new uncertainties in the behavior of these carbonaceous materials in the presence of external molecules. Although the coronene residue has lesser binding propensity with small molecules, it is again greater compared to curved convex surface, proving the above point that the curvature is inversely proportional to binding affinity.

6 Outlook

Graphene has become one of the most studied molecules in the last few years due to its limitless possibilities and wide range of applications. Functionalization has been proven to be an excellent way to fine-tune the properties of graphene and graphene-based compounds to make them useful for specific applications. In addition to graphene, graphene analogues like CNTs, fullerenes, graphane etc. and inorganic 2D materials like boron nitride, layered chalcogenides, borocarbonitrides, and MXenes also have important applications. The functionalization of graphene is either covalent or noncovalent in nature. Covalent functionalization is usually irreversible, while the noncovalent modes of interactions are reversible and weaker. The reversibility of functionalization is dependent on many factors like binding affinity, type of the interacting functional groups, solvent effects, the distance of separation between graphene and the functional group, and the size and geometry of functional groups.

Functionalization based on noncovalent interactions is via a cation-π, anion-π, π-π, π_{cation}-π, XH-π, lone pair-π, nonpolar gas-π, and σ_{hole}-π and π_{hole}-π type of linkage. While the cation-π interactions are the strongest among those listed, π-π interactions are the weakest. The cycloadditions are another important mode of graphene functionalization. Covalent functionalization—although irreversible—is very often applied on graphene and graphene oxide, the well-known additions being the addition of hydrogen-forming graphane, addition of halogens (mainly fluorine) leading to the formation of halographene, and addition of oxygen to form graphene epoxide. Graphene also binds to biomolecules to form biologically active substances. This interaction with biomolecules like nucleobases and amino acids is noncovalent in nature. Considering the aforementioned biomolecules, the interaction is mostly π-π or π_{cation}-π, while the other types of interactions rarely occur. Finally, a comparative understanding on the binding ability and effect of curvature on 0D fullerenes, 1D CNTs, and 2D graphene was considered.

In our opinion, chemical space has expanded significantly with the entry of carbon-based clusters in the last four decades. While irreversible functionalization of graphene will lead to new materials, the reversible functionalization provides a handle to fine-tune and modulates a number of physicochemical properties of these materials.

References

[1] V. Georgakilas, M. Otyepka, A.B. Bourlinos, V. Chandra, N. Kim, K.C. Kemp, P. Hobza, R. Zboril, K.S. Kim, Functionalization of graphene: covalent and non-covalent approaches, derivatives and applications, Chem. Rev. 112 (2012) 6156.

[2] K.S. Novoselov, A.K. Geim, S.V. Morozov, D. Jiang, Y. Zhang, S.V. Dubonos, I.V. Grigorieva, A.A. Firsov, Electric field effect in atomically thin carbon films, Science 306 (2004) 666.

[3] S. Guo, S. Dong, Graphene nanosheet: synthesis, molecular engineering, thin film, hybrids, and energy and analytical applications, Chem. Soc. Rev. 40 (2011) 2644.

[4] M.J. Allen, V.C. Tung, R.B. Kaner, Honeycomb carbon: a review of graphene, Chem. Rev. 110 (2010) 132.

[5] K. Anish, J. Mohammad, N.E.A. Bernaurdshaw, Graphene Functionalization Strategies: From Synthesis to Applications, Springer, 2019.

[6] A.B. Bourlinos, V. Georgakilas, R. Zboril, T.A. Steriotis, A.K. Stubos, Liquid-phase exfoliation of graphite towards solubilized graphenes, Small 5 (2009) 1841.

[7] C.E. Hamilton, J.R. Lomeda, Z. Sun, J.M. Tour, A.R. Barron, High-yield organic dispersions of unfunctionalized graphene, Nano Lett. 9 (2009) 3460.

[8] Y. Hernandez, V. Nicolosi, M. Lotya, F.M. Blighe, Z. Sun, S. De, I. McGovern, B. Holland, M. Byrne, Y.K. Gun'Ko, High-yield production of graphene by liquid-phase exfoliation of graphite, Nat. Nanotechnol. 3 (2008) 563.

[9] R. Khan, R. Nakagawa, B. Campeon, Y. Nishina, A simple and robust functionalization of graphene for advanced energy devices, ACS Appl. Mater. Interfaces 12 (2020) 12736.

[10] R. Khan, Y. Nishina, Covalent functionalization of carbon materials with redox-active organic molecules for energy storage, Nanoscale 13 (2021) 36.

[11] W. Yu, L. Sisi, Y. Haiyan, L. Jie, Progress in the functional modification of graphene/graphene oxide: a review, RSC Adv. 10 (2020) 15328.

[12] D.V. Kosynkin, A.L. Higginbotham, A. Sinitskii, J.R. Lomeda, A. Dimiev, B.K. Price, J.M. Tour, Longitudinal unzipping of carbon nanotubes to form graphene nanoribbons, Nature 458 (2009) 872.

[13] A. Sinitskii, A. Dimiev, D.A. Corley, A.A. Fursina, D.V. Kosynkin, J.M. Tour, Kinetics of diazonium functionalization of chemically converted graphene nanoribbons, ACS Nano 4 (2010) 1949.

[14] W. Du, H. Geng, Y. Yang, Y. Zhang, X. Rui, C.C. Li, Pristine graphene for advanced electrochemical energy applications, J. Power Sources 437 (2019) 226899.

[15] A. Fasolino, J. Los, M.I. Katsnelson, Intrinsic ripples in graphene, Nat. Mater. 6 (2007) 858.

[16] J.O. Sofo, A.S. Chaudhari, G.D. Barber, Graphane: a two-dimensional hydrocarbon, Phys. Rev. B 75 (2007) 153401.

[17] D.C. Elias, R.R. Nair, T. Mohiuddin, S. Morozov, P. Blake, M. Halsall, A.C. Ferrari, D. Boukhvalov, M. Katsnelson, A. Geim, Control of graphene's properties by reversible hydrogenation: evidence for graphane, Science 323 (2009) 610.

[18] V. Nagarajan, R. Chandiramouli, A novel approach for detection of NO_2 and SO_2 gas molecules using graphane nanosheet and nanotubes—a density functional application, Diam. Relat. Mater. 85 (2018) 53.

[19] D. Umadevi, G.N. Sastry, Graphane versus graphene: a computational investigation of the interaction of nucleobases, aminoacids, heterocycles, small molecules (CO_2, H_2O, NH_3, CH_4, H_2), metal ions and onium ions, Phys. Chem. Chem. Phys. 17 (2015) 30260.

[20] R. Zhang, Y. Zhang, Q. Zhang, et al., Growth of half-meter long carbon nanotubes based on Schulz Flory distribution, ACS Nano 7 (2013) 6156.

[21] S. Iijima, Helical microtubules of graphitic carbon, Nature 354 (1991) 56.

[22] R. Rao, C.L. Pint, A.E. Islam, R.S. Weatherup, S. Hofmann, E.R. Meshot, F. Wu, C. Zhou, N. Dee, P.B. Amama, Carbon nanotubes and related nanomaterials: critical advances and challenges for synthesis toward mainstream commercial applications, ACS Nano 12 (2018) 11756.

[23] D. Umadevi, G.N. Sastry, Quantum mechanical study of physisorption of nucleobases on carbon materials: graphene versus carbon nanotubes, J. Phys. Chem. Lett. 2 (2011) 1572.

[24] D. Umadevi, G.N. Sastry, Impact of the chirality and curvature of carbon nanostructures on their interaction with aromatics and amino acids, ChemPhysChem 14 (2013) 2570.

[25] A. Hirsch, Functionalization of single-walled carbon nanotubes, Angew. Chem. Int. 41 (2002) 1853.

[26] H.W. Kroto, A. Allaf, S. Balm, C60: buckminsterfullerene, Chem. Rev. 91 (1991) 1213.

[27] S. Higashibayashi, H. Sakurai, Synthesis of sumanene and related buckybowls, Chem. Lett. 40 (2011) 122.
[28] X. Li, F. Kang, M. Inagaki, Buckybowls: corannulene and its derivatives, Small 12 (2016) 3206.
[29] B.M. Schmidt, D. Lentz, Syntheses and properties of buckybowls bearing electron-withdrawing groups, Chem. Lett. 43 (2014) 171.
[30] U. DevaáPriyakumar, G. Narahariá Sastry, Ring closure synthetic strategies toward buckybowls: benzannulation versus cyclopentannulation, J. Chem. Soc. Perkin Trans. 2 (2002) 94.
[31] T. Dinadayalane, S. Deepa, A.S. Reddy, G.N. Sastry, Density functional theory study on the effect of substitution and ring annelation to the rim of corannulene, J. Organomet. Chem. 69 (2004) 8111.
[32] S. Higashibayashi, S. Onogi, H.K. Srivastava, G.N. Sastry, Y.T. Wu, H. Sakurai, Stereoelectronic effect of curved aromatic structures: favoring the unexpected endo conformation of benzylic-substituted sumanene, Angew. Chem. 125 (2013) 7455.
[33] S. Higashibayashi, R. Tsuruoka, Y. Soujanya, U. Purushotham, G.N. Sastry, S. Seki, T. Ishikawa, S. Toyota, H. Sakurai, Trimethylsumanene: enantioselective synthesis, substituent effect on bowl structure, inversion energy, and electron conductivity, Bull. Chem. Soc. Jpn. 85 (2012) 450.
[34] M.A. Hussain, D. Vijay, G.N. Sastry, Buckybowls as adsorbents for CO_2, CH_4, and C_2H_2: binding and structural insights from computational study, J. Comput. Chem. 37 (2016) 366.
[35] U.D. Priyakumar, M. Punnagai, G.K. Mohan, G.N. Sastry, A computational study of cation–π interactions in polycyclic systems: exploring the dependence on the curvature and electronic factors, Tetrahedron 60 (2004) 3037.
[36] U.D. Priyakumar, G.N. Sastry, Tailoring the curvature, bowl rigidity and stability of heterobuckybowls: theoretical design of synthetic strategies towards heterosumanenes, J. Mol. Graph. Model. 19 (2001) 266.
[37] U. Purushotham, G.N. Sastry, Conjugate acene fused buckybowls: evaluating their suitability for p-type, ambipolar and n-type air stable organic semiconductors, Phys. Chem. Chem. Phys. 15 (2013) 5039.
[38] A.M. Rice, E.A. Dolgopolova, N.B. Shustova, Fulleretic materials: buckyball-and buckybowl-based crystalline frameworks, Chem. Mater. 29 (2017) 7054.
[39] G.N. Sastry, Computational studies on siblings en-route to fullerenes: study of curved polycyclic aromatic hydrocarbons, J. Mol. Struct. THEOCHEM 771 (2006) 141.
[40] G.N. Sastry, U.D. Priyakumar, The role of heteroatom substitution in the rigidity and curvature of buckybowls. A theoretical study, J. Chem. Soc. Perkin Trans. 2 (2001) 30.
[41] G.N. Sastry, H.S.P. Rao, P. Bednarek, U.D. Priyakumar, Effect of substitution on the curvature and bowl-to-bowl inversion barrier of bucky-bowls. Study of mono-substituted corannulenes ($C_{19}XH_{10}$, X = B^-, N^+, P^+ and Si), Chem. Commun. 10 (2000) 843.
[42] D. Vijay, H. Sakurai, V. Subramanian, G.N. Sastry, Where to bind in buckybowls? The dilemma of a metal ion, Phys. Chem. Chem. Phys. 14 (2012) 3057.
[43] H. Sakurai, T. Daiko, T. Hirao, A synthesis of sumanene, a fullerene fragment, Science 301 (2003) 1878.
[44] U.D. Priyakumar, G.N. Sastry, Theory provides a clue to accomplish the synthesis of sumanene, $C_{21}H_{12}$, the prototypical C3v-buckybowl, Tetrahedron Lett. 42 (2001) 1379.
[45] M. Chhowalla, H.S. Shin, G. Eda, L.-J. Li, K.P. Loh, H. Zhang, The chemistry of two-dimensional layered transition metal dichalcogenide nanosheets, Nat. Chem. 5 (2013) 263.
[46] X. Huang, Z. Zeng, H. Zhang, Metal dichalcogenide nanosheets: preparation, properties and applications, Chem. Soc. Rev. 42 (2013) 1934.
[47] R. Mas-Balleste, C. Gomez-Navarro, J. Gomez-Herrero, F. Zamora, 2D materials: to graphene and beyond, Nanoscale 3 (2011) 20.
[48] K.S. Novoselov, D. Jiang, F. Schedin, T. Booth, V. Khotkevich, S. Morozov, A.K. Geim, Two-dimensional atomic crystals, Proc. Natl. Acad. Sci. 102 (2005) 10451.
[49] K. Raidongia, A. Gomathi, C. Rao, Synthesis and characterization of nanoparticles, nanotubes, nanopans, and graphene-like structures of boron nitride, Isr. J. Chem. 50 (2010) 399.
[50] C. Rao, A. Nag, Inorganic analogues of graphene, Eur. J. Inorg. Chem. 2010 (2010) 4244.
[51] C. Rao, H. Ramakrishna Matte, U. Maitra, Graphene analogues of inorganic layered materials, Angew. Chem. Int. Ed. 52 (2013) 13162.
[52] X. Song, J. Hu, H. Zeng, Two-dimensional semiconductors: recent progress and future perspectives, J. Mater. Chem. C 1 (2013) 2952.
[53] Q.H. Wang, K. Kalantar-Zadeh, A. Kis, J.N. Coleman, M.S. Strano, Electronics and optoelectronics of two-dimensional transition metal dichalcogenides, Nat. Nanotechnol. 7 (2012) 699.

[54] M. Xu, T. Liang, M. Shi, H. Chen, Graphene-like two-dimensional materials, Chem. Rev. 113 (2013) 3766.
[55] R. Frindt, Single crystals of MoS_2 several molecular layers thick, J. Appl. Phys. 37 (1966) 1928.
[56] P. Joensen, R. Frindt, S.R. Morrison, Single-layer MoS_2, Mater. Res. Bull. 21 (1986) 457.
[57] H. Ramakrishna Matte, A. Gomathi, A.K. Manna, D.J. Late, R. Datta, S.K. Pati, C. Rao, MoS2 and WS_2 analogues of graphene, Angew. Chem. Int. Ed. 49 (2010) 4059.
[58] B.K. Miremadi, S.R. Morrison, The intercalation and exfoliation of tungsten disulfide, J. Appl. Phys. 63 (1988) 4970.
[59] D. Yang, R. Frindt, Li-intercalation and exfoliation of WS_2, J. Phys. Chem. Solids 57 (1996) 1113.
[60] C. Chiritescu, D.G. Cahill, N. Nguyen, D. Johnson, A. Bodapati, P. Keblinski, P. Zschack, Ultralow thermal conductivity in disordered, layered WSe_2 crystals, Science 315 (2007) 351.
[61] X. Wang, Y. Gong, G. Shi, W.L. Chow, K. Keyshar, G. Ye, R. Vajtai, J. Lou, Z. Liu, E. Ringe, Chemical vapor deposition growth of crystalline monolayer $MoSe_2$, ACS Nano 8 (2014) 5125.
[62] A. Nagashima, N. Tejima, Y. Gamou, T. Kawai, C. Oshima, Electronic dispersion relations of monolayer hexagonal boron nitride formed on the Ni (111) surface, Phys. Rev. B 51 (1995) 4606.
[63] Y. Lin, J.W. Connell, Advances in 2D boron nitride nanostructures: nanosheets, nanoribbons, nanomeshes, and hybrids with graphene, Nanoscale 4 (2012) 6908.
[64] X. Li, L. Basile, M. Yoon, C. Ma, A.A. Puretzky, J. Lee, J.C. Idrobo, M. Chi, C.M. Rouleau, D.B. Geohegan, Revealing the preferred interlayer orientations and stackings of two-dimensional bilayer gallium selenide crystals, Angew. Chem. Int. Ed. 127 (2015) 2750.
[65] A. Kuhn, A. Chevy, R. Chevalier, Crystal structure and interatomic distances in GaSe, Phys. Status. Solidi. A 31 (1975) 469.
[66] N. Kumar, K. Moses, K. Pramoda, S.N. Shirodkar, A.K. Mishra, U.V. Waghmare, A. Sundaresan, C. Rao, Borocarbonitrides, $B_xC_yN_z$, J. Mater. Chem. A 1 (2013) 5806.
[67] C.N.R. Rao, M. Chhetri, Borocarbonitrides as metal-free catalysts for the hydrogen evolution reaction, Adv. Mater. 31 (2019) 1803668.
[68] C. Rao, K. Gopalakrishnan, Borocarbonitrides, $B_xC_yN_z$: synthesis, characterization, and properties with potential applications, ACS Appl. Mater. Interfaces 9 (2017) 19478.
[69] M. Naguib, M. Kurtoglu, V. Presser, J. Lu, J. Niu, M. Heon, L. Hultman, Y. Gogotsi, M.W. Barsoum, Two-dimensional nanocrystals produced by exfoliation of Ti_3AlC_2, Adv. Mater. 23 (2011) 4248.
[70] M.W. Barsoum, The $M_{n+1}AX_n$ phases: a new class of solids: thermodynamically stable nanolaminates, Prog. Solid State Chem. 28 (2000) 201.
[71] M. Naguib, V.N. Mochalin, M.W. Barsoum, Y. Gogotsi, 25th anniversary article: MXenes: a new family of two-dimensional materials, Adv. Mater. 26 (2014) 992.
[72] P. Arranz-Mascarós, M.L. Godino-Salido, R. López-Garzón, C. García-Gallarín, I. Chamorro-Mena, F.J. López-Garzón, E. Fernández-García, M.D. Gutiérrez-Valero, Non-covalent functionalization of graphene to tune its band gap and stabilize metal nanoparticles on its surface, ACS Omega 5 (2020) 18849.
[73] A.S. Mahadevi, G.N. Sastry, Cooperativity in noncovalent interactions, Chem. Rev. 116 (2016) 2775.
[74] A.S. Mahadevi, G.N. Sastry, Cation $-\ \pi$ interaction: its role and relevance in chemistry, biology, and material science, Chem Rev. 113 (2013) 2100.
[75] M.A. Hussain, A.S. Mahadevi, G.N. Sastry, Estimating the binding ability of onium ions with CO_2 and π systems: a computational investigation, Phys. Chem. Chem. Phys. 17 (2015) 1763.
[76] B. Liu, L.E. López-González, M. Alamri, E.F. Velázquez-Contrera, H. Santacruz-Ortega, J.Z. Wu, Cation–π interaction assisted molecule attachment and Photocarrier transfer in rhodamine/graphene Heterostructures, Adv. Mater. Interfaces 7 (2020) 2000796.
[77] G. Zhao, H. Zhu, Cation–π interactions in graphene-containing Systems for Water Treatment and beyond, Adv. Mater. 32 (2020) 1905756.
[78] E.M. Pérez, N. Martín, π–π interactions in carbon nanostructures, Chem. Soc. Rev. 44 (2015) 6425.
[79] C.A. Hunter, J.K. Sanders, The nature of. Pi.-Pi. Interactions, J. Am. Chem. Soc. 112 (1990) 5525.
[80] J. Yang, H. Miao, Y. Wei, W. Li, Y. Zhu, π–π interaction between self-assembled perylene diimide and 3D graphene for excellent visible-light photocatalytic activity, Appl. Catal. B Environ. 240 (2019) 225.
[81] N.J. Singh, S.K. Min, D.Y. Kim, K.S. Kim, Comprehensive energy analysis for various types of π-interaction, J. Chem. Theory Comput. 5 (2009) 515.
[82] A. Frontera, P. Gamez, M. Mascal, T.J. Mooibroek, J. Reedijk, Putting anion–π interactions into perspective, Angew. Chem. Int. Ed. 50 (2011) 9564.

[83] M. Majumder, A.K. Thakur, Graphene and Its Modifications for Supercapacitor Applications, Surface Engineering of Graphene, vol. 113, Springer, 2019.
[84] K.R. Maiyelvaganan, S. Kamalakannan, M. Prakash, Adsorption of ionic liquids on carbonaceous surfaces: the effect of curvature on selective anion$\cdots\pi$ and cation$\cdots\pi$ interactions, Appl. Surf. Sci. 495 (2019), 143538.
[85] B. Saha, P.K. Bhattacharyya, Anion$\cdots\pi$ interaction in oxoanion-graphene complex using coronene as model system: a DFT study, Comput. Theor. Chem. 1147 (2019) 62.
[86] J.W. Bloom, R.K. Raju, S.E. Wheeler, Physical nature of substituent effects in XH/π interactions, J. Chem. Theory Comput. 8 (2012) 3167.
[87] S.J. Grabowski, P. Lipkowski, Characteristics of XH$\cdots\pi$ interactions: ab initio and QTAIM studies, J. Phys. Chem. A 115 (2011) 4765.
[88] Y. Li, Z. Chen, XH/π (X = C, Si) interactions in graphene and silicene: weak in strength, strong in tuning band structures, J. Phys. Chem. Lett. 4 (2013) 269.
[89] E.E. de Moraes, M.Z. Tonel, S.B. Fagan, M.C. Barbosa, Density functional theory study of π-aromatic interaction of benzene, phenol, catechol, dopamine isolated dimers and adsorbed on graphene surface, J. Mol. Model. 25 (2019) 1.
[90] M. Egli, R.V. Gessner, Stereoelectronic effects of deoxyribose O4′on DNA conformation, Proc. Natl. Acad. Sci. U. S. A. 92 (1995) 180.
[91] T.J. Mooibroek, P. Gamez, J. Reedijk, Lone pair–π interactions: a new supramolecular bond? CrystEngComm 10 (2008) 1501.
[92] E.J. Stollar, J.L. Gelpí, S. Velankar, A. Golovin, M. Orozco, B.F. Luisi, Unconventional interactions between water and heterocyclic nitrogens in protein structures, Proteins 57 (2004) 1.
[93] B.W. Gung, Y. Zou, Z. Xu, J.C. Amicangelo, D.G. Irwin, S. Ma, H.-C. Zhou, Quantitative study of interactions between oxygen lone pair and aromatic rings: substituent effect and the importance of closeness of contact, J. Organomet. Chem. 73 (2008) 689.
[94] S. Güryel, M. Alonso, B. Hajgató, Y. Dauphin, G. Van Lier, P. Geerlings, F. De Proft, A computational study on the role of noncovalent interactions in the stability of polymer/graphene nanocomposites, J. Mol. Model. 23 (2017) 1.
[95] P. Hobza, O. Bludský, H. Selzle, E. Schlag, A binitio second-and fourth-order Mo/ller–Plesset study on structure, stabilization energy, and stretching vibration of benzene$\cdots$ X (X = He, ne, Ar, Kr, Xe) van der Waals molecules, J. Chem. Phys. 97 (1992) 335.
[96] P. Tarakeshwar, K.S. Kim, E. Kraka, D. Cremer, Structure and stability of fluorine-substituted benzene-argon complexes: the decisive role of exchange-repulsion and dispersion interactions, J. Chem. Phys. 115 (2001) 6018.
[97] D.G. Hwang, E. Jeong, S.G. Lee, Density functional theory study of CH_4 and CO_2 adsorption by fluorinated graphene, Carbon. Lett. 20 (2016) 81.
[98] G. Cavallo, P. Metrangolo, R. Milani, T. Pilati, A. Priimagi, G. Resnati, G. Terraneo, The halogen bond, Chem. Rev. 116 (2016) 2478.
[99] P. Politzer, J.S. Murray, T. Clark, G. Resnati, The σ-hole revisited, Phys. Chem. Chem. Phys. 19 (2017) 32166.
[100] X. Yang, F. Yang, R.-Z. Wu, C.-X. Yan, D.-G. Zhou, P.-P. Zhou, X. Yao, Linear σ-hole$\cdots$ CO$\cdots$ σ-hole intermolecular interactions between carbon monoxide and dihalogen molecules XY (X, Y = cl, Br), J. Mol. Graph. Model. 76 (2017) 419.
[101] F.-L. Yang, X. Yang, R.-Z. Wu, C.-X. Yan, F. Yang, W. Ye, L.-W. Zhang, P.-P. Zhou, Intermolecular interactions between σ-and π-holes of bromopentafluorobenzene and pyridine: computational and experimental investigations, Phys. Chem. Chem. Phys. 20 (2018) 11386.
[102] Y.-H. Zhang, Y.-L. Li, J. Yang, P.-P. Zhou, K. Xie, Noncovalent functionalization of graphene via π-hole$\cdots\pi$ and σ-hole$\cdots\pi$ interactions, Struct. Chem. 31 (2020) 97.
[103] X. Zou, S. Wei, J. Jasensky, M. Xiao, Q. Wang, C.L. Brooks III, Z. Chen, Molecular interactions between graphene and biological molecules, J. Am. Chem. Soc. 139 (2017) 1928.
[104] D. Li, W. Zhang, X. Yu, Z. Wang, Z. Su, G. Wei, When biomolecules meet graphene: from molecular level interactions to material design and applications, Nanoscale 8 (2016) 19491.
[105] N. Varghese, U. Mogera, A. Govindaraj, A. Das, P.K. Maiti, A.K. Sood, C. Rao, Binding of DNA nucleobases and nucleosides with graphene, ChemPhysChem 10 (2009) 206.
[106] P.A. Bozkurt, Sonochemical green synthesis of ag/graphene nanocomposite, Ultrason. Sonochem. 35 (2017) 397.

[107] M. Khan, M.N. Tahir, S.F. Adil, H.U. Khan, M.R.H. Siddiqui, A.A. Al-warthan, W. Tremel, Graphene based metal and metal oxide nanocomposites: synthesis, properties and their applications, J. Mater. Chem. A 3 (2015) 18753.
[108] S.P. Lonkar, V. Pillai, A. Abdala, Solvent-free synthesis of ZnO-graphene nanocomposite with superior photocatalytic activity, Appl. Surf. Sci. 465 (2019) 1107.
[109] B.P. Tarasov, A.A. Arbuzov, S.A. Mozhzhuhin, A.A. Volodin, P.V. Fursikov, M.V. Lototskyy, V.A. Yartys, Hydrogen storage behavior of magnesium catalyzed by nickel-graphene nanocomposites, Int. J. Hydrogen. Energ. 44 (2019) 29212.
[110] C. Wang, D. Astruc, Recent developments of metallic nanoparticle-graphene nanocatalysts, Prog. Mater. Sci. 94 (2018) 306.
[111] A. Criado, M. Melchionna, S. Marchesan, M. Prato, The covalent functionalization of graphene on substrates, Angew. Chem. Int Ed. 54 (2015) 10734.
[112] X. Xu, J. Chen, X. Luo, J. Lu, H. Zhou, W. Wu, H. Zhan, Y. Dong, S. Yan, J. Qin, Poly (9, 9′-diheylfluorene carbazole) functionalized with reduced graphene oxide: convenient synthesis using nitrogen-based nucleophiles and potential applications in optical limiting, Chem–Eur. J. 18 (2012) 14384.
[113] P.P. Li, Y. Chen, J. Zhu, M. Feng, X. Zhuang, Y. Lin, H. Zhan, Charm-bracelet-type poly (N-vinylcarbazole) functionalized with reduced graphene oxide for broadband optical limiting, Chem–Eur. J. 17 (2011) 780.
[114] C. Yuan, W. Chen, L. Yan, Amino-grafted graphene as a stable and metal-free solid basic catalyst, J. Mater. Chem. 22 (2012) 7456.
[115] S. Niyogi, E. Bekyarova, M.E. Itkis, H. Zhang, K. Shepperd, J. Hicks, M. Sprinkle, C. Berger, C.N. Lau, W.A. Deheer, Spectroscopy of covalently functionalized graphene, Nano Lett. 10 (2010) 4061.
[116] A.B. Bourlinos, K. Safarova, K. Siskova, R. Zbořil, The production of chemically converted graphenes from graphite fluoride, Carbon 50 (2012) 1425.
[117] M.Z. Flores, P.A. Autreto, S.B. Legoas, D.S. Galvao, Graphene to graphane: a theoretical study, Nanotechnology 20 (2009) 465704.
[118] F. Karlický, R. Zbořil, M. Otyepka, Band gaps and structural properties of graphene halides and their derivates: a hybrid functional study with localized orbital basis sets, J. Chem. Phys. 137 (2012) 034709.
[119] R. Zbořil, F. Karlický, A.B. Bourlinos, T.A. Steriotis, A.K. Stubos, V. Georgakilas, K. Šafářová, D. Jančík, C. Trapalis, M. Otyepka, Graphene fluoride: a stable stoichiometric graphene derivative and its chemical conversion to graphene, Small 6 (2010) 2885.
[120] S. Ryu, M.Y. Han, J. Maultzsch, T.F. Heinz, P. Kim, M.L. Steigerwald, L.E. Brus, Reversible basal plane hydrogenation of graphene, Nano Lett. 8 (2008) 4597.
[121] P. Chandrachud, B.S. Pujari, S. Haldar, B. Sanyal, D. Kanhere, A systematic study of electronic structure from graphene to graphane, J. Phys. Condens. Mat. 22 (2010) 465502.
[122] R. Balog, B. Jørgensen, L. Nilsson, M. Andersen, E. Rienks, M. Bianchi, M. Fanetti, E. Lægsgaard, A. Baraldi, S. Lizzit, Bandgap opening in graphene induced by patterned hydrogen adsorption, Nat. Mater. 9 (2010) 315.
[123] H. Gao, L. Wang, J. Zhao, F. Ding, J. Lu, Band gap tuning of hydrogenated graphene: H coverage and configuration dependence, J. Phys. Chem. C 115 (2011) 3236.
[124] Q. Sun, X. Wang, B. Li, Y. Wu, Z. Zhang, X. Zhang, X. Zhao, X. Liu, Acetic acid assistant hydrogenation of graphene sheets with ferromagnetism, Chem. Res. Chin. Univ. 34 (2018) 344.
[125] L. Yuan, C. Zhang, X. Zhang, M. Lou, F. Ye, C.R. Jacobson, L. Dong, L. Zhou, M. Lou, Z. Cheng, Photocatalytic hydrogenation of graphene using Pd nanocones, Nano Lett. 19 (2019) 4413.
[126] J.T. Robinson, J.S. Burgess, C.E. Junkermeier, S.C. Badescu, T.L. Reinecke, F.K. Perkins, M.K. Zalalutdniov, J.W. Baldwin, J.C. Culbertson, P.E. Sheehan, Properties of fluorinated graphene films, Nano Lett. 10 (2010) 3001.
[127] R.R. Nair, W. Ren, R. Jalil, I. Riaz, V.G. Kravets, L. Britnell, P. Blake, F. Schedin, A.S. Mayorov, S. Yuan, Fluorographene: a two-dimensional counterpart of Teflon, Small 6 (2010) 2877.
[128] F. Withers, M. Dubois, A.K. Savchenko, Electron properties of fluorinated single-layer graphene transistors, Phys. Rev. B 82 (2010) 073403.
[129] S.-H. Cheng, K. Zou, F. Okino, H.R. Gutierrez, A. Gupta, N. Shen, P. Eklund, J. Sofo, J. Zhu, Reversible fluorination of graphene: evidence of a two-dimensional wide bandgap semiconductor, Phys. Rev. B 81 (2010) 205435.
[130] S.O. Olanrele, Z. Lian, C. Si, B. Li, Halogenation of graphene triggered by heteroatom doping, RSC Adv. 9 (2019) 37507.
[131] S. Hajian, P. Khakbaz, B. Narakathu, S. Masihi, M. Panahi, D. Maddipatla, V. Palaniappan, R. Blair, B. Bazuin, M. Atashbar, Humidity sensing properties of halogenated graphene: a comparison of fluorinated

graphene and chlorinated graphene, in: 2020 IEEE International Conference on Flexible and Printable Sensors and Systems (FLEPS), IEEE, 2020, p. 1.
[132] J.E. Johns, M.C. Hersam, Atomic covalent functionalization of graphene, Acc. Chem. Res. 46 (2013) 77.
[133] A. Barinov, O.B. Malcioglu, S. Fabris, T. Sun, L. Gregoratti, M. Dalmiglio, M. Kiskinova, Initial stages of oxidation on graphitic surfaces: photoemission study and density functional theory calculations, J. Phys. Chem. C 113 (2009) 9009.
[134] D.C. Sorescu, K.D. Jordan, P. Avouris, Theoretical study of oxygen adsorption on graphite and the (8, 0) single-walled carbon nanotube, J. Phys. Chem. B 105 (2001) 11227.
[135] M.Z. Hossain, J.E. Johns, K.H. Bevan, H.J. Karmel, Y.T. Liang, S. Yoshimoto, K. Mukai, T. Koitaya, J. Yoshinobu, M. Kawai, Chemically homogeneous and thermally reversible oxidation of epitaxial graphene, Nat. Chem. 4 (2012) 305.
[136] M.H.A. Kudus, M.R. Zakaria, H.M. Akil, F. Ullah, F. Javed, Oxidation of graphene via a simplified hummers' method for graphene-diamine colloid production, J. King Saud Univ. Sci. 32 (2020) 910.
[137] M.P. Araújo, O. Soares, A. Fernandes, M. Pereira, C. Freire, Tuning the surface chemistry of graphene flakes: new strategies for selective oxidation, RSC Adv. 7 (2017) 14290.
[138] D. Mandal, Pericyclic Reactions: Introduction, Classification and the Woodward–Hoffmann Rules, Elsevier, 2018.
[139] V. Georgakilas, Functionalization of Graphene, Wiley-VCH, 2014.
[140] X. Zhang, W.R. Browne, B.L. Feringa, Preparation of dispersible graphene through organic functionalization of graphene using a zwitterion intermediate cycloaddition approach, RSC Adv. 2 (2012) 12173.
[141] S. Sarkar, E. Bekyarova, S. Niyogi, R.C. Haddon, Diels–Alder chemistry of graphite and graphene: graphene as diene and dienophile, J. Am. Chem. Soc. 133 (2011) 3324.
[142] J. Yuan, G. Chen, W. Weng, Y. Xu, One-step functionalization of graphene with cyclopentadienyl-capped macromolecules via Diels–Alder "click" chemistry, J. Mater. Chem. 22 (2012) 7929.
[143] T.A. Strom, E.P. Dillon, C.E. Hamilton, A.R. Barron, Nitrene addition to exfoliated graphene: a one-step route to highly functionalized graphene, Chem. Commun. 46 (2010) 4097.
[144] X. Xu, Q. Luo, W. Lv, Y. Dong, Y. Lin, Q. Yang, A. Shen, D. Pang, J. Hu, J. Qin, Functionalization of graphene sheets by polyacetylene: convenient synthesis and enhanced emission, Macromol. Chem. Phys. 212 (2011) 768.
[145] H. He, C. Gao, General approach to individually dispersed, highly soluble, and conductive graphene nanosheets functionalized by nitrene chemistry, Chem. Mater. 22 (2010) 5054.
[146] L.-H. Liu, M. Yan, Functionalization of pristine graphene with perfluorophenyl azides, J. Mater. Chem. 21 (2011) 3273.
[147] S. Vadukumpully, J. Gupta, Y. Zhang, G.Q. Xu, S. Valiyaveettil, Functionalization of surfactant wrapped graphene nanosheets with alkylazides for enhanced dispersibility, Nanoscale 3 (2011) 303.
[148] H. Ismaili, D. Geng, A.X. Sun, T.T. Kantzas, M.S. Workentin, Light-activated covalent formation of gold nanoparticle–graphene and gold nanoparticle–glass composites, Langmuir 27 (2011) 13261.
[149] C.K. Chua, M. Pumera, Covalent chemistry on graphene, Chem. Soc. Rev. 42 (2013) 3222.
[150] X. Zhong, J. Jin, S. Li, Z. Niu, W. Hu, R. Li, J. Ma, Aryne cycloaddition: highly efficient chemical modification of graphene, Chem. Commun. 46 (2010) 7340.
[151] C. Bingel, Cyclopropanierung von fullerenen (Cyclopropylation of fullerenes), Chem. Ber. 126 (1993) 1957.
[152] S.P. Economopoulos, G. Rotas, Y. Miyata, H. Shinohara, N. Tagmatarchis, Exfoliation and chemical modification using microwave irradiation affording highly functionalized graphene, ACS Nano 4 (2010) 7499.
[153] X. Wang, G. Shi, An introduction to the chemistry of graphene, Phys. Chem. Chem. Phys. 17 (2015) 28484.
[154] X. Duan, S. Indrawirawan, H. Sun, S. Wang, Effects of nitrogen-, boron-, and phosphorus-doping or codoping on metal-free graphene catalysis, Catal. Today 249 (2015) 184.
[155] D. Wei, Y. Liu, Y. Wang, H. Zhang, L. Huang, G. Yu, Synthesis of N-doped graphene by chemical vapor deposition and its electrical properties, Nano Lett. 9 (2009) 1752.
[156] D. Umadevi, G.N. Sastry, Metal ion binding with carbon nanotubes and graphene: effect of chirality and curvature, Chem. Phys. Lett. 549 (2012) 39.
[157] M. Treier, C.A. Pignedoli, T. Laino, R. Rieger, K. Müllen, D. Passerone, R. Fasel, Surface-assisted cyclodehydrogenation provides a synthetic route towards easily processable and chemically tailored nanographenes, Nat. Chem. 3 (2011) 61.
[158] X. Wang, L. Zhi, N. Tsao, Ž. Tomović, J. Li, K. Müllen, Transparent carbon films as electrodes in organic solar cells, Angew. Chem. Int. Ed. 47 (2008) 2990.

CHAPTER

8

Interaction of amino acids, peptides, and proteins with two-dimensional carbon materials

Kanagasabai Balamurugan[a,*] *and Venkatesan Subramanian*[a,b]

[a]Centre for High Computing, CSIR—Central Leather Research Institute, Chennai, India

[b]Academy of Scientific and Innovative Research (AcSIR), CSIR-CLRI Campus, Chennai, India

1 Introduction

In the last four decades, we have witnessed unprecedented developments in nanoscience and nanotechnology which lead to numerous opportunities in various branches of science and technology such as chemistry, physics, life science including biotechnology, material science, and engineering combined with possible solutions. Among the different types of novel materials discovered at the nanoscale, carbon-based nanomaterials have received intensive attention of researchers. This carbon-based superfamily includes fullerenes, carbon nanotubes, carbon nanohorns, diamonds, and graphenes [1–3]. The unique feature of elemental carbon is responsible for the formation of zero-dimensional (0D, fullerene), one-dimensional (1D, carbon nanotube), two-dimensional (2D, graphene), and three-dimensional (3D, graphite) novel materials. The exceptional electrical, electronic, and mechanical properties and chemical reactivities of these nanomaterials are completely responsible for the aforementioned opportunities in various fields in general and in particular for a series of applications in biology ranging from diagnostic devices to drug delivery vehicles [4–9].

The applications of these nanomaterials and engineered nanoparticles in the industrial products increase the exposure of these materials to various life forms in the environment including human beings. Therefore, in addition to the exciting applications of these new

*Present address: Structural Bioinformatics, BIOTEC TU Dresden, Tatzberg 47-51, Dresden 01307, Germany.

https://doi.org/10.1016/B978-0-12-819514-7.00004-X

materials in different fields, the studies on the interaction of these materials with biological molecules call for significant attention of researchers [10–12]. Creation of safety profiles of these nanomaterials is in progress. Based on the experience gained from the carbon nanotube, the rules for the employment of graphene (G) have been summarized by Kostarelos and coworkers [13].

These nanomaterials and nanoparticles interact with the biomolecules like DNA, proteins, membranes, organelles as well as cells and in turn have their effect on the biological systems. These interactions lead to the formation of protein coronas, particle wrapping, intracellular uptake, and biocatalytic processes that could have biocompatible or bioadverse outcomes [14]. As these are aliens to biological systems, they are capable of inducing changes in the biological systems, which must be carefully assessed to ensure safety. Understanding the nano-biointerface enhances the knowledge base for developing a relationship between the properties of nanomaterial such as size, shape, curvature, surface chemistry, and roughness of the nanomaterials with the structure and function of the various biomolecular systems.

Two important contributions in science fueled the research in the area of nanoscience and nanotechnology. They are: (i) the invention of scanning tunneling microscope (STM) in 1981 by Gerd Binnig and Heinrich Rohrer, which enabled the visualization of individual atoms and bonds. The aforementioned scientists received the Nobel Prize in Chemistry in the year 1986 for their contributions toward the development of STM and its applications. This technique enabled to explore atomistic details of nanomaterials; and (ii) the discovery of fullerene in 1985 by Harry Kroto, Richard Smalley, and Robert Curl [1,15,16]. They won the Nobel Prize in Chemistry in 1996 for the discovery of fullerene [17]. Initially, fullerene was not viewed as a potential candidate in nanoscience and technology. It was viewed as a new allotrope of carbon when graphite and diamond are the only known allotropes of carbon. Subsequently, the era of the carbon allotropes began followed by the discovery of carbon nanotube (CNT) in 1991 by Iijima and graphene by Andre Geim and Konstantin Novoselov in 2004 [2,3]. Both Andre Geim and Konstantin Novoselov received Nobel Prize for their discovery [18]. The fullerenes, carbon nanotubes, and graphenes are found to possess extraordinary physical, chemical, and electronic properties, which are also appreciably different from one another and thus touted to have phenomenal applications in various fields. It is well known that carbon is the element in the periodic table that provides the basis for the life on the earth. Interestingly, the carbon-based nanomaterials form the essential part of the nanoscience and nanotechnology. The importance of these materials can be understood from scientific literature that numerous studies have been made in the past on these materials and their applications in various fields of science and technology.

1.1 Graphene 2D materials

Graphene is a two-dimensional sheet of sp^2-hybridized carbon, and it is the thinnest material known so far with one-atom thickness. It can be stacked together to form 3D graphite, rolled to form a one-dimensional (1D) (single-walled (SWCNT) and multiwalled carbon nanotubes (MWCNT)) and wrapped to form zero-dimensional (0D) fullerene. The structural relationship between the graphenes and carbon nanotubes (CNTs) is presented in Fig. 8.1. Thus, graphene is an important material, which is interrelated to various forms of carbon nanomaterials. The extended π-conjugation in graphene gives rise to their excellent thermal, mechanical, and electrical properties, which makes it as a material with extraordinary application potential. The isolation of single-layer graphene by Geim and coworkers led to enormous research interest in the field of nanomaterials. Graphene is one of the few materials in the world that is transparent, conductive, strongest, and flexible—all at the same time. The properties mentioned previously make graphene a wonder molecule owing to its phenomenal application potential in electrical, electronics, energy, and biological fields.

The extraordinary optical properties of CNT and graphene made them suitable for molecular and cellular imaging applications in the area of biomedical applications. Thus, these materials received the widespread attention of researchers. Both CNT- and graphene-based field-effect transistors (FETs) have been extensively explored to develop sensitive chemical- and biosensors because of their functionalizable surface and highly sensitive electrical properties. These carbon-based FETs are useful in detecting biomolecules such as nucleic acids (NAs), proteins, and growth factors [19–23]. The covalently functionalized CNTs can be used as atomic force microscopy (AFM) tips. It was demonstrated that nanotube tips with the capability of chemical and biological discrimination could be developed with the required functionality [24].

The pristine graphene possesses remarkable and unique properties such as chemical mechanical, thermal, optical, and electronic properties. Graphene is one of the world's thinnest materials but at the same time the toughest 2D materials with a tensile strength of 1Tpa. Graphene is light in terms of weight and flexible in nature. Graphene has high current density and intrinsic mobility. It has lower resistivity than any other materials known so far at room temperature. Graphene is a perfect thermal conductor, and it conducts heat in all directions (isotropic conductor). Graphene possesses extraordinary optical properties whereby it absorbs 2–3% of white light but still transparent to human eye, which makes it potential application as a transparent conductor. Graphene is an inert material and does not readily interact with other atoms. It can, however, adsorb other atoms and molecules on its surface, which can lead to changes in their electronic properties, thus making their potential application as sensors. Graphene can also be functionalized with chemical groups leading to graphene derivatives such as graphene oxide and fluorinated graphene. Thus, the extraordinary electronic and physicochemical properties of the carbon-based 2D materials made them a potential candidate for the suitable exploitation in the area of bimolecular application.

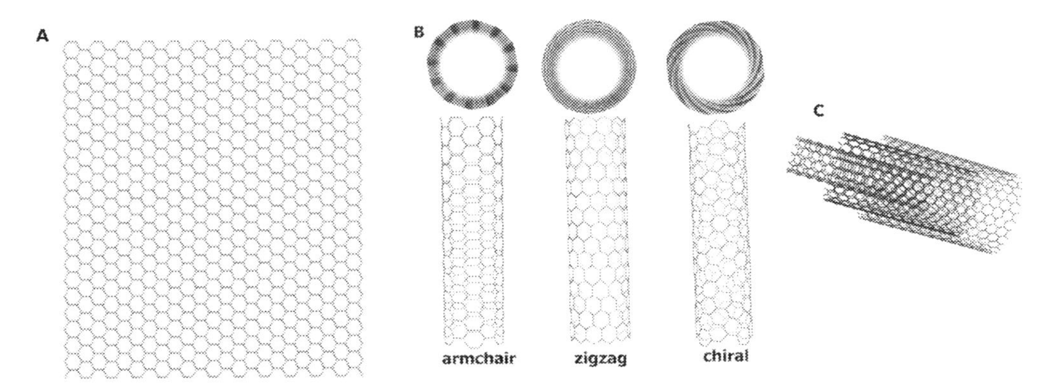

FIG. 8.1 Structure of carbon nanomaterials (A) graphene; (B) armchair, zigzag, and chiral SWCNTs; and (C) MWCNT.

The pristine graphene sheet is shown to induce cytotoxicity through the depletion of the mitochondrial membrane potential and intracellular increase in reactive oxygen species (ROS) [25]. Functionalization of the graphene with biocompatible polymers, drugs, and other molecules reduces the cellular toxicity of the graphene when compared to its pristine form. It is found from a previous study that administration of graphene oxide (GO) in mice induces chronic toxicity and lung granuloma death [26]. Polyethylene glycosylation (PEG) of GO reduces the toxic effects in mice, and similarly, no severe toxicity was measured in vivo upon the administration of GO as a component in injectable hydrogels for tissue engineering [27,28]. The different applications of graphene and graphene oxide in biology are illustrated in Fig. 8.2. On the contrary, the toxicity of these materials can be exploited for antimicrobial applications [29]. Recent results elicit that graphene-based materials like GO and reduced GO (rGO) have strong antibacterial activity. Thus, various factors should be taken into consideration on assessing the toxicity of the carbon nanomaterials and due care should be taken to adopt suitable methods for assessing the toxicity. Leszczynski and coworkers have made seminal contributions to the in silico toxicity assessment and prediction of the nanomaterials and nanoparticles [30,31]. Surface reactivity of SWCNTs on chlorination and the changes in the reactivity due to Stone-Wales defects have been illustrated [32,33].

1.2 Protein structure

The different hierarchical levels of protein structure are primary, secondary, tertiary, and quaternary structures. The sequence of the protein forms the primary structure, and the secondary structure is mainly formed through hydrogen bonds (H-bonds) between backbone atoms [34]. The α-helix and β-sheet are the widely occurring stable secondary structural elements. On the contrary, random coils, loops, and turns do not have a stable secondary structure in proteins. Deconstruction of various secondary structural elements of proteins is illustrated in Fig. 8.3.

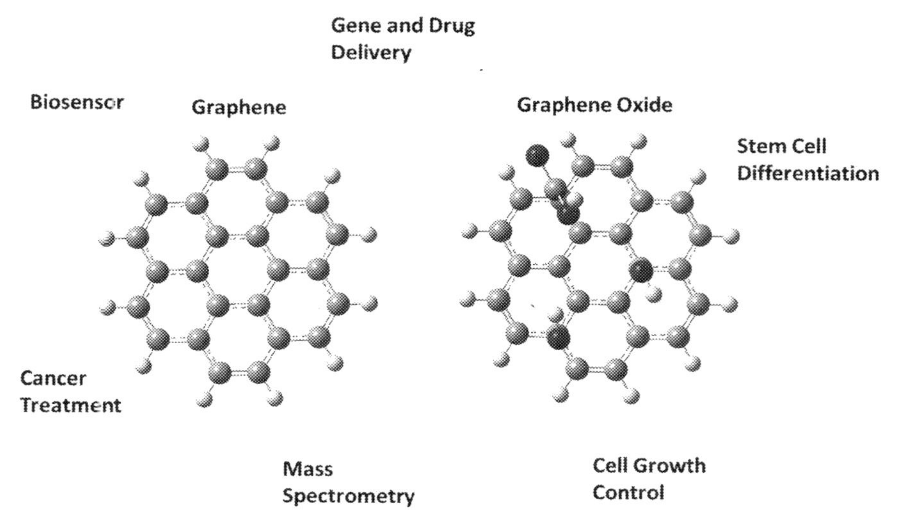

FIG. 8.2 Applications of graphene and graphene oxide in biology.

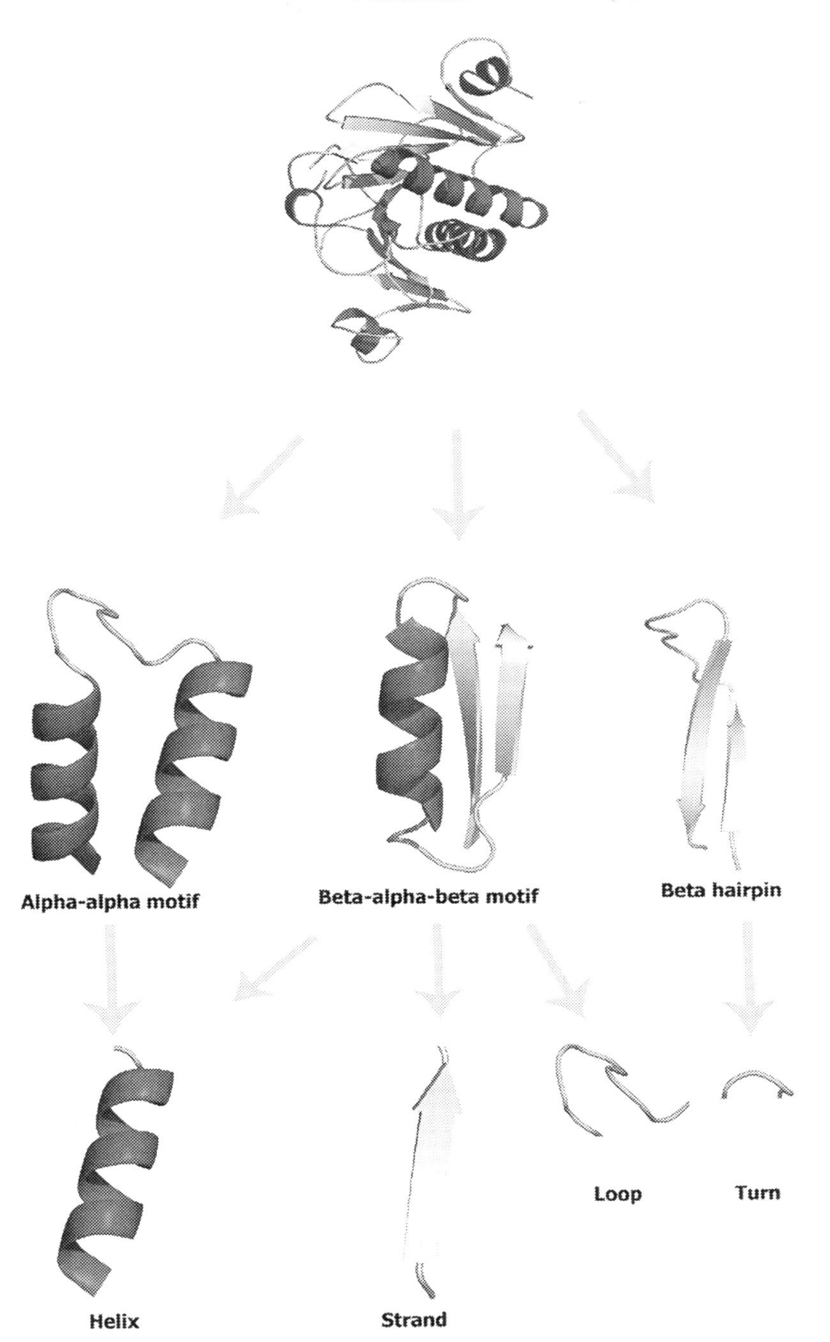

FIG. 8.3 Deconstruction of protein structure into super secondary and secondary structural elements (PDB ID IAV5).

Tertiary structure of the protein is described as the organization of secondary structures together at a whole level of one polypeptide chain. The organization of two or more polypeptide chains together forms a higher level of protein organization, which is referred to as quaternary structure [35]. Protein folding is the process by which a protein structure assumes its functional shape or conformation. It is the physical process, by which a polypeptide folds into its characteristic and functional three-dimensional structure from random coils.

A protein fold is defined by the way the secondary structure elements of the structure are arranged relative to each other in space. In this context, it is important to mention the Ramachandran plot, which is based on the simple idea of avoidance of steric overlap between nonbonded atoms, and it is immensely useful for understanding the structure and folding of proteins [36].

1.3 Protein-nanointerface

Due to the potential biological application of nanomaterials, it is imperative to understand the interaction of nanomaterials with biomolecules [37]. The structure–function relationship of proteins is highly important. The interactions between hierarchical organization of protein at different levels with carbon nanomaterials provide associated structural/conformational changes combined with energies. These details are significantly important due to their biological relevance [38]. In this chapter, the findings from investigations on the interaction of nanomaterials with biological molecules will be summarized.

2 Computational simulation methods

Experiment and theory are the two important paradigms of science. With the explosion of computer power, the computational modeling has emerged as the third paradigm of the scientific research. The computer simulation methods have been employed to gain insight into atomic scale details of nanomaterials. Several computational methods have been developed in the last 50 years, which enabled the simulation of systems at various levels of accuracies and approximations. Due to the importance of density functional theory (DFT) and the development of computational methods in quantum chemistry to solve problems in various branches of science, the Nobel Prize in Chemistry for the year 1998 was awarded to J.A. Pople and W. Kohn [39]. On the similar lines, the Nobel Prize in Chemistry for the year 2013 was awarded jointly to Martin Karplus, Michael Levitt, and Arieh Warshel "for the development of multiscale models for complex chemical systems" which shows the importance of these methods in the scientific research [40]. These awards reinforce the importance of computational modeling in science. It is the choice of the user to apply the best possible method for the problem under study. The selection of the appropriate method is of paramount importance for getting reliable results. The different levels of theoretical rigor associated with the length and timescales are schematically presented in Fig. 8.4.

Quantum mechanical methods are accurate, and these methods are usually employed to study systems at atomic level. However, they are computationally very expensive and can be applied for a small system typically up to ~200 atoms employing Hartree-Fock and

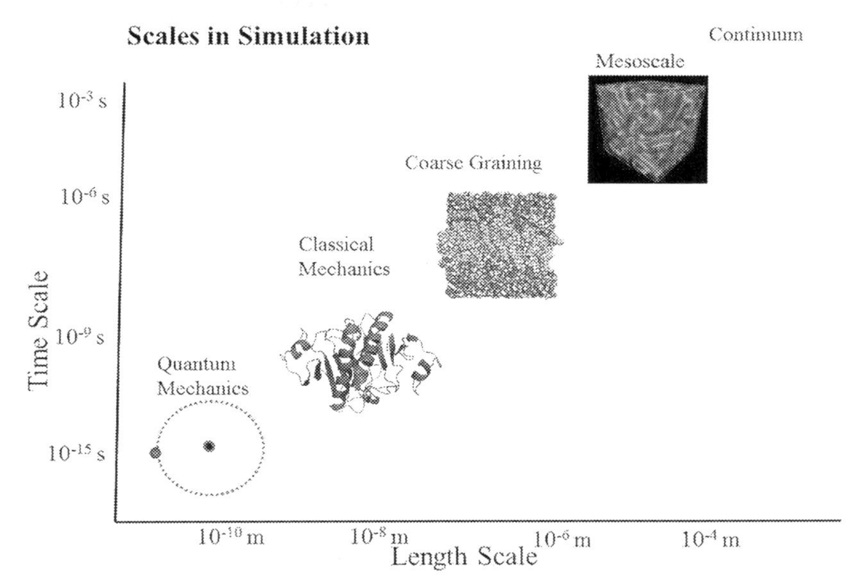

FIG. 8.4 Various scales of simulation in computational modeling.

DFT-based methods [41,42]. The quantum dynamics of these systems is possible over a very short timescale, of the order of picoseconds. In most of the QM studies reported in this chapter, the hybrid generalized gradient approximation (GGA) and meta-GGAs DFT functionals such as ωB97X-D, M05-2X, M06-2X, and ab initio MP2 method are applied to understand the nano-biointerface [43–46].

Atomistic simulation methods such as molecular dynamics and Monte Carlo simulations are based on classical mechanics. The principles of statistical mechanics are also applied to make necessary approximations in the development of these methods. These techniques can be used for simulating systems with a few thousands to several thousands of atoms at the larger timescale of nanoseconds to microseconds. In most of the studies discussed in this chapter, all atomistic classical molecular dynamics simulations were used as the preferred methods to understand the nanoprotein interface. In very few cases, enhanced sampling method like replica exchange molecular dynamics (REMD) was applied.

Coarse-grained (CG) methods are similar to the atomistic simulation. In this, each particle represents a collection of atoms. Each coarse-grained particle represents three to four heavy atoms so that the complexity of the system is reduced considerably, and thus, the systems with larger length scales (millions of atoms) can be simulated to longer timescale of μs to ms. The main disadvantage of this method is the absence of detailed atomistic motions and interactions.

Mesoscale methods are particle-based methods in which each particle accounts from 10 to 100 atoms. These are suitable to study the systems in the mesoscopic scale (μm). Solvent effects are usually handled by dissipative and random force to each particle. Mesoscale simulation techniques include dissipative particle dynamics and lattice Boltzmann methods. Continuum methods describe a system by partial differential equations and thus are useful when the discreteness of matter can be approximated by a continuum. These methods employ finite element techniques to obtain approximate solutions to the differential equations.

3 Interaction of amino acids with 2D materials

Amino acids are the basic building blocks of the proteins, and these are joined together by means of peptide bonds to form polypeptides. Amino acids are classified as aliphatic, aromatic, positively charged, negatively charged, and polar based on the nature of their side chains. The interaction of amino acids with the two-dimensional carbon materials has gained widespread research interest owing to their relevance in biological applications. Especially, the interaction of aromatic (Phe, Trp, and Tyr) and positively charged (His, Arg, and Lys) amino acids with the carbon nanomaterials has gained much attention due to their delicate noncovalent interactions such as π-π and cation-π interactions [47].

Sastry and coworkers studied the impact of chirality and curvature of carbon nanostructures on interaction with aromatic amino acids. They showed that the π-π and CH-π mode of interactions are possible between the aromatic amino acid and carbon nanosurface apart from the OH-π and NH-π interactions [48]. It was also shown that the order of interaction of amino acids with carbon nanomaterials varies as Trp > His > Tyr > Phe. The curvature of the nanomaterial plays an important role in determining the magnitude of interaction where the planar graphene sheet is favored over the curved nanotubes. The chirality of nanotube plays a very subtle role in determining interaction magnitude. Dinadayalane and coworkers investigated the interaction of various conformers of amino acid histidine and proline with the graphene surface. It is illustrated that the histidine possesses better binding energy with the graphene when compared to proline. The binding energy of the amino acids with graphene is reduced in the aqueous phase [49].

Sanyal and coworkers investigated the interaction of aromatic amino acids with the graphene and graphene oxide sheets and concluded that the aromatic amino acids interact with the graphene and graphene oxide (GO) in two distinct modes [50]. In the case of graphene, the aromatic amino acids interact through π-π stacking interactions, whereas hydrogen bonding is responsible for the interaction of amino acids with GO. The magnitude of interaction energy is higher in the case of GO when compared to graphene. The order of interaction varies between graphene and GO sheets. In the case of graphene, the order of interaction is Trp > Tyr > His > Phe. The same for GO is His > Trp > Tyr > Phe. The significant observations mentioned earlier are predominately arising due to the presence of OH groups in the GO surface. Schematic representation of graphene and graphene oxide with aromatic amino acids is given in Fig. 8.5. Sastry and coworkers investigated the interaction of amino acids with the graphene and graphene sheets [51]. They found that the aromatic amino acids interact with the graphane sheets through the CH-π mode of interaction. In this mode of interaction, the aromatic ring is parallel to the graphane sheet. The role of π-π interaction is evident from the interaction of aromatic amino acids with graphene. The interaction between graphene and aromatic amino acid is more favorable when compared to graphane-aromatic amino acid interactions. The order of interaction for graphene is Trp > Tyr > His > Phe. The trend for graphane is Trp > Tyr ~ His > Phe. Dinadayalane and coworkers explored the interactions of various conformers of tyrosine with graphene.

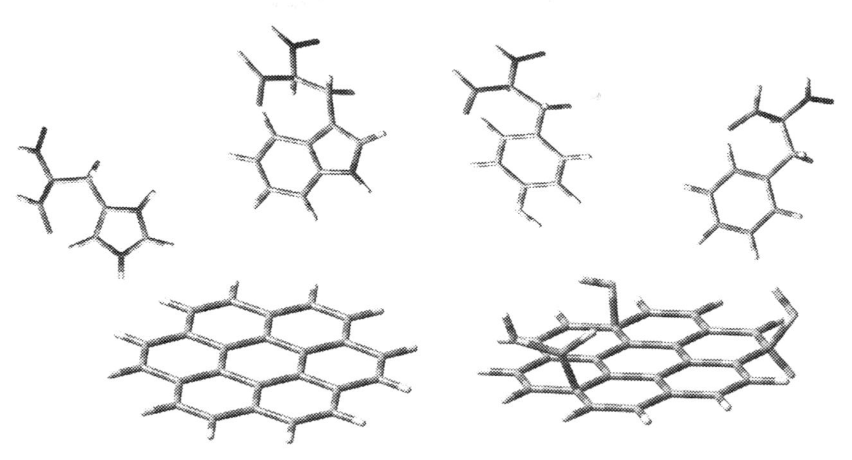

FIG. 5 Schematic representation of graphene and graphene oxide with aromatic amino acids.

Their results show that the aromatic amino acid tyrosine prefers the bent orientation over the parallel orientation on the surface of the graphene sheet [52]. Apart from theoretical studies, experimental estimation of binding thermodynamics between various amino acids with GO has been carried out using isothermal titration calorimetry by De and coworkers [53]. The results illustrated that the interaction of GO with amino acids is mainly driven by the electrostatic and π-π interactions with variable enthalpy and entropy values.

Bryce and coworkers investigated the efficiency of Trp, Tyr, and Val in the exfoliation of graphene flakes using theoretical and experimental methods [54]. They observed that Trp was found to be the most favorable for exfoliation. Carzola et al. investigated the interaction of Gly, Pro, and Hyp (most predominant amino acids in collagen) with the graphane, graphene and Ca-doped graphene by ab initio molecular dynamics simulation [55]. They found that these amino acids have weak interaction with graphane and graphene. The doping of Ca on the graphene sheets significantly strengthens the interaction of amino acids. The observation described previously is important in terms of application of 2D carbon materials in the area of collagen-based biomaterials.

4 Interaction of peptides with 2D materials

4.1 Interaction of oligopeptides with 2D materials

It is necessary to understand the interaction of the oligopeptides with 2D material surfaces to assess the impact of the 2D materials on the structure of proteins. Comer and coworkers analyzed the conformational equilibria of alanine dipeptide on the graphene and graphene oxide surfaces using molecular dynamics simulation and free energy calculations [56]. They found that the dipeptide is stabilized with the aid of amide-π interactions contrary to the nonplanar α-helix and β-sheet conformation in the aqueous phase. These results show that the peptides are adsorbed onto the graphene and graphene oxide surfaces. These are trapped in a metastable state in contrast to their more stable form in the aqueous phase. As the dipeptides are considered to be the simplest form of the protein backbone structure, the aforementioned study provides valuable information on the impact of graphene/graphene oxide surface on the protein backbone. Chen and coworkers probed the molecular-level interaction of peptides such as cecropin P1 and MSI-78(C1) with the graphene using sum frequency generation (SFG) vibrational spectroscopy and MD approaches [57]. They found that the peptide sequence determines the orientation of the peptide adsorbed on a graphene surface.

Leng and coworkers investigated the interaction of RGD (Arg-Gly-Asp) peptides with the pristine and defective graphene and graphene oxide [58]. The results illustrated that strongest binding energy was observed when RGD was parallel to graphene surfaces. In this mode, all three functional groups of RGD, such as NH_3^+, COO^-, and guanidine, interact with graphene. The interaction of $NH_3^+\cdots\pi$ was stronger than that of guanidine $-NH_2\cdots\pi$ and $COO^-\cdots\pi$. The monovacancy further enhances the interaction with RGD peptides through the dangling C-atoms. Graphene oxide (OH, epoxy, and mixed OH/epoxy groups interaction with RGD) possesses a stronger interaction with RGD when compared to pristine and defective graphene. Morales and coworkers investigated the interaction of oligopeptide onto the carbon surfaces using MD simulation [59]. They found that amino acid residues such as tryptophan, isoleucine, and histidine significantly contribute to the interaction with carbon surface. Walsh and coworkers investigated the interaction of tryptophan-containing oligopeptides with carbon nanotube and graphite surfaces [60]. Findings showed that the tryptophan residues significantly enhanced the interaction with the carbon surfaces.

4.2 Interaction of α-helical peptides with 2D materials

The α-helical structure is one of the most important secondary structural elements in the protein structural organization. The backbone hydrogen bonds present in the α-helices stabilize the structure and act as skeletal structural component in terms of protein structure. It is important to understand what kind of structural changes an α-helical structure undergoes on interaction with carbon nanostructures. Our previous studies on the interaction of polyalanine-based α-helical peptides with carbon nanotubes have revealed the decrease in the helicity of the peptide due to the loss of backbone hydrogen bonding and concomitant electrostatic energy [61]. This reduction in the electrostatic energy was compensated by the gain in the van der Waals interaction energy between the helix and carbon nanosurface. The effect of the amino acid sequence on the interaction pattern of the α-helix with the carbon nanostructure has also been investigated using the soluble *N*-ethylmaleimide-sensitive factor attachment protein receptor (SNARE) protein. The results showed that the stability of the α-helical region depends on the innate propensity of the amino acid to form α-helical structure. The regions of amino acids with higher α-helical propensity remain stable, whereas the regions with low α-helical propensity undergo changes in the helical content.

Furthermore, we have investigated the effect of curvature of the carbon nanosurfaces on the stability of α-helical peptides [62]. In this study, we have explored the interactions of polyalanine-based α-helix with carbon nanotubes of different diameters as well as graphene sheets. Results clearly elicited that the extent of α-helix breakage is inversely proportional to the curvature of carbon nanostructures. The helix breaking tendency is minimum for the

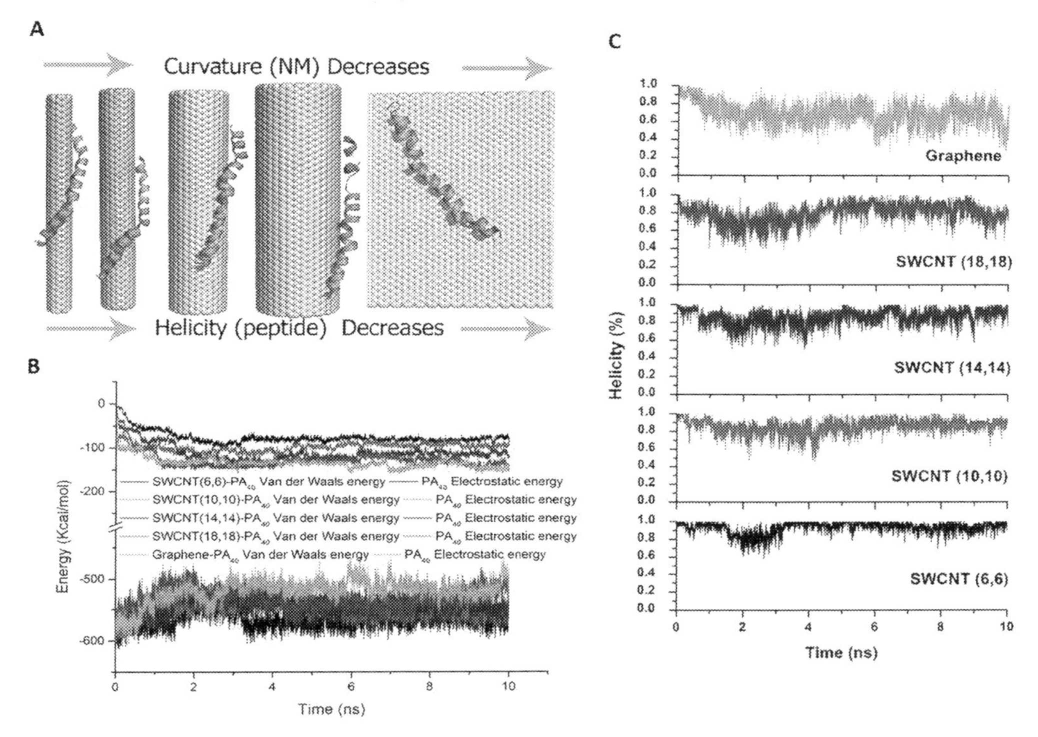

FIG. 8.6 Curvature-dependent helicity loss in carbon nanomaterials. (A) Snapshot of PA_{40} helix interaction with nanomaterials, (B) energetics of the interaction energies during simulation, and (C) helical content during the course of simulation. *Reprint permission from K. Balamurugan, E.R. Azhagiya Singam, V. Subramanian. Effect of curvature on the α-helix breaking tendency of carbon based nanomaterials, J. Phys. Chem. C 115 (2011) 8886–8892.*

carbon nanotube having the highest curvature and maximum for the planar graphene sheet. The energetics behinds this observation is clearly illustrated by the analysis of the interacting complexes. The results show that there is a significant decrease in the electrostatic interaction energy of the α-helix. However, the overall stability of the complex is attained by the gain in van der Waals interaction energy. These results are illustrated in Fig. 8.6.

In another study, we have investigated the interaction of cytoplasmic α-helical protein complexin with graphene oxide, reduced graphene oxide, and graphene [63]. Results showed that adsorption of protein on graphene oxide was highly selective and mediated through electrostatic interactions such as hydrogen bond and salt bridge interactions. Further, both van der Waals and π-π stacking interactions were the major driving forces for the adsorption of protein onto reduced graphene oxide and graphene. The stability of α-helix on the carbon surfaces was found to be in the order of graphene oxide > reduced graphene oxide > graphene. The conformational stability of the protein on graphene oxide was attributed to the extensive hydration of graphene oxide surface. Wei and coworkers investigated the adsorption, conformational changes, and the dimerization of the designed de novo α-helical peptides onto the graphene surfaces [64]. Findings elucidated that the two chains are mostly dimeric and keep α-helical structure in solution. Both unfolding and assembling occurred to form an amorphous dimer on the surface of graphene. Overall results from various studies showed that the graphene surface significantly reduces the α-helical content of the protein on adsorption, which is determined by various factors such as amino acid content, aromaticity of amino acids, charged residues, α-helix propensity, and functionalization of graphene surface. In addition, solvation plays an additional role in the interaction process.

4.3 Interaction of β-sheet peptides with 2D materials

The β-sheets can be classified as parallel and antiparallel β-sheets based on the directionality of N- and C-terminals of the interacting peptides. These structures are stabilized by the hydrogen-bonding interactions between the backbone atoms of the peptides. In general, antiparallel β-sheets are found to be more stable when compared to the parallel β-sheets. It is of scientific interest to understand the interaction of β-sheets with carbon nanomaterials due to its implications in neurodegenerative diseases. We investigated the interaction of amyloid-beta ($A\beta_{1-40}$) peptide with graphene oxide and reduced graphene oxide sheets using classical MD simulation [65]. The results illustrated that graphene oxide and reduced graphene oxide could inhibit the conversion of α-helix to toxic β-sheet aggregates. The reduced graphene oxide is shown to be more efficient in blocking the α-helix to toxic β-sheet. The mechanism of Aβ inhibition by graphene oxide is illustrated in Fig. 8.7.

Gupta and coworkers explored the stability of the hepcidin peptide in the solution phase as well as at the interface of graphene and graphene oxide [66]. Hepcidin is a cysteine-rich cationic peptide that has a simple hairpin-like β-sheet structure stabilized by four disulfide bonds. The results showed that hepcidin peptide adsorbed onto the graphene-based nanomaterial surface and formed a stable complex by destabilization of β-sheet content of the peptide. Among various nanomaterials, graphene is observed to be more destabilizing

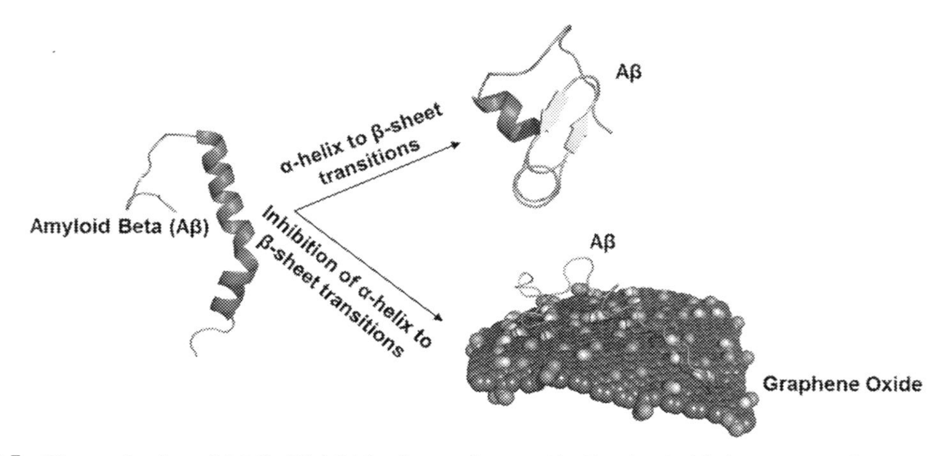

FIG. 8.7 The mechanism of Aβ fibril inhibition by graphene oxide. *Reprinted with the permission from G. Gianese, V. Rosato, F. Cleri, M. Celino, P. Morales, Atomic-scale modeling of the interaction between short polypeptides and carbon surfaces, J. Phys. Chem. B 113 (2009) 12105–12112.*

propensity for β-sheet structure when compared to the graphene oxide counterpart. This observation partially explains the different toxicity profiles of graphene analogues. Liu and coworkers carried out the investigation on the interaction of prion proteins with carbon nanotubes and graphenes [67]. The computational simulation of the $PrP_{127-147}$ with carbon nanotube and graphene using REMD methodology showed that the presence of nanomaterials reduces the aggregation of proteins. The carbon nanomaterials decrease the interpeptide interaction in proteins and thus inhibit the β-sheet formation. It is noteworthy to mention that the contribution of π-π interactions from the aromatic amino acids significantly inhibits the aggregation of proteins.

Dasgupta and coworkers illustrated the inhibition of amyloid fibril formation in human serum albumin in the presence of graphene oxide using various experimental methods [68]. Cheng and coworkers investigated the adsorption mechanism of amyloid beta-fibril to the graphene nanosheets [69]. They carried out the simulation of $A\beta_{17-42}$ pentamer with the graphene sheet in different orientations. Results revealed that graphene has a strong capability to interact with the amyloid fibril initially through the side chains of the amyloid peptide followed by the collapse of the secondary structure (β-sheet) which leads to the dissociation of the fibril structure. The interaction is primarily driven through the π-π interactions of the graphene surface further mediated by the solvent-induced interactions in the dissociation of the amyloid fibril. Thus, graphene could induce the damage of amyloid fibril. Yang and coworkers explored the inhibitory potential of fluorinated graphene quantum dots toward the human islet amyloid polypeptide (hIAPP) fibrillation using experimental methods [70]. Their results pointed out that the aggregation morphologies of hIAPP were observed to change from the entangled long fibrils to short thin fibrils. Thus, it is well established by both computational and experimental studies that the graphene-based nanomaterials can significantly destabilize the β-sheet structure and inhibit the amyloid-like fibril formation in amyloid/prion proteins.

4.4 Interaction of graphene-binding peptides with 2D materials

Several graphene-binding peptides (GBPs) have been designed which could interact with graphene with high affinity to enable exfoliation in water. Naik and coworkers made seminal contributions in the area of peptide-based graphene exfoliation by employing both experimental and docking/classical MD simulations [71]. They clearly illustrated the preferential binding of peptides obtained to the edges or on the plane of the graphene. These results can be used not only for the separation of the nanomaterials but also to fine-tune the electronic properties of the 2D materials for various applications. In continuation of the aforementioned study, they have also explored the effect of the quality of graphene as well the number of layers of graphene in modulating the graphene-peptide interaction in the case of GBPs and CBPs using classical MD simulation [72]. They also investigated the interaction of the dodecamer graphene-binding peptide GAMHLPWHMGTL with graphene and graphite using experimental and classical MD simulations [73].

Walsh and coworkers have reviewed how the functionalization of the GBPs can be efficiently applied for the exfoliation of graphene sheets minimizing the sheet damage in an aqueous medium [74]. It was shown that the fatty acid modification of the GBPs remarkably enhances the exfoliation of graphene along with reduced sheet damage. MD simulation results shed light on the molecular mechanism behind the sheet protection properties of the fatty acid group.

5 Interaction of proteins with 2D materials

5.1 Interaction of different proteins/enzymes with 2D materials

Recently, numerous studies have been undertaken to unravel the interaction of proteins/enzymes with graphene and its derivatives with multifarious objectives to monitor: (i) protein–protein interactions, (ii) drug delivery systems, (iii) toxicity of graphene and its derivatives, and (iv) bioelectronic devices including sensors. Graphene oxide was used as a quencher for the fluorescence assay of amino acids, peptides, and proteins [75]. Zhou et al. had investigated the interaction of villin headpiece as a model protein with graphene and phosphorene 2D materials using classical molecular dynamics simulation approach [76]. They showed that the graphene surfaces cause a severe disruption in the protein structure when compared to a weaker interaction of phosphorene to protein. Wei and coworkers have studied the atomic-level adsorption of peptides onto the surface of graphene using classical MD simulation [64]. Without using any cross-linking reagents, horseradish peroxidase (HRP) and lysozyme molecules were immobilized onto GO with the aid of noncovalent mode of interaction. Similarly, adsorption of oxalate oxidase on chemically reduced GO has been investigated.

Lee et al. reported the adsorption of heparin on the surface of graphene with the help of hydrophobic interaction between the two systems [77]. Changes in the conformation of blood proteins on interaction with nanosurfaces are the first step in the initiation of immunological response caused by the nanomaterials. In this context, the interaction of graphene with serum protein has been investigated [78]. The adsorption of fibronectin onto the surface of the graphene sheet was investigated by Floriano and coworkers by employing classical MD simulation [79]. Zhang and coworkers investigated the interaction of fibrinopeptide-A with graphene sheets of different functionalization and surface defects with the aid of classical MD simulation [80]. Their results illustrated that the addition of functional groups in graphene enhances the interaction of fibrinopeptide. The introduction of defects in the graphene surface also increases the interaction of the peptide with the nanosurface. Raffaini et al. have investigated the surface topology effects of the carbon nanomaterials on protein adsorption [81]. The results showed that the interaction propensity of protein varies as concave inner surface of nanotube > sheet > convex outer surface of the nanotube. Zhou and coworkers have investigated the translocation of $A\beta_{42}$ protein through the graphene-MoS_2 2D hetero structure by applying classical MD simulation [82]. It is found that the protein can successfully pass through the hybrid nanopore due to the variation in the van der Waals interaction potential of the graphene and MoS_2 materials. This study opens up the possibility of using these kinds of hybrid 2D platforms of different sizes, shapes, and thicknesses for the protein sequencing applications. Classical molecular dynamics and quantum transport simulations have been applied to investigate the passage of angiotensin peptides through the graphene nanopore [83].

It has been experimentally illustrated that the individual graphene and GO sheets could serve as an ideal solid substrate for enzyme immobilization [84,85]. It was demonstrated that enzyme molecules could be directly immobilized on GO without using any coupling reagent due to the intrinsic surface functional groups of GO, π-π stacking, and/or hydrophobic interactions [77,86]. It is possible to gain further insight into immobilized enzymes with the help of AFM [87]. Molecular dynamics simulations were conducted to investigate the noncovalent adsorption of protein/enzyme onto GO. Results indicated that the π-π stacking between GO and aromatic residues of the enzymes plays a key role in the immobilization of enzymes. Experimental studies have shown that GO can act as a most potent inhibitor of α-chymotrypsin when compared to all other available artificial inhibitors. Li and coworkers investigated the interaction of α-chymotrypsin with graphene and GO sheets using classical MD simulation [88]. They found that the α-chymotrypsin interacts with graphene through its hydrophobic part. On the contrary, the protein interacts with GO through the cationic and hydrophobic residues where the active site is in close contact with the GO surface. In the case of GO, the active site of protein undergoes a significant deformation which in turn leads to the loss of enzymatic activity. However, the interaction between α-chymotrypsin and graphene/GO does not alter the structure of α-chymotrypsin. Similarly, the interaction between trypsin and graphene/GO has been investigated. It is found that the activity of trypsin is not affected by the presence of these materials. Peng and coworkers carried out experimental and molecular dynamics studies on the interaction of the polyethylene-glycosylated GO (PEG-GO) surface with trypsin and α-chymotrypsin [89]. They showed that the GO selectively enhances the thermostability of trypsin but not α-chymotrypsin. More surprisingly, the enzymatic assay showed that the presence of GO and PEG-GO increases the enzymatic activity of trypsin and inhibits the activity of α-chymotrypsin. This study illustrated the versatile and contrasting role of GO with two different enzymes that are closely related evolutionarily.

Using classical molecular dynamics simulation methods, Na and coworkers have explored the interaction of glucose oxidase enzyme with the graphene surface [90]. They showed that the glucose oxidase can be immobilized on to the surface of the graphene by modifying the key hydrophobic residues without compromising the enzymatic active site. These findings would be useful in designing the graphene-based platform for the glucose sensing application. Yang and coworkers studied the molecular mechanism of lipase immobilization as well as lid opening in the graphene-based nanosupports [91]. Salient findings showed that hydrophobic surface increases lipase activity due to the opening of the helical lid present on the lipase. They also illustrated that the open and active form of lipase can be achieved and fine-tuned with an optimized activity through the chemical reduction of graphene oxide.

6 Summary

There are a significant number of experimental and theoretical studies on the various aspects of the interaction of carbon-based nanomaterials with amino acids, peptides, and proteins. We discussed important findings from various techniques such as electronic structure calculations and statistical mechanics-based molecular dynamics simulations. The electronic structure calculations have been carried out to unravel the interaction of amino acids with graphene and GO. The overall results illustrated that the aromatic amino acids and positively charged amino acids dominate the interaction process. In the case of aromatic amino acids, the order of interaction is in general observed to be Trp > Tyr > His > Phe.

Results from the study on the interaction of alanine dipeptide with graphene and graphene oxide showed that the presence of graphene-backbone amide interaction. In addition, the trapping of the oligopeptide in the metastable stable state is found from the findings. Subsequent investigations have elicited the role of various amino acids in the interaction process.

The structural changes induced by carbon nanomaterials on the secondary structural elements of the proteins like α-helixes and β-sheets have been explored. The results showed that the carbon nanomaterials induce loss of α-helical content. The loss in helicity is directly proportional to the planarity of the carbon nanomaterial, i.e., highly curved CNT induces minimal helix breakage, whereas the planar graphene causes a considerable helical disruption. The stability of α-helix on the carbon surfaces is in the order of graphene oxide > reduced graphene oxide > graphene. The interaction of β-sheet peptides such as Aβ and Hepcidin revealed that the graphene-based materials have the potential to inhibit β-sheet formation. The graphene is found to be most effective in β-sheet inhibition when compared to graphene oxide. Usefulness of graphene-binding peptides in exfoliation of graphene has been demonstrated. The exploration on the interaction of proteins with the carbon nanostructures elucidated that these materials are capable of inducing conformational changes in the protein structure based on various factors such as nature of protein, curvature of the nanomaterial, and solvation.

The investigation of graphene oxide with the enzymes such as α-chymotrypsin and trypsin showed that GO can act as an inhibitor in some enzymes. The GO inhibits the activity of α-chymotrypsin, whereas the same stabilizes the trypsin without affecting its enzymatic activity. Thus, there is a huge potential of graphene-based materials in life science applications by suitably tailoring them according to the envisaged applications.

Acknowledgment

The authors thank the Council of Scientific and Industrial Research (CSIR), New Delhi; the Board of Research in Nuclear Sciences (BRNS), Mumbai; and the Department of Science and Technology (DST), New Delhi, for financial support. The authors like to thank Prof. Swapan K. Ghosh and Dr. C.N. Patra from BARC, Mumbai, for their support. The authors wish to thank the CSIR-Central Leather Research Institute, Adyar, Chennai, India, for the institutional support.

References

[1] H.W. Kroto, J.R. Heath, S.C. O'Brien, R.F. Curl, R.E. Smalley, C_{60}: buckminsterfullerene, Nature 318 (1985) 162–163.
[2] S. Iijima, Helical microtubules of graphitic carbon, Nature 354 (1991) 56–58.
[3] K. Novoselov, A.S. Geim, V. Morozov, D. Jiang, Y. Zhang, S.V. Dubonos, I.V. Grigorieva, A.A. Firsov, Electric field effect in atomically thin carbon films, Science 306 (2004) 666–669.
[4] P.M. Ajayan, Nanotubes from carbon, Chem. Rev. 99 (1999) 1787–1800.
[5] M. Dresselhaus, G. Dresselhaus, P. Eklund, R. Saito, Carbon nanotubes, Phys. World 11 (1998) 33.
[6] Y. Lin, S. Taylor, H.P. Li, K.A.S. Fernando, L.W. Qu, W. Wang, L.R. Gu, B. Zhou, Y.P. Sun, Advances toward bio applications of carbon nanotubes, J. Mater. Chem. 14 (2004) 527–541.

[7] E. Katz, I. Willner, Biomolecule-functionalized carbon nanotubes: applications in nanobioelectronics, Chem. Phys. Chem. 5 (2004) 1085–1104.
[8] L. Lacerda, A. Bianco, M. Prato, K. Kostarelos, Carbon nanotubes as nanomedicines: from toxicology to pharmacology, Adv. Drug Deliv. Rev. 58 (2006) 1460–1470.
[9] R.H. Baughman, C.X. Cui, A.A. Zakhidov, Z. Iqbal, J.N. Barisci, G.M. Spinks, G.G. Wallace, A. Mazzoldi, D. De Rossi, A.G. Rinzler, O. Jaschinski, S. Roth, M. Kertesz, Carbon nanotube actuators, Science 284 (1999) 1340–1344.
[10] G. Oberdorster, E. Oberdorster, J. Oberdorster, Nanotoxicology: an emerging discipline evolving from studies of ultrafine particles, Environ. Health Perspect. 113 (2005) 823–839.
[11] V.L. Colvin, The potential environmental impact of engineered nanomaterials, Nat. Biotechnol. 21 (2003) 1166–1170.
[12] L.K. Limbach, Y.C. Li, R.N. Grass, T.J. Brunner, M.A. Hintermann, M. Muller, D. Gunther, W. Stark, Oxide nanoparticle uptake in human lung fibroblasts: effects of particle size, agglomeration, and diffusion at low concentrations, J. Environ. Sci. Technol. 39 (2005) 9370–9376.
[13] C. Bussy, H. Ali-Boucetta, K. Kostarelos, Safety considerations for graphene: lessons learnt from carbon nanotubes, Acc. Chem. Res. 46 (2013) 692–701.
[14] A.E. Nel, L. Madler, D. Velegol, T. Xia, E.M.V. Hoek, P. Somasundaran, F. Klaessig, V. Castranova, M. Thompson, Understanding biophysicochemical interactions at the nano–bio interface, Nat. Mater. 8 (2009) 543–557.
[15] G. Binnig, H. Rohrer, Scanning tunneling microscopy, IBM J. Res. Dev. 30 (1986) 4.
[16] http://www.nobelprize.org/nobel_prizes/physics/laureates/1986/press.html.
[17] http://www.nobelprize.org/nobel_prizes/chemistry/laureates/1996/.
[18] http://www.nobelprize.org/nobel_prizes/physics/laureates/2010/.
[19] M. Labib, E.H. Sargent, S.O. Kelley, Electrochemical methods for the analysis of clinically relevant biomolecules, Chem. Rev. 116 (2016) 9001–9090.
[20] Y. Du, S. Dong, Nucleic acid biosensors: recent advances and perspectives, Anal. Chem. 89 (2017) 189–215.
[21] S.K. Krishnan, E. Singh, P. Singh, M. Meyyappan, H.S. Nalwa, A review on graphene-based nanocomposites for electrochemical and fluorescent biosensors, RSC Adv. 9 (2019) 8778–8881.
[22] E. Vermisoglou, D. Panáček, K. Jayaramulu, M. Pykal, I. Frébort, M. Kolář, M. Hajdúch, R. Zbořil, M. lOtyepka, Human virus detection with graphene-based materials, Biosens. Bioelectron. 166 (2020) 112436.
[23] A. Hashem, M.A.M. Hossain, A.R. Marlinda, M.A. Mamun, K. Simarani, M.R. Johan, Nanomaterials based electrochemical nucleic acid biosensors for environmental monitoring: a review, Appl. Surf. Sci. 4 (2021) 100064.
[24] S. Stanislaus, W.E. Joselevich, A.T. Woolley, C.L. Cheung, C.M. Lieber, Covalently functionalized nanotubes as nanometre-sized probes in chemistry and biology, Nature 394 (1998) 52–55.
[25] Y. Li, Y. Liu, F. Yujian, W. Taotao, L.L. Guyadera, G. Gao, L. Ru-Shi, C. Yan-Zhong, C. Chen, The triggering of apoptosis in macrophages by pristine graphene through the MAPK and TGF-beta signaling pathways, Biomaterials 33 (2012) 402–411.
[26] K. Wang, J. Ruan, H. Song, J. Zhang, Y. Wo, S. Guo, D. Cui, Biocompatibility of graphene oxide, Nanoscale Res. Lett. 6 (2011) 1–8.
[27] K. Yang, S. Zhang, G. Zhang, X. Sun, S.T. Lee, Z. Liu, Graphene in mice: ultrahigh *in vivo* tumor uptake and efficient photothermal therapy, Nano Lett. 10 (2010) 3318–3323.
[28] A. Sahu, W.I. Choi, G. Tae, A stimuli-sensitive injectable graphene oxide composite hydrogel, Chem. Commun. 48 (2012) 5820–5822.
[29] C.M. Santos, M.C.R. Tria, R.A.M.V. Vergara, F. Ahmed, R.C. Advincula, D.F. Rodrigues, Antimicrobial graphenepolymer (PVK-GO) nanocomposite films, Chem. Commun. 47 (2011) 8892–8894.
[30] M. Turabekova, B. Rasulev, M. Theodore, J. Jackman, D. Leszczynska, J. Leszczynski, Immunotoxicity of nanoparticles: a computational study suggests that CNTs and C60 fullerenes might be recognized as pathogens by toll-like receptors, Nanoscale 6 (2014) 3488–3495.
[31] T. Puzyn, B. Rasulev, A. Gajewicz, X. Hu, T.P. Dasari, A. Michalkova, H.M. Hwang, A. Toropov, D. Leszczynska, J. Leszczynski, Using nano-QSAR to predict the cytotoxicity of metal oxide nanoparticles, Nat. Nanotechnol. 6 (2011) 175–178.
[32] S. Saha, T.C. Dinadayalane, J.S. Murray, D. Leszczynska, J. Leszczynski, Surface reactivity for chlorination on chlorinated (5,5) armchair SWCNT: a computational approach, J. Phys. Chem. C 116 (2012) 22399–22410.
[33] T.C. Dinadayalane, J.S. Murray, M.C. Concha, P. Politzer, J. Leszczynski, Reactivities of sites on (5,5) single-walled carbon nanotubes with and without a stone-wales defect, J. Chem. Theory Comput. 6 (2010) 1351–1357.

[34] L. Pauling, R.B. Corey, H.R. Branson, The structure of proteins: two hydrogen-bonded helical configurations of the polypeptide chain, Proc. Natl. Acad. Sci. 37 (1951) 205–211.
[35] E.J. Stollar, D.P. Smith, Uncovering protein structure, Essays Biochem. 64 (2020) 649–680.
[36] G.N. Ramachandran, C. Ramakrishnan, V. Sasisekharan, Stereochemistry of polypeptide chain configurations, J. Mol. Biol. 7 (1963) 95–99.
[37] K. Kostarelos, K.S. Novoselov, Materials science. Exploring the interface of graphene and biology, Science 344 (2014) 261–263.
[38] S. Wei, X. Zou, J. Tian, H. Huang, W. Guo, Z. Chen, Control of protein conformation and orientation on graphene, J. Am. Chem. Soc. 141 (2019) 20335–20343.
[39] http://www.nobelprize.org/nobel_prizes/chemistry/laureates/1998/.
[40] http://www.nobelprize.org/nobel_prizes/chemistry/laureates/2013/.
[41] J.C. Slater, A simplification of the Hartree-Fock method, Phys. Rev. 81 (1951) 385.
[42] P. Hohenberg, W. Kohn, Inhomogeneous electron gas, Phys. Rev. A: Atom. Mol. Opt. Phys. 136 (1964) B864.
[43] J. Chai, M. Head-Gordon, Long-range corrected hybrid density functionals with damped atom-atom dispersion corrections, Phys. Chem. Chem. Phys. 10 (2008) 6615–6620.
[44] Y. Zhao, N.E. Schultz, D.G. Truhlar, Design of density functionals by combining the method of constraint satisfaction with parametrization for thermochemistry, thermochemical kinetics, and noncovalent interactions, J. Chem. Theory Comput. 2 (2006) 364–382.
[45] Y. Zhao, D.G. Truhlar, The M06 suite of density functionals for main group thermochemistry, thermochemical kinetics, noncovalent interactions, excited states, and transition elements: two new functionals and systematic testing of four M06-class functionals and 12 other functionals, Theor. Chem. Accounts 120 (2006) 215–241.
[46] M. Head-Gordon, J A. Pople, M.J. Frisch, MP2 energy evaluation by direct methods, Chem. Phys. Lett. 153 (1988) 503–506.
[47] A.S. Mahadevi, G.N. Sastry, Cation–π interaction: its role and relevance in chemistry, biology, and material science, Chem. Rev. 113 (2013) 2100–2138.
[48] D. Umadevi, G.N. Sastry, Impact of the chirality and curvature of carbon nanostructures on their interaction with aromatics and amino acids, Chem Phys Chem 14 (2013) 2570–2578.
[49] D. Daggag, T. Dorlus, T. Dinadayalane, Binding of histidine and proline with graphene: DFT study, Chem. Phys. Lett. 730 (2019) 147–152.
[50] H. Vovusha, S. Sanyal, B. Sanyal, Interaction of nucleobases and aromatic amino acids with graphene oxide and graphene flakes, J. Phys. Chem. Lett. 4 (2013) 3710–3718.
[51] D. Umadevi, G.N. Sastry, Graphane versus graphene: a computational investigation of the interaction of nucleobases, aminoacids, heterocycles, small molecules (CO_2, H_2O, NH_3, CH_4, H_2), metal ions and onium ions, Phys. Chem. Chem. Phys. 17 (2015) 30260–30269.
[52] D. Daggag, J. Lazare, T. Dinadayalane, Conformation dependence of tyrosine binding on the surface of graphene: Bent prefers over parallel orientation, Appl. Surf. Sci. 483 (2019) 178–186.
[53] S. Pandit, M. De, Interaction of amino acids and graphene oxide: trends in thermodynamic properties, J. Phys. Chem. C 121 (2017) 600–608.
[54] F.A. Alkathiri, C. McCallion, A.P. Golovanov, J. Burthem, A. Pluen, R.A. Bryce, Solvation of pristine graphene using amino acids: a molecular simulation and experimental analysis, J. Phys. Chem. C 123 (2019) 30234–30244.
[55] C. Cazorla, Ab initio study of the binding of collagen amino acids to graphene and A-doped (a = H, ca) graphene, Thin Solid Films 518 (2010) 6951–6961.
[56] H. Poblete, I. Miranda-Carvajal, J. Comer, Determinants of alanine dipeptide conformational equilibria on graphene and hydroxylated derivatives, J. Phys. Chem. B 121 (2017) 3895–3907.
[57] X. Zou, S. Wei, J. Jasensky, M. Xiao, Q. Wang, C.L. Brooks III, Z. Chen, Molecular interactions between graphene and biological molecules, J. Am. Chem. Soc. 139 (1928 – 1936) 2017.
[58] Y. Guo, X. Lu, J. Weng, Y. Leng, Density functional theory study of the interaction of arginine-glycine-aspartic acid with graphene, defective graphene, and graphene oxide, J. Phys. Chem. C 117 (2013) 5708–5717.
[59] G. Gianese, V. Rosato, F. Cleri, M. Celino, P. Morales, Atomic-scale modeling of the interaction between short polypeptides and carbon surfaces, J. Phys. Chem. B 113 (2009) 12105–12112.
[60] S.M. Toma'sio, T.R. Walsh, Modeling the binding affinity of peptides for graphitic surfaces, influences of aromatic content and interfacial shape, J. Phys. Chem. C 113 (2009) 8778–8785.

[61] K. Balamurugan, R. Gopalakrishnan, S. Sundar Raman, V. Subramanian, Exploring the changes in the structure of α-helical peptides adsorbed onto a single walled carbon nanotube using classical molecular dynamics simulation, J. Phys. Chem. B 114 (2010) 14048–14058.
[62] K. Balamurugan, E.R. Azhagiya Singam, V. Subramanian, Effect of curvature on the α-helix breaking tendency of carbon based nanomaterials, J. Phys. Chem. C 115 (2011) 8886–8892.
[63] L. Baweja, K. Balamurugan, V. Subramanian, A. Dhawan, Hydration patterns of graphene-based nanomaterials (GBNMs) play a major role in the stability of a helical protein: a molecular dynamics simulation study, Langmuir 29 (2013) 14230–14238.
[64] L. Ou, Y. Luo, G. Wei, Atomic-level study of adsorption, conformational change, and dimerization of an α-helical peptide at graphene surface, J. Phys. Chem. B 115 (2011) 9813–9822.
[65] L. Baweja, K. Balamurugan, V. Subramanian, A. Dhawan, Effect of graphene oxide on the conformational transitions of amyloid beta peptide: a molecular dynamics simulation study, J. Mol. Graph. Model. 61 (2015) 175–185.
[66] K.P. Singh, L. Baweja, O. Wolkenhauer, Q. Rahman, S.K. Gupta, Impact of graphene-based nanomaterials (GBNMs) on the structural and functional conformations of hepcidin peptide, J. Comput. Aid. Mol. Des. 32 (2018) 487–496.
[67] S. Zhou, Y. Zhu, X. Yao, H. Liu, Carbon nanoparticles inhibit the aggregation of prion protein as revealed by experiments and atomistic simulations, J. Chem. Inf. Model. 59 (2019) 1909–1918.
[68] S. Bag, R. Mitra, S. Dasgupta, S. Dasgupta, Inhibition of human serum albumin fibrillation by two-dimensional nanoparticles, J. Phys. Chem. B 121 (2017) 5474–5482.
[69] N. Zhang, X. Hu, P. Guan, K. Zeng, Y. Cheng, Adsorption mechanism of amyloid fibrils to graphene nanosheets and their structural destruction, J. Phys. Chem. C 123 (2019) 897–906.
[70] M. Yousaf, H. Huang, P. Li, C. Wang, Y. Yang, Fluorine functionalized graphene quantum dots as inhibitor against hIAPP amyloid aggregation, ACS Chem. Neurosci. 8 (2017) 1368–1377.
[71] S.N. Kim, Z. Kuang, J.M. Slocik, S.E. Jones, Y. Cui, B.L. Farmer, M.C. McAlpine, R.R. Naik, Preferential binding of peptides to graphene edges and planes, J. Am. Chem. Soc. 133 (2011) 14480–14483.
[72] S.S. Kim, Z. Kuang, Y.H. Ngo, B.L. Farmer, R.R. Naik, Biotic–abiotic interactions: factors that influence peptide – graphene interactions, ACS Appl. Mater. Interfaces 7 (2015) 20447–20453.
[73] J. Katoch, S.N. Kim, Z. Kuang, B.L. Farmer, R.R. Naik, S.A. Tatulian, M. Ishigami, Structure of a peptide adsorbed on graphene and graphite, Nano Lett. 12 (2012) 2342–2346.
[74] T.R. Walsh, M.R. Knecht, Biomolecular material recognition in two dimensions: peptide binding to graphene, h-BN, and MoS_2 nanosheets as unique bioconjugates, Bioconjug. Chem. 30 (2019) 2727–2750.
[75] S. Li, A.N. Aphale, I.G. Macwan, P.K. Patra, W.G. Gonzalez, J. Miksovska, R.M. Leblanc, Graphene oxide as a quencher for fluorescent assay of amino acids, peptides, and proteins, ACS Appl. Mater. Interfaces 4 (2012) 7069–7075.
[76] Z. Guanghong, X. Zhou, Q. Huang, H. Fang, R. Zhou, Adsorption of villin headpiece onto graphene, carbon nanotube, and C60: effect of contacting surface curvatures on binding affinity, J. Phys. Chem. C 115 (2011) 23323–23328.
[77] D.Y. Lee, Z. Khatun, J.H. Lee, Y.K. Lee, Blood compatible graphene/heparin conjugate through noncovalent chemistry, Biomacromolecules 12 (2011) 336–341.
[78] T. Xiaofang, F. Liangzhu, Z. Jing, Y. Kai, Z. Shuai, L. Zhuang, P. Rui, Functionalization of graphene oxide generates a unique interface for selective serum protein interactions, ACS Appl. Mater. Interfaces 5 (2013) 1370–1377.
[79] N. Dragneva, O. Rubel, W.B. Floriano, Molecular dynamics of fibrinogen adsorption onto graphene, but not onto poly(ethylene glycol) surface, increases exposure of recognition sites that trigger immune response, J. Chem. Inf. Model. 56 (4) (2016) 706–720.
[80] M. Wang, Q. Wang, X. Lu, K. Wang, L. Fang, F. Ren, G. Lu, H. Zhang, Interaction behaviors of fibrinopeptide-a and graphene with different functional groups: a molecular dynamics simulation approach, J. Phys. Chem. B 121 (2017) 7907–7915.
[81] G. Raffaini, F. Ganazzoli, Surface topography effects in protein adsorption on nanostructured carbon allotropes, Langmuir 29 (2013) 4883–4893.
[82] B. Luan, R. Zhou, Single-file protein translocations through graphene – MoS_2 heterostructure nanopores, J. Phys. Chem. Lett. 9 (2018) 3409–3415.
[83] Q. Wanzhi, S. Efstratios, Detection of protein conformational changes with multilayer graphene nanopore sensors, ACS Appl. Mater. Interfaces 6 (2014) 16777–16781.

[84] J. Zhang, F. Zhang, H. Yang, X. Huang, H. Liu, J. Zhang, S. Guo, Graphene oxide as a matrix for enzyme immobilization, Langmuir 26 (2010) 6083–6085.
[85] Y. Wang, Z. Li, J. Wang, J. Li, Y. Lin, Graphene and graphene oxide: biofunctionalization and applications in biotechnology, Trends Biotechnol. 29 (2011) 205–212.
[86] F. Zhang, B. Zheng, J. Zhang, X. Huang, H. Liu, S. Guo, J. Zhang, Horseradish peroxidase immobilized on graphene oxide: physical properties and applications in phenolic compound removal, J. Phys. Chem. C 114 (2010) 8469–8473.
[87] P. Laaksonen, M. Kainlauri, T. Laaksonen, A. Shchepetov, H. Jiang, J. Ahopelto, M.B. Linder, Interfacial engineering by proteins: exfoliation and functionalization of graphene by hydrophobins, Angew. Chem. Int. Ed. 49 (2010) 4946–4949.
[88] X. Sun, Z. Feng, T. Hou, Y. Li, Mechanism of graphene oxide as an enzyme inhibitor from molecular dynamics simulations, ACS Appl. Mater. Interfaces 6 (10) (2014) 7153–7163.
[89] K. Yao, P. Tan, Y. Luo, L. Feng, L. Xu, Z. Liu, Y. Li, R. Peng, Graphene oxide selectively enhances thermostability of trypsin, ACS Appl. Mater. Interfaces 7 (2015) 12270–12277.
[90] I. Baek, H. Choi, S. Yoon, S. Na, Effects of the hydrophobicity of key residues on the characteristics and stability of glucose oxidase on a graphene surface, ACS Biomater. Sci. Eng. 6 (1899-1908) 2020.
[91] M. Mathesh, B. Luan, T.O. Akanbi, J.K. Weber, J. Liu, C.J. Barrow, R. Zhou, W. Yang, Opening lids: modulation of lipase immobilization by graphene oxides, ACS Catal. 6 (2016) 4760–4768.

CHAPTER

9

Structures, properties, and applications of nitrogen-doped graphene

Tandabany Dinadayalane, Jovian Lazare, Nada F. Alzaaqi, Dinushka Herath, Brittany Hill, and Allea E. Campbell

Department of Chemistry, Clark Atlanta University, Atlanta, GA, United States

1 Introduction

Some common morphologies of graphene-based nanomaterials [1] include two-dimensional (2D) graphene nanosheets (GNSs), one-dimensional (1D) graphene nanoribbons (GNRs) [2–4], and zero-dimensional (0D) graphene quantum dots (GQDs) [5,6]. The bandgap of pristine graphene is zero, which limits its use in electronics and must be opened. Common approaches to tuning the properties are by size and concentration of edge effects. An example of tuning the bandgap by size effect can be demonstrated: GNRs with widths less than 10 nm showed semiconductor characteristics (small bandgap), whereas those having more than 10 nm showed very weak gate-voltage dependence as a result of an even smaller bandgap [7]. An example of tuning by the edge is when a higher concentration of zigzag edge produces a smaller bandgap compared to a lower concentration [6]. The shape of the graphene system shows an impact on the electronic properties based on the edge effects [8].

There are other ways to tune the properties of graphene. Doping is a popular approach by either physisorption or chemisorption (by gases [9,10], metals [11,12], or organic molecules [13,14]) [15–17]. In the case of N-doped graphene systems, doping reaction mechanisms result in various configurations [18]. The formation mechanism is not clearly understood, especially in thermally treated graphene [18]. Hu et al. highlighted that synthesizing nitrogen-doped carbon materials with a high doping level (e.g., over 10% N content) and controllable doping contents are very challenging [19]. Different configurations of nitrogen doping have been used to control the properties that enable various applications. For example, controlling the dopants can achieve n- or p-type semiconductors (see Fig. 9.1) [20] and/or allow half-metallicity, and/or modify catalytic and sensing properties [21,22]. Nitrogen-doped

Theoretical and Computational Chemistry, Volume 21
https://doi.org/10.1016/B978-0-12-819514-7.00010-5

(N-doped) graphene has attracted significant interest because of its extraordinary properties and various applications [23]. The desired properties and applications can be controlled by the morphology.

Some common configurations are pyridinic, pyrrolic, and graphitic (quaternary) N as shown in Scheme 9.1. Pyridinic N is bonded to two carbon atoms on the edge ring location or near the defect site and has one p electron donated to the π-system. Pyrrolic N is like pyridinic N configuration except for two distinctions: (1) nitrogen is in a 5-membered ring

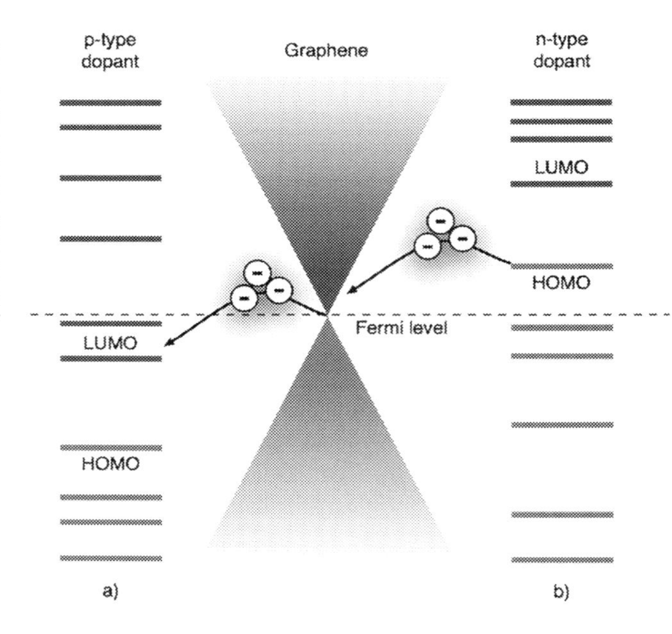

FIG. 9.1 Scheme of the relative position of the highest occupied molecular orbital (HOMO) and lowest unoccupied molecular orbital (LUMO) of an adsorbate to the Fermi level of graphene for (A) p-type and (B) n-type dopants. *This figure was reproduced from H. Pinto, A. Markevich, Electronic and electrochemical doping of graphene by surface adsorbates. Beilstein J. Nanotechnol. 5 (2014) 1842–1848. (© 2014 H. Pinto et al., distributed under the terms of the Creative Commons Attribution 2.0 International License, https://creativecommons.org/licenses/by/2.0).*

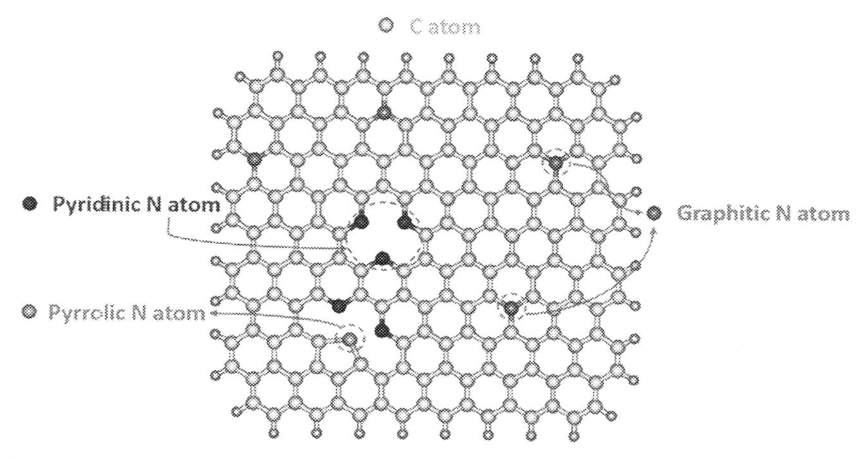

SCHEME 9.1 Different configurations of nitrogen doping with graphene model.

rather than a 6-membered ring with sp^3 hybridization at the nitrogen site instead of sp^2 and (2) pyrrolic N donates two p electrons rather than one to the π-system [24,25]. Quaternary N or graphitic N can be found at the edge location terminated by hydrogen or in the bulk region [26]. The two forms (pyridinic and quaternary N) are of interest for high-performance supercapacitors and electrocatalytic reactions including oxygen reduction reactions (ORRs) and oxygen evolution reactions (OERs) [22,27].

Nitrogen (N)-doped carbon materials are some of the most efficient metal-free catalysts [28]. Nitrogen doping in graphene helps to improve the activity of oxygen reduction reaction slightly; however, nitrogen doping at the edge of graphene does not contribute noticeably to the enhancement of activity [29]. Nitrogen-doped graphene systems have been investigated for the electrochemical sensor applications due to their capacity to improve electrical conductivity [26]. It has been reported that spin density and charge distributions near the doped regions are influenced by dopants and are responsible for activated regions involved in catalytic reactions on the graphene surface [22,30,31]. These distinct characteristics were attributed to different electronic properties. Besides various catalytic applications, N-doped graphene further enhances Raman scattering-derived molecular sensing in comparison with graphene [32]. Other applications may be enabled in medicine since N-doped graphene materials show reduced cytotoxicity and antioxidant effect depending on the content of nitrogen [33]. In fact, the N-doped graphene materials demonstrate less toxicity in comparison with graphene oxide (GO) [34].

2 Synthesis and characterization techniques

Two types of methods for obtaining N-doped graphene are direct synthesis and postsynthesis. Direct synthesis has a high potential for even distribution of nitrogen doping on graphene. Some of the common direct synthesis methods include chemical vapor deposition (CVD), segregation, solvothermal, hydrothermal, and arc discharge [26,35–40]. Postsynthesis primarily involves doping of the surface region. Some common postsynthesis methods include thermal treatment, plasma treatment, ion bombardment, and hydrazine treatment [40–44]. Generally, the characterization techniques of X-ray photoelectron spectroscopy (XPS) [42,45], Raman [46,47], and scanning tunneling microscopy (STM) are used to confirm the N-doped graphene. Other characterization techniques also exist and will be discussed in brief [48].

2.1 Direct synthesis and postsynthesis for N-doped graphene

2.1.1 Direct synthesis by CVD

A metal catalyst such as Cu or Ni was used as a substrate in the CVD approach [39,46,49]. A carbon gas is usually mixed with nitrogen gas at high temperatures on the substrate and leads to dissociation and recombination into N-doped graphene [39,50]. Some examples of the gas mixtures are C_2H_2/NH_4 and CH_4/NH_3 [39,46,49]. Acetonitrile [51] or pyridine [52] could be used as a precursor for obtaining N-doped graphene. When acetonitrile precursor was used, the number of N-doped graphene layers could be controlled by adjusting the flow time [51]. To control the nitrogen content, the flow rate or ratio of carbon to nitrogen gas source can be adjusted. Wei et al. reported that CH_4/NH_3 ratio increased from 1:1 to 4:1 and this led to a decrease in nitrogen atom content from 8.9% to 1.2% [39]. The percentage of nitrogen atom content was normally found to vary between 4% and 9% on the basis of the CVD approach, but nitrogen content as high as 16% was reported [46]. On Cu catalyst, a 1:1 ratio was used for CH_4/NH_3 precursor resulting in graphitic N being the main configuration [39]. When 5:1 ratio of CH_4/NH_3 precursor was used on Ni catalyst, pyridinic and pyrrolic N configuration types were reported to be predominant [49]. Pyridinic N becomes the main configuration type in N-doped graphene by utilizing C_2H_2/NH_4 as a precursor on Cu catalyst [46]. Other sequences of configurations can also be controlled by choice of catalyst and growth temperature [53–56]. By using CH_4/NH_3 precursor at the high temperature, mainly pyridinic N was obtained; but lowering the temperature led to a gradual increase in pyrrolic N with increased nitrogen content [57]. In a number of studies, graphitic N was reported [39,47,58].

2.1.2 *Direct synthesis by segregation growth*

In the segregation growth approach, nitrogen-containing boron layers and carbon-containing nickel layers were consecutively placed onto SiO_2/Si substrate by electron beam evaporation mechanism [59]. A vacuum annealing process took place where nickel trapped boron atoms while carbon and nitrogen atoms diffused and segregated on the nickel surface to form N-doped graphene. Though thick multilayers were observed sometimes in the segregation growth approach, thin layers of N-doped graphene were reported to be common and consistently observed on a large scale. Variation of the size of the boron or nickel films controls the nitrogen atom content which was generally 0.3–2.9%. By inserting nitrogen species onto the site-specific area of substrate, doping can be done in a selective area. Pyridinic and pyrrolic N types in N-doped graphene are the main configurations in the segregation approach.

2.1.3 *Direct synthesis by solvothermal approach*

Solvothermal synthesis is a chemical reaction that occurs in nonaqueous solutions at relatively high temperatures. The solvothermal approach has been used for gram-scale production of N-doped graphene at approximately 300° C. Around 1–6 layers were commonly obtained using this approach. Two different mixtures lithium nitride (Li_3N) with tetrachloromethane (CCl_4) and cyanuric chloride ($N_3C_3Cl_3$) with Li_3N and CCl_4 were involved in this process. The synthetic process with the latter mixture resulted in higher nitrogen atom content of 16.4% in comparison with the process with the first mixture that resulted in 4.5% nitrogen content. Pyridinic and pyrrolic N-doped graphene systems were predominantly produced in the process with the second mixture, and graphitic N was predominant in the reaction with the first mixture [60].

2.1.4 *Direct synthesis by hydrothermal approach*

Hydrothermal synthesis is a chemical reaction that occurs in aqueous solutions above the boiling point of water [61]. These types of reactions have become popular for simple N-doped graphene synthesis. For example, by using hexamethylenetetramine (HMTA) as a single source of carbon and nitrogen, N-doped graphene was prepared in aqueous solutions in a single-step reaction. Nitrogen atom content in this reaction was reported as 1.68% [62]. Hydrophilic and hydrophobic N-doped graphene quantum dots (N-GQDs) were synthesized from exfoliation and disintegration of graphite flakes using one-pot hydrothermal treatment [63]. Furthermore, a similar one-pot hydrothermal treatment of citric acid and dicyandiamide was employed to produce N-GQDs [64]. Two-step synthesis is also commonly used. Agusu et al. obtained N-doped graphene by a hydrothermal reduction of GO and urea mixed in deionized H_2O with a 3:1 weight ratio of urea and GO preceding a treatment of 190 °C for 12 hours [65]. Similarly, Mageed et al. reduced GO in NH_3 solution and later treated samples to 200 °C under N_2 gas to obtain N-doped reduced GO, possessing 5.98% and 15.85% of nitrogen and oxygen atom contents, respectively [37]. Hydrothermal synthesis of N-doped graphene using amino acids with different acidities can be used to control the nitrogen content, doping configuration, and morphology [35,66]. Aqueous solutions of amino acids (e.g., glycine) were used for producing N-doped graphene [67].

2.1.5 Direct synthesis by discharge

In the arc discharge approach, a carbon source such as graphite is usually evaporated at high temperatures [68,69]. By additionally using NH_3 or pyridine vapor, N-doped graphene was successfully synthesized [70,71]. Nanodiamond as a carbon source yielded higher hydrogen content than that of graphite. The nitrogen atom content in N-doped graphene was reported from 0.5% to 1.5%, and two or three layers were primarily observed. However, single-layer N-doped graphene was found occasionally. The scale of production is normally less than 1 μm in the arc discharge approach [70].

2.1.6 Postsynthesis by thermal treatment

Thermal treatment involves high temperature conditions (usually $\geq$800 °C) to produce N-doped graphene [40,72]. Such high temperatures were achieved by electrical annealing to produce N-doped graphene nanoribbons [41]. The nitrogen atom content in N-doped graphene thermal treatment was from as low as 1.1% at 1100 °C [40], to as high as 2.8% at 800 °C and 900 °C [72]. The nitrogen contents were considered to be low and the reasons were ascribed to: (1) high annealing temperatures causing C—N bonds to break and (2) a low number of defects in the high-quality graphene [73]. In thermal treatment methods, N-doping tends to occur at the defect sites and the edge of graphene [72]. Pyridinic and pyrrolic N configurations were predominant in the thermal treatment methods. Although annealing of NH_3 on graphene introduced N-doping in the defect sites, annealing N^+ on graphene in N_2 did not produce chemical doping at the defect sites [40].

Graphene oxide (GO) can be used instead of pristine graphene for the thermal treatment process. For example, annealing of melamine and GO between 700 °C and 1000 °C yielded N-doped graphene [45]. The layer distribution was reported to depend on the synthesis conditions. For example, the number of layers of N-doped graphene produced is dependent on the temperature and number of GO layers used in the annealing process of the melamine-GO mixture. Thermal treatment of a single layer of GO with melamine produced single-layer N-doped graphene. However, using few GO layers resulted in few-layer N-doped graphene. The nitrogen content in N-doped graphene was influenced by both temperature and ratio of melamine to GO used. The highest nitrogen content that has been observed by this method is 10.1% from a thermal condition of less than 700 °C and a mass ratio of 0.2 GO to melamine [74]. Using an annealing temperature of 500 °C in NH_3 atmosphere produced only 5% nitrogen content in N-doped graphene. Previous studies indicated that temperature severely affects the nitrogen content. Thermal treatments involving high temperatures decrease the number of oxygen functional groups that are believed to be responsible for the formation of C-N bond(s). The reactivity of nitrogen with GO is reduced upon the decomposition of oxygen functional groups [45,74]. Lowering the ratio of GO to melamine at 800 °C impacted increased N-doping suggesting that the amount of melamine can be used at high temperatures as an additional control parameter for maximizing the nitrogen content. Faisal et al. reported N-doped graphene with nitrogen atom content as high as 9.22% by using GO and uric acid [27]. Interestingly, the additional use of uric acid yielded a high percentage of pyridinic-N and graphitic-N rather than pyridinic- and pyrrolic-N which are usually predominant in the thermal treatment methods.

Pyrolysis is an irreversible high-temperature thermal decomposition in an inert atmosphere. Pyrolysis methods were regularly used in the synthesis of N-doped carbon nanotubes (CNTs) [54–56]. Recently, this method has become a common facile approach in preparing N-doped graphene [75–77]. N-doped graphene-like nanoflakes were achieved via facile one-step/one-pot pyrolysis of carbon nanoflakes with biomass-derived citric acid and dicyandiamide. Gu et al. reported nitrogen atom content as high as 6.2% [75], whereas Wang et al. achieved nitrogen atom content as high as 9.2% for N-doped graphene [77].

2.1.7 *Postsynthesis by plasma treatment*

Plasma treatment is another method that can be used to synthesize N-doped graphene by placing carbon material into a nitrogen plasma. In this treatment, carbon atoms are replaced by nitrogen atoms. This method was applied to carbon nanotubes (CNTs) [78–80], graphene [42,81,82], and GO in producing nitrogen-doped carbon materials [83]. NH_3 plasma can also be used on mechanically exfoliated graphene to produce N-doped graphene. The exposure time and plasma strength were controlling factors for nitrogen atom content between 3.0% and 8.5%. One drawback in this method was reported to the formation of oxygen species in the process [42,82]. Formation of reactive carbons at defect sites was blamed for not obtaining high-quality N-doped graphene [42]. Significant exposure to N_2 plasma diminished the electrocatalytic activity [82]. Recent experiments used microwave plasma for large-scale synthesis [84] and for fast, low-temperature enhancement to the CVD approach [85].

2.1.8 *Postsynthesis by ion bombardment*

The ion bombardment method is also known as ion implantation. By using an ion gun, N^+ ions are irradiated into graphene. Graphene with two nitrogen atoms doped in the same sublattice with metaconfiguration (as shown in Fig. 9.2A) was achieved and confirmed by STM. Single nitrogen atoms in sublattices were also observed (Fig. 9.2) [43]. Furthermore, at sufficiently low ion implantation energies, a low percentage of vacancy-related defect sites was observed [86]. Cress et al. showed that only the top layer of graphene was doped using this process on bilayer graphene [87].

2.1.9 *Postsynthesis by hydrazine treatment*

N-doped graphene can also be synthesized from GO using hydrazine (N_2H_4) hydrate treatment [44]. This method is widely used by the reduction of GO in a mixture of NH_3 and N_2H_4 solution. Nitrogen content of a maximum of 5% can be achieved at 80 °C. If the reaction temperature is improperly controlled (reaction temperature of 160 °C or more), then the

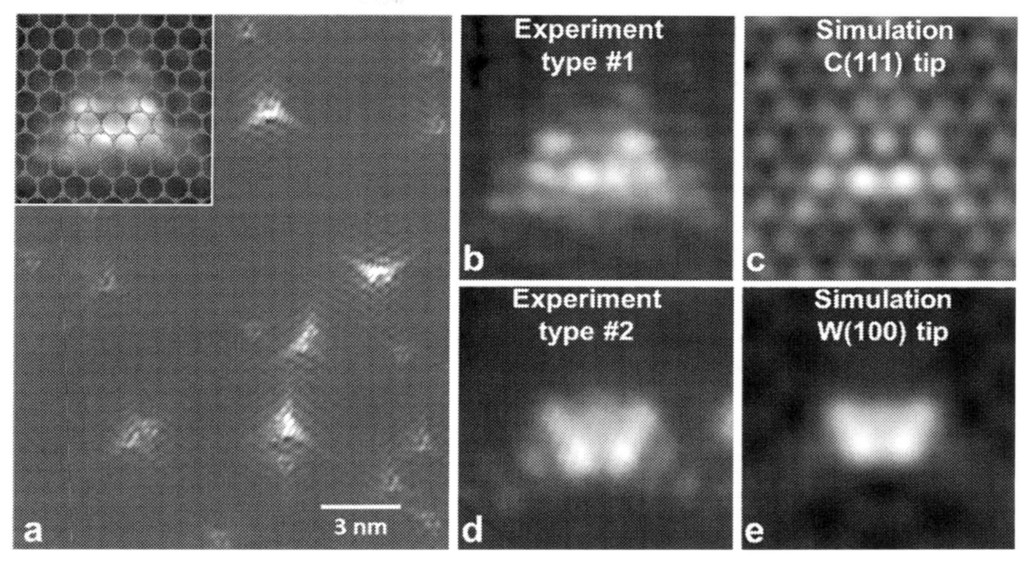

FIG. 9.2 (A) STM image of a sample doped with a higher amount of N defects; the *inset* shows an overlay of the graphene lattice with two substitutional N defects on a zoomed metadefect. (B–E) Experiment to theory comparison of the two types of the most frequently observed atomically resolved contrasts. *This was reproduced from M. Telychko, P. Mutombo, M. Ondráček, P. Hapala, F.C. Bocquet, J. Kolorenč, M. Vondráček, P. Jelínek, M. Švec, Achieving high-quality single-atom nitrogen doping of graphene/SiC(0001) by ion implantation and subsequent thermal stabilization. ACS Nano 8 (7) (2014) 7318–7324 with permission. Copyright 2014 American Chemical Society.*

nitrogen atom content reduces to around 4%. Temperature also controls the morphology. Reaction temperatures of 120 °C or less lead to flat N-doped graphene, whereas higher temperatures lead to more agglomeration. The use of ultrasonication can also be used in the presence of hydrazine to produce N-doped graphene. Agglomeration can also occur after N-doping. Only ~1% of nitrogen atom is obtained with pyridinic and pyrrolic nitrogen configurations. The very low N atom content and the configurations suggest that doping sites are at only the edges or defects [88].

2.2 Experimental characterizations of N-doped graphene

2.2.1 Characterization by X-ray photoelectron spectroscopy (XPS)

X-ray photoelectron spectroscopy (XPS) is commonly used to confirm nitrogen doping in graphene. Peaks observed around 400 and 285 eV represent N1s and C1s, respectively. By using the peak intensities, the ratio of N1s and C1s can be used to determine the percentage of N atom content. Deconvolution of N1s spectrum results in several peaks that can be used to determine the content of different configurations. Peaks in the ranges of 398.1–399.3, 399.8–401.2, and 401.1–402.7 eV are assigned to pyridinic-N, pyrrolic-N, and quaternary-N, respectively (see Fig. 9.3) [42]. The variation in environment of N-doped graphene may attribute to different peak positions of different configurations [24]. The peak positions are affected by the charge of atoms including nitrogen and neighboring carbon atoms as well

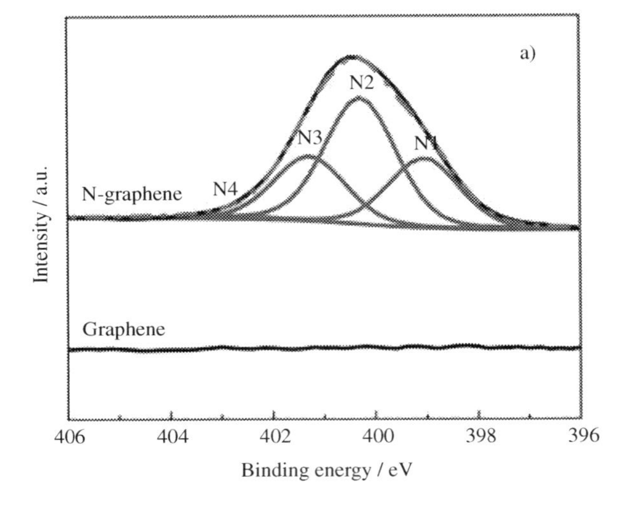

FIG. 9.3 Core-level high-resolution N1s XPS spectra of graphene and N-graphene. Labels N1, N2, N3, and N4 represent pyridinic-N, pyrrolic-N, quaternary-N, and N-oxides of pyridinic-N, respectively [42]. *Republished with permission of the Royal Society of Chemistry, from Y. Shao, S. Zhang, M.H. Engelhard, G. Li, G. Shao, Y. Wang, J. Liu, I.A. Aksay, Y. Lin, Nitrogen-doped graphene and its electrochemical applications. J. Mater. Chem. 20 (35) (2010) 7491–7496; permission conveyed through Copyright Clearance Center, Inc.*

as charge distribution after ionization. In addition to three well-known N-doped configurations, N-oxides of pyridinic-N configuration were also observed. The peak of 402.8 eV is assigned to this type of configuration [42,45].

The C1s peaks are used in assessing configurations of GO. Sharp peaks at ~284.5 eV are assigned to C=C bond (sp^2). The peaks around 286.2, 287.8, and 289.2 eV correspond to sp^3 C—O bonds, carbonyls (C=O), and carboxylates (O=C—O). Fig. 9.4 shows that the annealing process removes C—O peaks [45]. Furthermore, new peaks were assigned to C-N groups of N-doped graphene [44,45,59,89]. Physisorbed carbon-oxygen groups on graphene can also be observed by XPS. For example, a new peak at 289 eV has been observed and attributed to the physisorbed carbon-oxygen [59,89]. Peaks corresponding to carbon atoms change after N-doping occurs [44,45,59,74,89]. Change in peak at higher energy in the C1s spectrum generally suggests the N-doping occurred in the graphene.

2.2.2 Characterization by Raman spectroscopy

Raman spectroscopy is also a common tool in assessing the synthesis of N-doped graphene. Peaks ranging from 1320 to 1350, 1570 to 1585, and 2640 to 2680 cm^{-1} correspond to D-, G-, and 2D-bands in N-doped graphene. Even if the D-band is not observed in graphene or N-doped graphene (defects needed for activation), the 2D-band is always observable. In some studies, D′ peak is also observed at $1602-1625\ cm^{-1}$ [46,60]. The G-band originates from the first-order Raman scattering related to the doubly degenerate E_{2g} phonons at the Brillouin zone. The 2D and D-bands come from the second-order double-resonance scattering process related to zone-boundary phonons. In the 2D mode, two zone-boundary phonons are involved (one phonon and one defect).

The defect-induced D′ peak originates from the double-resonance scattering process. Defects are introduced to graphene in the N-doping process. The relative intensity ratio of D- and G-bands (I_D/I_G) can be used to assess the defects. The Tuinstra-Koenig (TK) relation, $L_a(\text{nm}) = (2.4 \times 10^{-10})\,\lambda_4(I_D/I_G)^{-1}$, can be used to determine crystallite size (L_a), where λ is the

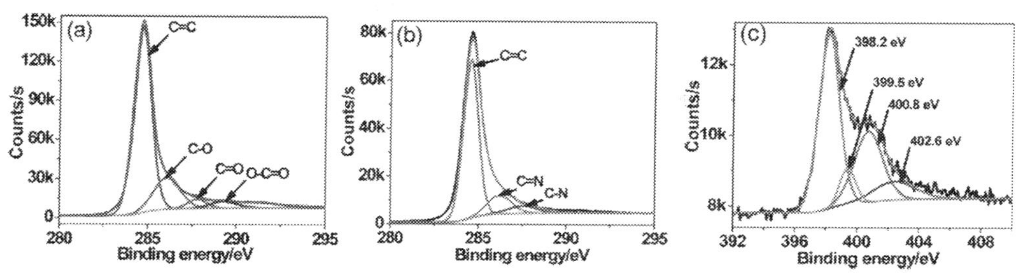

FIG. 9.4 High-resolution C1s spectra of the (A) GO, (B) graphene prepared by annealing GO in Ar at 800 °C, and (C) N-graphene prepared by annealing GO/melamine (1:5) at 800 °C for 30 min. *This was reproduced from Z.-H. Sheng, L. Shao, J.-J. Chen, W.-J. Bao, F.-B. Wang, X.-H. Xia, Catalyst-free synthesis of nitrogen-doped graphene via thermal annealing graphite oxide with melamine and its excellent electrocatalysis. ACS Nano 5 (6) (2011) 4350–4358 with permission. Copyright 2011 American Chemical Society.*

Raman excitation wavelength. Due to the inversely proportional relationship, more defects lead to smaller L_a. Furthermore, more defects are indicated by a higher ratio of I_D/I_G and higher content of N-doping. I_{2D}/I_G was suggested to be used to characterize the nitrogen atom content [59,89]. One report noted that the increase in G-band intensity is almost linear with the incremental increase in graphene layers [90]. Relation to thickness of graphene sheets was also determined based on I_G/I_{2D} ratio [91,92].

Homogeneous layers can be properly characterized by I_{2D}/I_G and I_D/I_G ratios. If the layers are inhomogeneous, then a discrete location cannot represent the entire N-doping environment. Luo et al. reported an uneven distribution due to inhomogeneous N-doping by high and very low I_D/I_G ratio depending on the location [46]. Raman spectral mapping is generally applied to avoid poor assessment of N-doping. This technique involves the assessment of a large scale of N-doped graphene [47]. Unlike I_D/I_G, lower ratio of I_{2D}/I_G means higher concentration of N-doping in the analysis of Raman spectral maps. Due to change in charge density [89,93], uneven layer distribution [91,94], and/or defects related to N-doping [95], an upward [47,89] or downward shift [60,74] of the G-band may be observed.

2.2.3 Characterization by STM

Another tool for assessing N-doping of graphene is scanning tunneling microscopy (STM). STM technique is widely used for investigating electronic properties of solid-state materials by probing the charge density at the Fermi level (conduction and valence bands). The conduction bands are probed when positive bias voltage (V_{bias}) is applied causing electrons to tunnel from the tip to the sample. In an opposite effect, the valence bands are probed when negative V_{bias} is applied causing electron tunneling from the sample to the tip [96]. The tip is very sharp and allows researchers to view atomic resolution images. STM-related electronic properties have been studied on N-doped graphene both experimentally and theoretically [43,47,60,97].

N-doped graphene displays some brighter features that allow us to determine N-doped locations that are distinct from pristine regions of the graphene surface [43,60]. This is clearly shown in Fig. 9.5 [43]. Quaternary N-doped graphene was revealed experimentally. Theoretical STM images showed that C atoms close to N atoms in quaternary N-doped graphene have the brightest feature due to an increase in charge density. By using line scans above the bright

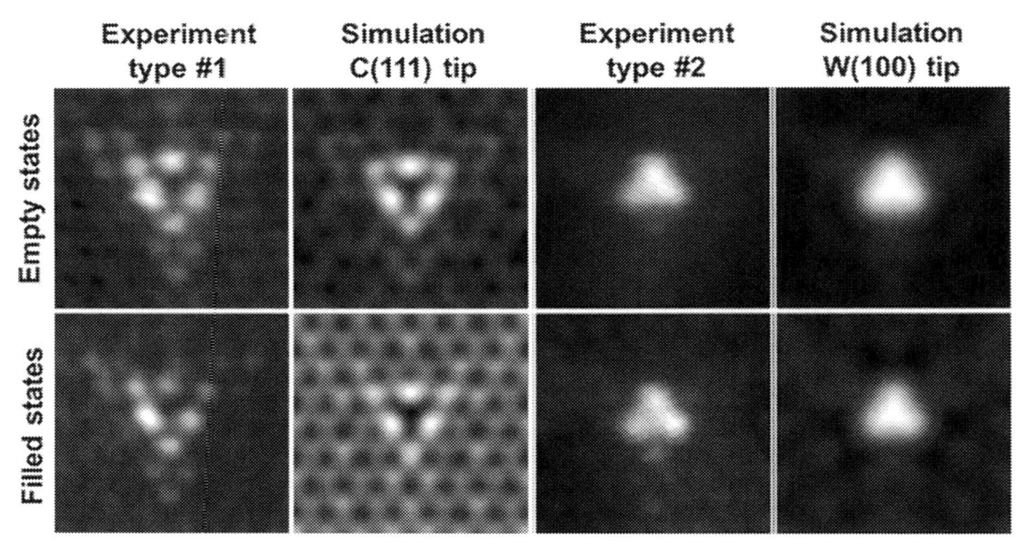

FIG. 9.5 Experimental and calculated STM current maps of the single N defects, empty and filled states, $1.7 \times 1.7\ nm^2$. STM simulations (+0.5 and −0.5 V, surface-tip distance 3.5 Å) for various tip models are matched to the two most frequently experimentally observed patterns (+0.4 and −0.4 V). *This was reproduced from M. Telychko, P. Mutombo, M. Ondráček, P. Hapala, F.C. Bocquet, J. Kolorenč, M. Vondráček, P. Jelínek, M. Švec, Achieving high-quality single-atom nitrogen doping of graphene/SiC(0001) by ion implantation and subsequent thermal stabilization. ACS Nano 8 (7) (2014) 7318–7324 with permission. Copyright 2014 American Chemical Society.*

regions, out-of-plane heights were also assessed to confirm the nitrogen substitution. In N-doped graphene, multiple nitrogen atoms prefer to be located in the same hexagonal sublattice [43]. Long-range STM shows that other than the same sublattice, N atoms are randomly doped in graphene [47].

2.2.4 Other characterization techniques

Besides the aforementioned techniques (XPS, Raman, and STM), transmission electron microscopy (TEM), atomic force microscopy (AFM), scanning electron microscopy (SEM), selected area electron diffraction (SAED), isotherm by Brunauer-Emmett-Teller (BET) method, X-ray diffraction (XRD), Fourier-transform infrared spectroscopy (FT-IR), X-ray absorption near-edge structure (XANES), and thermogravimetric analysis (TGA) are some other characterization techniques. SEM, TEM, XRD, and BET are used to find microstructural differences [48]. High-resolution transmission electron microscopy (HRTEM) and AFM are generally used for studying the number of graphene layers by looking closely at cross sections and edges or interlayer distance [52,53,70,71,88]. Based on diffraction patterns, SAED has also been employed to understand the crystalline nature. For example, hexagonal diffraction pattern in a location indicates well-ordered crystallinity [45,52], whereas ring-like diffraction patterns of a location suggest distortion that is usually observed after doping [49]. FT-IR and XANES are used for assessing C—N bonding configurations [48]. Tables 9.1 and 9.2 provide the list of information of synthesis methods, configurations of doped nitrogen, experimental characterization techniques employed, and the percentage of nitrogen in N-doped graphene reported in the literature.

TABLE 9.1 Summary of direct synthesis methods, configuration of doped nitrogen, characterization techniques used for N-doped graphene reported with percentage of nitrogen (N at%).

Method	N at%	Configuration	Characterization techniques	Ref.
CVD	8.9	Graphitic (main), pyridinic, and pyrrolic	TEM, SEM, Raman, and XPS	[39]
CVD	0–16	Pyridinic	Raman, ToF-SIMS, XPS, and RDE voltammetry	[46]
CVD	0.2–0.4	Graphitic (main)	STM, Raman, and XPS	[47]
CVD	4	Pyridinic, pyrrolic	AFM, TEM, SEM, Raman, XRD, and XPS	[49]
CVD	9	Graphitic, pyridinic, and pyrrolic	TEM, Raman, XRD, and XPS	[51]
CVD	2.4	Graphitic and pyridinic	AFM, SAED, HRTEM, Raman, and XPS	[52]
CVD	0–10	Graphitic and pyridinic	HRTEM, SEM, Raman, XPS, and EELS	[53]
CVD	0.7–4.6	Pyridinic and pyrrolic	Raman and XPS	[57]
CVD	~0.4	Graphitic (main), pyridinic, and nitrilic	XAS, XES, XPS, and STM	[58]
Segregation growth	0.3–2.9	Pyridinic and pyrrolic	XPS, optical microscope, and Raman	[59]
Solvothermal	4.5, 16.4	Graphitic (main for 16.4%), pyridinic, and pyrrolic (main for 4.5)	STM, Raman, TEM, HRTEM, and XPS	[60]
Hydrothermal	1, 3, 8.9	Pyridinic, pyrrolic, and graphitic	SEM, XRD, Raman, XPS, and N-physisorption by pressure sorption analyzer	[35]
Hydrothermal	5.98	Not reported	SEM, TEM, EDX (energy dispersion X-ray spectroscopy), FT-IR, and UV-Vis	[37]
Hydrothermal	1.68	Graphitic, pyridinic, pyrrolic, oxidized, and chemisorbed	SEM, TEM, HRTEM, Raman, XRD, and XPS	[62]
Hydrothermal	18.13	Pyridinic and pyrrolic	TEM, HRTEM, AFM, FT-IR, UV-Vis for photoluminescence, XPS, Raman, and LCSM	[64]
Hydrothermal	13.82	Pyrrolic (main)	SEM, XRD, FT-IR, XPS, and Raman	[66]
Hydrothermal	1.91	Pyrrolic (main), pyridinic, and graphitic	SEM, TEM, HRTEM, XRD, FT-IR, Raman, SAED, and XPS	[67]
Arc discharge	0.5–1.5	Not reported	XPS, TEM, AFM, Raman, DFT, and EELS	[70]
Arc discharge	Not reported	Not reported	TEM, Raman, XRD, AFM, BET, and FESEM	[71]

ToF-SIMS—Time-of-flight secondary ion mass spectrometry; RDE—Rotating disk electrode; LCSM—Laser confocal scanning microscopy; XES—X-ray emission spectroscopy; XAS—X-ray adsorption spectroscopy; FESEM—Field emission scanning electron microscopy; EELS—Electron energy loss spectroscopy; BET—Brunauer-Emmett-Teller method; HRTEM—High-resolution transmission electron microscopy; AFM—Atomic force microscopy; SEM—Scanning electron microscopy; SAED—selected area electron diffraction.

TABLE 9.2 Summary of postsynthesis methods, characterization techniques used for N-doped graphene reported with the percentage of nitrogen (N at%) and configuration reported in the literature.

Method	N at%	Configuration	Characterization techniques	Ref.
Thermal	1.1	n/a	Raman, atomic emission spectroscopy (AES), XPS, and SEM	[40]
Thermal	2.0–2.8	Pyridinic, pyrrolic, and graphitic	XPS, cyclic voltammograms (CVs), SEM, TEM, and Raman	[72]
Thermal	10.1	Pyridinic and graphitic	TEM, HRTEM, AFM, XPS, and Raman	[45]
Thermal	5	Pyridinic and graphitic	AFM, XPS, and Raman	[74]
Thermal	1.5, 6.8, 9.2	Pyridinic (41–47%), graphitic (19–32%), and pyrrolic (35–14%)	HRTEM, FESEM, SAED, TGA, XRD, Raman, BET, XPS, and cyclic voltammograms (CVs)	[27]
Plasma	8.5	Pyridinic, pyrrolic, and graphitic	Cyclic voltammograms (CVs), XPS, and Raman	[42]
Plasma	0.11–1.35	Pyridinic, pyrrolic, and graphitic	XPS, TEM, and cyclic voltammograms (CVs)	[82]
Ion bombardment	0.1–1.6	Graphitic	STM and XANES	[43]
Ion bombardment	~1	Predominantly substitutionally incorporated graphene	Electron energy loss (EEL) spectra and atomic resolution high-angle annular dark-field (HAADF) images	[86]
Hydrazine	5	Graphitic, pyridinic, and pyrrolic	XRD, ^{13}C NMR, SEM, TEM, and XPS	[13]
Hydrazine-assisted ultrasonication	1.04	Pyridinic and pyrrolic	XPS, cyclic voltammograms (CVs), SEM, TEM, and AFM	[88]

3 Computational studies on N-doped graphene

In many reported computational studies [15,36,98–101], density-functional theory (DFT) calculations were used to investigate the stabilities and electronic properties of nitrogen-doped graphene systems. DFT calculations were performed for N-doped graphene models using cluster and periodic approaches. Both approaches allow scientists to make use of the advantages of both formalisms in order to obtain a thorough understanding of the structures, properties, and reactivities of the materials of interest. The main advantage of the periodic approach is the description of the system closer to realistic and avoiding the typical drawbacks of small clusters related to artificial boundaries as the case of cluster approach. On the contrary, the small size of the cluster and its "molecular" nature make this type of calculation computationally less demanding. In fact, very large, and consequently computationally expensive, slabs would be required to describe the local changes in the active sites within the periodic approach [98].

3.1 Spin densities of N-doped graphene systems

For the N-doped graphene systems [98], charge and spin density on each atom were estimated by using the Mulliken population analysis [102]. The results of the calculations using the periodic approach indicated that the doping of nitrogen as graphitic type in graphene network caused a local redistribution of charges and spin densities on C atoms adjacent to N. In each case, the N atom has one more electron than C and is more electronegative compared to carbon. The N atom exhibited a considerable negative charge. The *ortho* C atoms have the largest compensating positive charge. However, the charges on *meta* and *para* C atoms are close to zero. Spin density due to the unpaired electron introduced with nitrogen doping is mainly delocalized onto the N and *ortho* C atoms. The results of Mulliken charge distribution obtained from the calculations of the cluster approach agree with the periodic results. The nitrogen atom always bears a significant negative charge, and the compensating positive charge is localized mainly on the *ortho* carbon atoms. Fig. 9.6 depicts the charge and spin density redistribution of N-doped graphene models of 1.02 and 3.06 at% of N-doping.

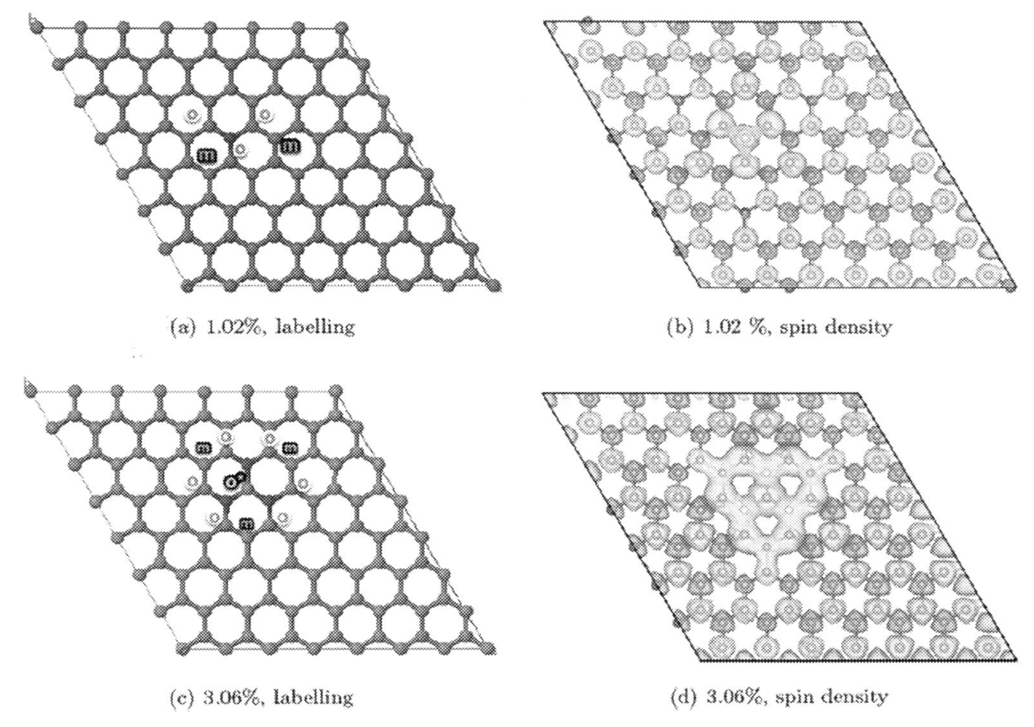

FIG. 9.6 Labeling scheme and spin density plots of N-doped graphene with doping amounts of 1.02 and 3.06 at%. Nitrogen atoms as *blue spheres* (*dark gray* in the print version). Isosurface values of 0.0005. *This was reproduced from C. Ricca, F. Labat, N. Russo, C. Adamo, E. Sicilia, Oxidation of Ethylbenzene to acetophenone with n-doped graphene: insight from theory. J. Phys. Chem. C 118 (23) (2014) 12275–12284 with permission. Copyright 2014 American Chemical Society.*

3.2 Formation energies of different configurations of n-doped graphene systems

Kresse et al. investigated different configurations of nitrogen-doped graphene such as graphitic N-doped graphene and pyridinic N-doped graphene (Fig. 9.7) on Cu(111) surface using the DFT method. All geometry optimization calculations were carried out using PBE-D2 as implemented in Vienna ab initio Simulation Package (VASP) [103,104]. The BSKAN code was used to compute the simulation of STM images [105,106]. The results from both experiments and theoretical calculations suggested that nitrogen dopants were generally pyridinic configuration, which preferably located in the edge of the graphene systems [107].

Theoretical studies are helpful to understand the structural, electronic, and magnetic properties of different N-doping configurations and positions within the rings of the graphene sheet. Combined experimental and computational studies revealed that doping of nitrogen

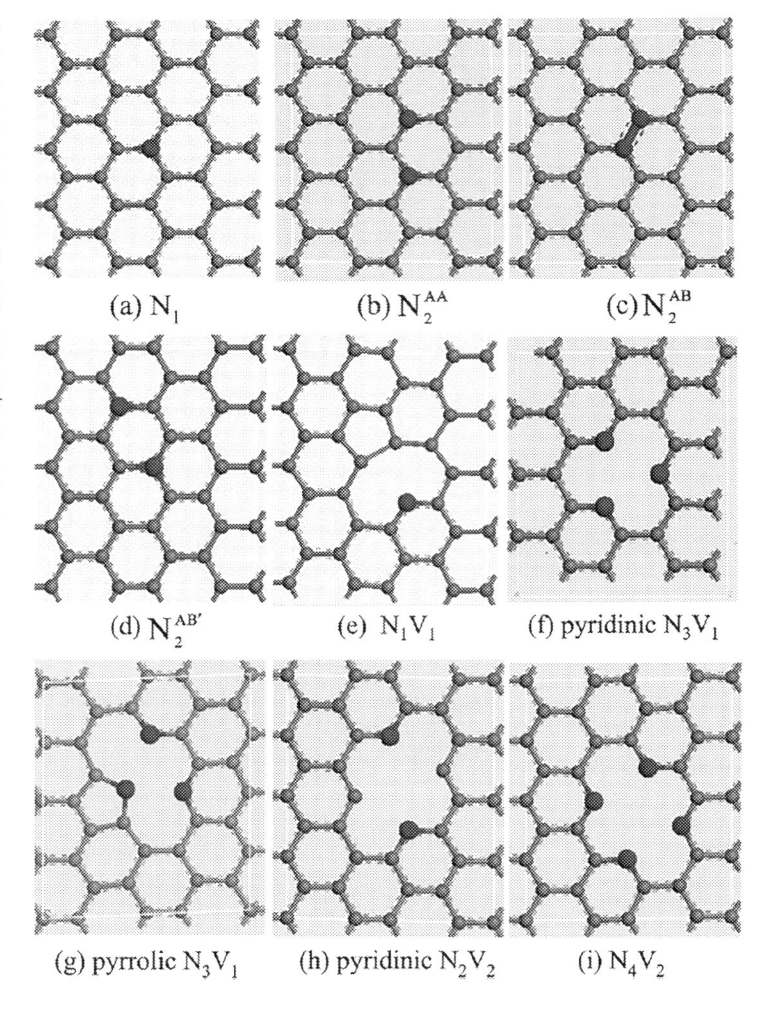

FIG. 9.7 Structures of graphene models with different N-doped configurations: (A) N_1, (B) N_2^{AA}, (C) N_2^{AB}, (D) $N_2^{AB'}$, (E) N_1V_1, (F) pyridinic N_3V_1, (G) pyrrolic N_3V_1, (H) pyridinic N_2V_2, and (I) pyridinic N_4V_2. The small *gray* and *large blue balls* (*dark gray* in the print version) represent carbon and nitrogen atoms, respectively. The superscripts B and B′ refer to, respectively, the first-nearest and third-nearest neighbors of a nitrogen atom represented by A. *Reproduced from Y.-X. Yu, Can all nitrogen-doped defects improve the performance of graphene anode materials for lithium-ion batteries? Phys. Chem. Chem. Phys. 15 (39) (2013) 16819–16827 with permission from the Royal Society of Chemistry.*

atoms into graphene could alter its physicochemical properties depending on the doping configuration within the sublattices [36]. A unique dopant was obtained within the same graphene sublattice (N_2^{AA}) and was characterized by STM images and Raman spectra. The controlled synthesis was carried out by varying the reaction temperature and time to manipulate the doping configuration [36]. The computational study using density-functional theory calculations highlighted that the electronic and adsorption properties of graphene can be changed significantly through substitutional doping with nitrogen and nitrogen decoration of vacancies [99]. The calculated formation energies of different N-doped graphene structures have shown that two nitrogen atoms substituted within the six-membered ring of graphene (N_2^{AA}, N_2^{AB}, $N_2^{AB'}$) are more preferred than nitrogen atoms placed in a defect configuration (Fig. 9.7 and Table 9.3), where B and B′ correspond to the first-nearest and third-nearest neighbors of a nitrogen atom represented by A. The formation energy (ΔE_f) of the defects was computed using Eq. (9.1) as reported in the previous studies.

$$\Delta E_f = E_{GN_x} + (x+y)\mu_C - (E_G + x\mu_N) \tag{9.1}$$

where E_{GN_x} and E_G are the energies of N-doped graphene and the perfect (pristine) graphene, respectively; x is the number of carbon atoms substituted by nitrogen atoms, $x+y$ is the number of carbon atoms removed from the perfect graphene nanosheet to form the defects, μ_C is the total energy of perfect graphene per atom, and μ_N is a half of the total energy of the N_2 molecule in the gas phase.

In the DFT study by Yu [99], the Perdew and Wang (PW91) exchange-correlation functional was used in conjunction with a basis set of double-numerical plus polarization (DNP) for the calculations of N-doped graphene [108,109]. The DMol^3 package was used for all DFT calculations [110]. Table 9.3 provides the formation energy values, the charge transfer to a nitrogen

TABLE 9.3 Simulated formation energies (ΔE_f), charge transfer to the nitrogen atom (Q_N), and C-N (C-C for pristine) distances in N-doped graphene nanosheets.

System	ΔE_f (eV)	$\lvert Q_N \rvert$ (e)	$d_{C\text{-}N}$ (Å)
Pristine	0.00	—	1.42 (C—C)
N_1	0.97	0.21	1.41
N_2^{AA}	2.28	0.30	1.41
N_2^{AB}	3.01	0.19	1.40
$N_2^{AB'}$	1.96	0.29	1.41
N_1V_1	4.90	0.24	1.32
Pyridinic N_3V_1	3.57	0.15	1.34
Pyrrolic N_3V_1	5.35	0.18	1.33, 1.39
Pyridinic N_2V_2	7.20	0.21	1.34
Pyridinic N_4V_2	3.82	0.13	1.33

Reproduced from Y.-X. Yu, Can all nitrogen-doped defects improve the performance of graphene anode materials for lithium-ion batteries? Phys. Chem. Chem. Phys. 15 (39) (2013) 16819–16827 with permission from the Royal Society of Chemistry.

atom (Q_N), and the C-N bond lengths in the N-doped graphene. The data listed in Table 9.3 reveal that the formation energies of both N-substituted graphene and N-decorated vacancies in graphene are lower than those of the single and double vacancies in graphene. The less value of the formation energy means that the structure is more stable. The calculated formation energies of the selected systems are found to follow the trend: $N_1 < N_2^{AB'} < N_2^{AA} < N_2^{AB} <$ pyridinic $N_3V_1 <$ pyridinic $N_4V_2 < N_1V_1 <$ pyrrolic $N_3V_1 <$ pyridinic N_2V_2. It was reported that nitrogen atoms could stabilize single- and double-vacancy formation in graphene [99]. As shown in Table 9.3, the C—N bond lengths for the N-substituted graphene systems (N_1, N_2^{AA}, N_2^{AB}, $N_2^{AB'}$) are about 1.40–1.42 Å, which are very close to the C—C bond length in pristine graphene. In the case of N-doped graphene with single or double vacancies, the C-N bond distances reduce to 1.31–1.34 Å, and to 1.39 Å for the two C—N bonds in the pyrrolic N_3V_1 defect. DFT study revealed that N_1V_1, pyridinic N_3V_1, and pyridinic N_4V_2 defects are nonmagnetic, but pyrrolic N_3V_1 and pyridinic N_2V_2 defects are spin-polarized. The single and double substitutions of carbon atoms by N-dopants in graphene produce nonmagnetic structures since the nitrogen atom forms three σ bonds and the remaining two electrons are paired up [99].

Lv et al. mentioned that sp^2-hybridized carbon atoms in graphene are arranged in two different triangular sublattices [36]. Physicochemical properties of the N-doped graphene could be changed depending on the doping configuration within sublattices. Scanning tunneling microscopy/spectroscopy STM/STS results and computational data reveal that as-synthesized NG sheets contain an abundant amount of two nitrogen atoms doped within the same graphene sublattice (N_2^{AA}). It should be noted that 80% dominance of N_2^{AA} structures was observed among all the identified defects as well as other types of more complex configurations. Lv et al. performed density-functional theory (DFT)-based STM simulations using the Tersoff-Hamann approach within the generalized gradient approximation (GGA) for exchange and correlation to confirm experimentally obtained N-doped graphene structures as shown in Fig. 9.8 [36].

Computational studies focused on understanding the nitrogen-doped regions in graphene. Bond lengths and electron densities are important to determine the active sites of the nitrogen-doped graphene. The C—C bond length of pristine graphene from calculations is 1.42 Å which is in very good agreement with the experimental data [62]. Doping of nitrogen atoms to a perfect hexagonal carbon ring did not disturb the geometrical structure of the ideal graphene and the nitrogen atom seems to reside in a planar sp^2 hybridized configuration. Previous studies provided an explanation that the nitrogen atom has a similar size to the carbon atom and donates electrons to the graphene π-electron system. The N—C bond length of N-doped graphene having graphitic configuration is about 1.41 Å, which is slightly shorter than the C—C bond length of the pristine graphene [31,111–114].

3.3 DFT studies of double-N-doped graphene

Herath and Dinadayalane investigated double nitrogen doping in a finite-sized graphene sheet of $C_{82}H_{24}$ [101]. DFT calculations were performed to examine 21 isomers (Scheme 9.2) using B3LYP and M06-2X hybrid functionals in conjunction with 6-31G(d) basis set. The relative stabilities of these isomers of N-doped graphene structures shown in Scheme 9.2 were analyzed to determine the positional preferences of the doping site. For comparison, B3LYP

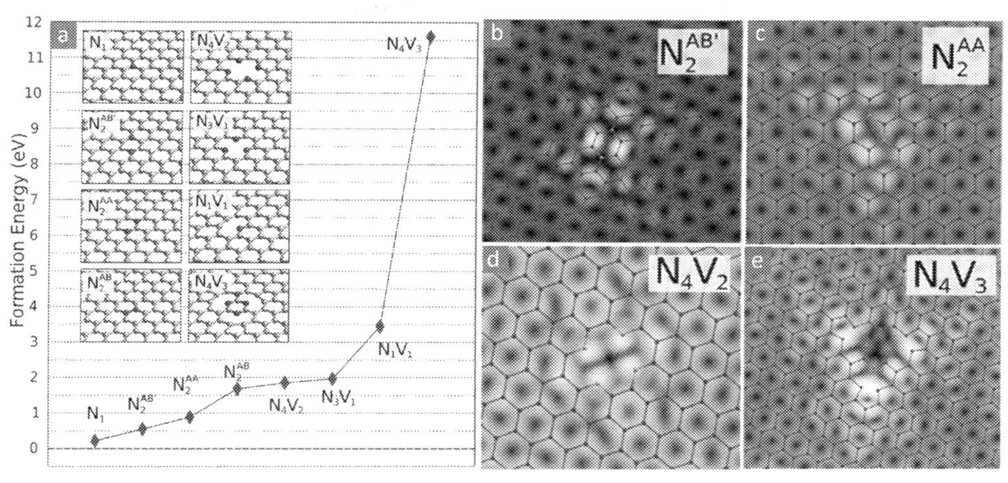

FIG. 9.8 Calculated formation energy, experimental, and simulated STM images of different N-doping configurations. (A) Formation energies of different N-doping configurations in NG sheets (as illustrated in *insets*) computed using *ab initio* calculations. (B–E) Simulated STM images depicting two different atomic configurations for double substitution of nitrogen dopants (B and C) and for two pyridine-like N-dopants (D–E). The biases are as follows: (B) −1.0 eV, (C) −1.0 eV, (D) −0.7 eV, and (E) −0.7 eV. The carbon and nitrogen atoms are illustrated using *gray and cyan balls* (*dark gray* in the print version), respectively. The superscript B′ is used to differentiate between two N atoms as being the first-nearest neighbors (N_2^{AB}) or third-nearest neighbors ($N_2^{AB'}$). *Reprinted by permission from Springer Nature. Scientific Reports, Ref. R Lv, Q. Li, A.R. Botello-Méndez, T. Hayashi, B. Wang, A. Berkdemir, Q. Hao, A.L. Elías, R. Cruz-Silva, H.R. Gutiérrez, Y.A. Kim, H. Muramatsu, J. Zhu, M. Endo, H. Terrones, J.-C. Charlier, M. Pan, M. Terrones, Nitrogen-doped graphene: beyond single substitution and enhanced molecular sensing. Sci. Rep. 2 (2012) 586–586, Copyright 2021.*

and M06-2X functionals were used to calculate the relative energies of 21 structures, and the results are shown in Fig. 9.9. Both functionals produce similar trends. The structure **2NG-19** is the most preferred in which the two nitrogen dopants are far in distance. On the contrary, the least favorable structure is **2NG-2** where the structure possesses 1,2-substitution of two nitrogen atoms, called ortho-like substitution [101]. This supports the experimental observation of two nitrogen atoms at the 1,3 (meta) and 1,4 (para) positions in a single six-membered ring of the graphene [36]. Doping of two nitrogen atoms at 1,4-positions (**2NG-3**) was reported to be slightly more preferred than 1,3- or 1,2-positions for the finite-sized graphene sheet. DFT results clearly indicate that N-N bond does not prefer to form in N-doped graphene as reported previously by Chaban and Prezhdo [115]. The doping energy of several two nitrogen-doped graphene configurations reduces when the two nitrogen dopants are placed away [31].

The electronic structure of monolayer graphene can be tuned by introducing electron-rich species to the system, such as nitrogen atoms. Many of the available theoretical studies provide knowledge on the effect of nitrogen doping in graphene on the energy gap between the highest occupied molecular orbital (HOMO) and the lowest unoccupied molecular orbital (LUMO). Early theoretical studies demonstrated that N-doping can modify the band structure of the zero bandgap graphene. The HOMO-LUMO bandgap opening was considered negligible at a low concentration of nitrogen dopants in graphene, whereas increasing the concentration of substituted nitrogen may effectively tune the bandgap [17,116,117].

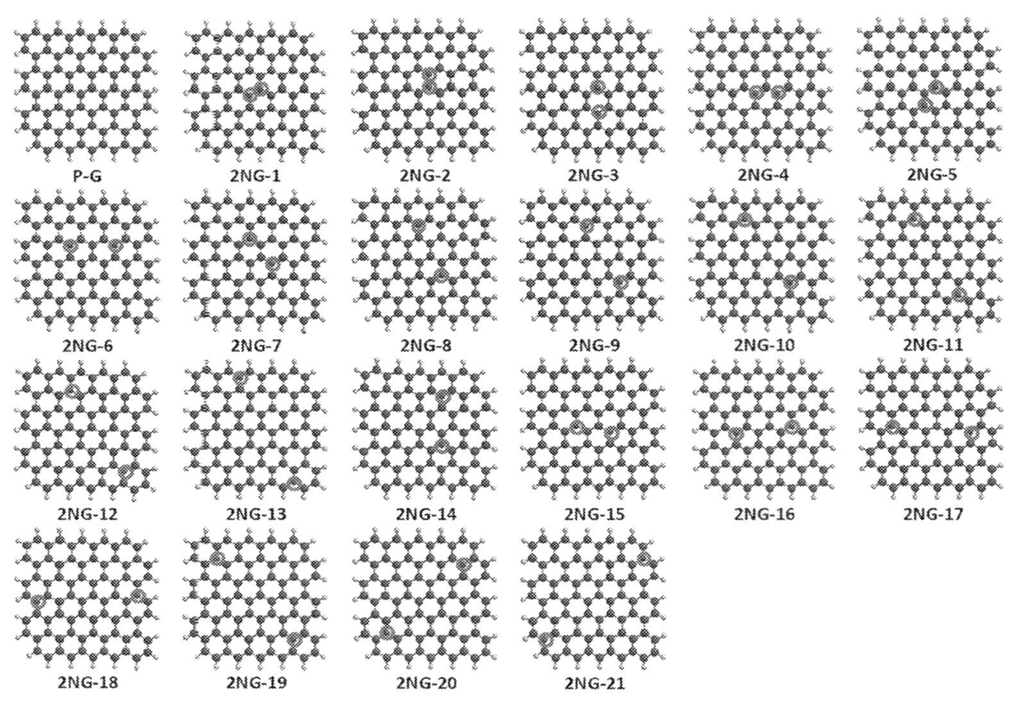

SCHEME 9.2 Structures of pristine graphene (P-G) and two nitrogen-doped graphene structures. *Red circles* (*light gray* in the print version) indicate the positions of nitrogen atoms in double-N-doped graphene. *Reprinted by permission from Springer Nature Customer Service Centre GmbH: Springer Nature, Journal of Molecular Modeling, ref. D. Herath, T. Dinadayalane, Computational investigation of double nitrogen doping on graphene. J. Mol. Model. 24 (1) (2018) 26, copyright Springer-Verlag GmbH Germany, part of Springer Nature 2017.*

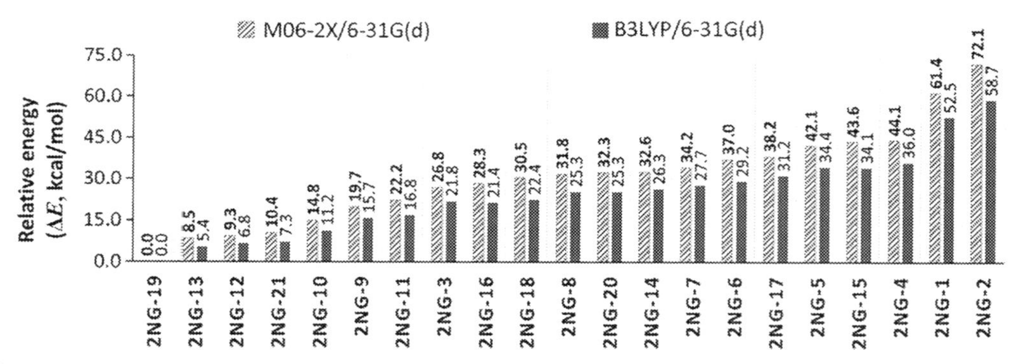

FIG. 9.9 Relative energies (ΔE, kcal/mol) of double-N-doped graphene obtained at the M06-2X/6-31(d) and the B3LYP/6-31(d) levels. See Scheme 9.2 for the naming of the structures. The graph is plotted with increasing order of relative energy. *Reprinted by permission from Springer Nature Customer Service Centre GmbH: Springer Nature, Journal of Molecular Modeling, ref. D. Herath, T. Dinadayalane, Computational investigation of double nitrogen doping on graphene. J. Mol. Model. 24 (1) (2018) 26, Copyright Springer-Verlag GmbH Germany, part of Springer Nature 2017.*

To calculate HOMO and LUMO energies, and the HOMO-LUMO energy gaps, Herath and Dinadayalane have used the TPSSh [118,119] functional that was developed by Tao, Perdew, Staroverov, Scuseria, and HSE06 [120,121] functional developed by Heyd, Scuseria, and Ernzerhof [101]. 6-31G(d) basis set was used to run single-point energy calculations with TPSSh and HSE06 functionals using B3LYP/6-31G(d)-level optimized geometries [101]. All calculations at the DFT level were carried out using NWCHEM ver.6.6 software program [122]. It is worth mentioning that the findings from the frontier molecular orbitals based on these both levels were varied from the pristine graphene.

The HOMO-LUMO energy gap is notably increased for the structures **2NG-19** and **2NG-13** compared to pristine graphene as depicted in Fig. 9.10. In general, double-N-doped structures yielded a higher HOMO-LUMO energy gap than the pristine graphene indicating that controlled, site-specific N-doping could open the bandgap of graphene [101]. Computational results showed that the location of the nitrogen doping in the network of the graphene layer played an important role in the electronic properties [101]. It could be possible to design experiments to produce some of the double-nitrogen-doped graphene structures reported. The knowledge on nitrogen-doped graphene will be useful in producing metal-free nitrogen-doped graphene-based catalysts for fuel cells and many other applications [42,101].

3.4 Thermal stability of N-doped graphene structures by increasing N concentration

Shi et al. used DFT, density-functional tight binding (DFTB), cluster expansion, and molecular dynamics approaches to investigate the thermal stability and electronic properties of a binary two-dimensional alloy of graphitic carbon and nitrogen [102]. They used the local spin density approximation (LSDA) with projector-augmented wave (PAW) potentials as implemented in VASP [103,104] to obtain total energies and band structures. The results of

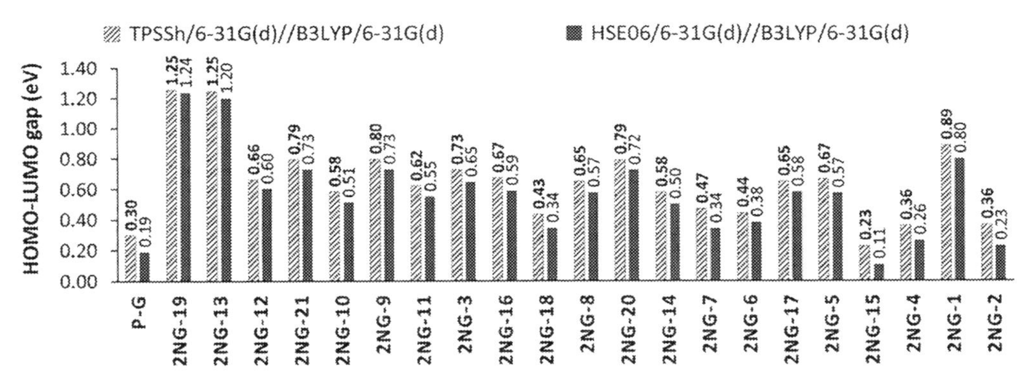

FIG. 9.10 HOMO-LUMO energy gaps (in eV) obtained for pristine graphene (P-G) and the double-nitrogen-doped graphene structures at TPSSh/6-31G(d) and HSE06/6-31G(d) levels using the B3LYP/6-31G(d)-level optimized geometries. *Reprinted by permission from Springer Nature Customer Service Centre GmbH: Springer Nature, Journal of Molecular Modeling, D. Herath, T. Dinadayalane, Computational investigation of double nitrogen doping on graphene. J. Mol. Model. 24 (1) (2018) 26, copyright Springer-Verlag GmbH Germany, part of Springer Nature 2017.*

phonon and MD calculations predicted that the percentage of N-doping >37.5% will lead to the unstable structures of graphene due to the strong repulsive interactions between nitrogen dopants. The largest achievable N-doping concentration of N-doped graphene is 37.5% [123]. The high-temperature ab initio MD simulations at 1500 K based on DFT reveal that the 8×8 supercell structures with nitrogen concentrations of 8.33%, 12.5%, and 16.7% remain stable. Thus, the N-doped graphene with these percentages of N concentration will most likely be very stable under the experimental conditions and may be realized easily. It should be noted that Deng et al. reported the experiments on nitrogen doping of graphene with N of 16.4% [60].

As shown in Fig. 9.11, the structures with N concentrations higher than 16.7% in graphene at 1500 K are unstable. This confirms the report of Xiang et al. for the structural stability of N-doping of 25% based on MD simulations at 500 K. N-doped graphene structure with N concentration of 33.3% was reported unstable at 1500, 1000 as well as 500 K as shown in Fig. 9.11. The breaking of the carbon-carbon (C—C) bond was reported to be responsible for the thermal decomposition of N-doped graphene structures. As the N concentration increases from 8.33% to 37.5%, the strain has a greater destabilizing effect on the C—C bonds and the strain was released by out-of-plane relaxations of carbon atoms as shown in Fig. 9.11B–E [123].

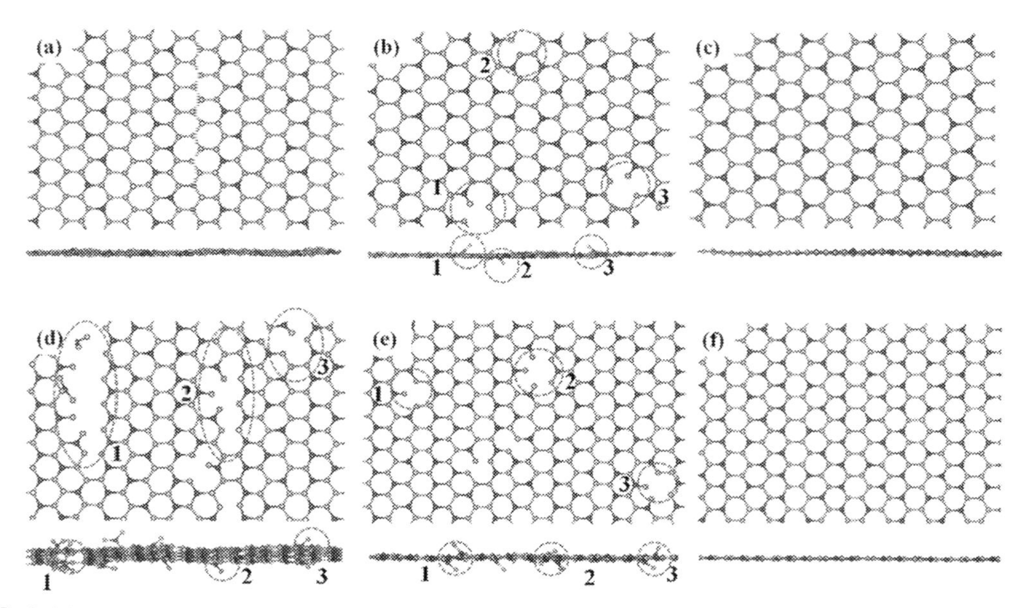

FIG. 9.11 *Top* and *side view* of snapshots of the selected low-energy 2D CN alloy structures as predicted by the cluster expansion after 10 ps of the MD simulation in the NVT ensemble using a 8×8 supercell for N concentrations of (A) 16.7% at 1500 K, (B) 25% at 1500 K, (C) 25% at 500 K, (D) 33.3% at 1500 K, (E) 33.3% at 500 K, and (F) 33.3% at 100 K. The *red circles* (*light gray* in the print version) mark the disrupted areas in each unstable configuration; the numbers are used to set the correspondence between the areas in the *top* and *side views*. *This was reproduced from Z. Shi, A. Kutana, B.I. Yakobson, How much N-doping can graphene sustain? J. Phys. Chem. Lett. 6 (1) (2015) 106–112 with permission. Copyright 2015 American Chemical Society.*

3.5 Magnetic properties of nitrogen-doped graphene

Magnetism can be induced in graphene due to the p_z electron imbalance in bipartite hexagonal sublattice, electron localization at the zigzag edge, and heteroatom doping [100,124]. Nitrogen doping was reported to be a promising approach to manipulate magnetism in graphene [31,100,124]. Błoński et al. reported the experimental evidence of ferromagnetic graphene achieved by controlled doping with graphitic, pyridinic, and chemisorbed nitrogen. Their results indicate that magnetic properties strongly depend on both the nitrogen concentration and the type of N-doping in graphene structures [100]. Hou et al. have also confirmed that nitrogen concentration and structural configuration play a critical role in the magnetic properties of N-doped graphene systems [31]. Nitrogen dopants at concentrations below 5 at.% in graphene are nonmagnetic [31,100]. Wang et al. using DFT calculations predicted nonmagnetic behavior for graphene doped with 2.1 at.% of nitrogen [125]. The computational study indicated the magnetic structure with a magnetic moment of ~0.2 μB per supercell of graphene in which N atoms substituted C atoms at para positions (the nitrogen content was 4.2 at.%) [100].

Babar and Kabir reported a systematic study [124] using spin-polarized density-functional theory calculations within the projector-augmented wave (PAW) formalism as implemented in the Vienna ab initio simulation package (VASP) [103,104]. Perdew-Burke-Ernzerhof (PBE) [126] and Heyd-Scuseria-Ernzerhof (HSE06) [120] functionals were used to calculate the electronic and magnetic properties. The computational study highlighted that the electronic and magnetic descriptions of the defects and nitrogen and/or boron doping in graphene were observed to be sensitive to the treatment of exchange-correlation energy as we compare the results from the PBE, hybrid HSE06, and SCAN meta-GGA functionals [124].

Babar and Kabir revealed that the long-range magnetic interactions were found to depend on the N concentration, defect type, sublattice, separation, and orientation. They mentioned that the magnetism in the graphitic N-doping is fundamentally very different from the vacancy-containing N-doping. Furthermore, they proposed the basis for the ferromagnetic order that originates from the direct exchange between the extended magnetic states of the graphitic defects. In the case of N-doped graphene systems, the trimerized pyridine, where three N atoms occupy the same graphene sublattice surrounding a vacancy, was reported to be energetically more favorable than the monomerized and dimerized pyridines. In the case of trimerized pyridine defect in graphene, the Bader analysis indicated a 0.9–1.12 e/N charge transfer to nitrogen from carbon. Figs. 9.12 and 9.13 show that both the electronic structure and the magnetism in the vacancy-containing N-doped graphene are fundamentally different from the graphitic N-doped graphene systems. The n-type band structure was observed in graphitic N-doped graphene, whereas p-type band structure was reported for vacancy-containing N-doped graphene (Fig. 9.12).

Vacancies in the same sublattice are coupled ferromagnetically. However, they prefer antiferromagnetic alignment when placed on the different sublattices. Single graphitic nitrogen defect is similar to the pristine vacancy. When two substitutional N atoms are placed on the same sublattice, the spin configuration is ferromagnetic. However, antiferromagnetic behavior was observed by placing two substitutional N atoms on different sublattices in graphene. DFT study predicted an experimental strategy to increase the concentration of the graphitic

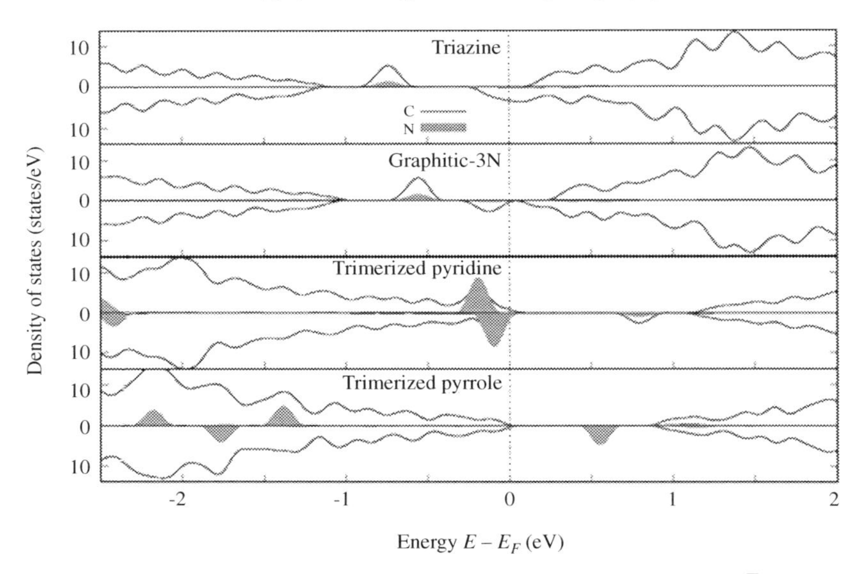

FIG. 9.12 Spin-polarized projected density of states calculated using the $(10 \times 5) - (\sqrt{3}, 3)a_0$ supercell for the graphitic and vacancy-containing N graphene systems. Substitutional N-doping opens up a gap at the Dirac point E_D. While the electronic structures are n-type for the graphitic defects, the vacancy-containing N-doped defects (e.g., pyridine and pyrrole) display p-type semiconducting behavior. The curves above or below the density of states' value = 0 line refer to the spin-up or spin-down components, respectively. *Reprinted with permission from R. Babar, M. Kabir, Ferromagnetism in nitrogen-doped graphene. Phys. Rev. B 99 (11) (2019) 115442. Copyright 2019 by the American Physical Society.*

defects to produce robust ferromagnetism with improved Curie temperature. A small amount of boron codoping was reported to further enhance the ferromagnetism of N-doped (graphitic type) graphene and this is in agreement with the experimental report [124].

3.6 Simulated X-ray photoelectron spectra (XPS) for N-doped graphene

The presence of nitrogen in N-doped graphene was confirmed by high-resolution C(1s) X-ray photoelectron spectroscopy (XPS). A peak at a binding energy of around 285.5 eV corresponds to the C-N bond in N-doped graphene. Here, we provide an example of how simulations complement experiments. The simulated spectra were helpful in elucidating doping configurations of samples by comparing them to experimental XPS spectra [100]. Furthermore, good agreement was shown between experimental XPS in Fig. 9.3 and simulated XPS in Fig. 9.14 for N1, N2, and N3 peak ranges. The values of XPS peaks for different types of N-doped graphene are given (as shown in Fig. 9.14): pyridinic nitrogen (at ~398.3 to ~398.5 eV), graphitic nitrogen (at ~401.0 to ~401.5 eV), chemisorbed N/N_2 (at ~404.5 to ~405.5 eV), and pyrrolic nitrogen (at ~400.0 eV).

The combined experimental and computational study reported that the presence of N/N_2 was also confirmed by thermogravimetric analysis (TGA) and evolved gas analysis (EGA) techniques with gas electron ionization mass spectrometry [100]. A small shift in

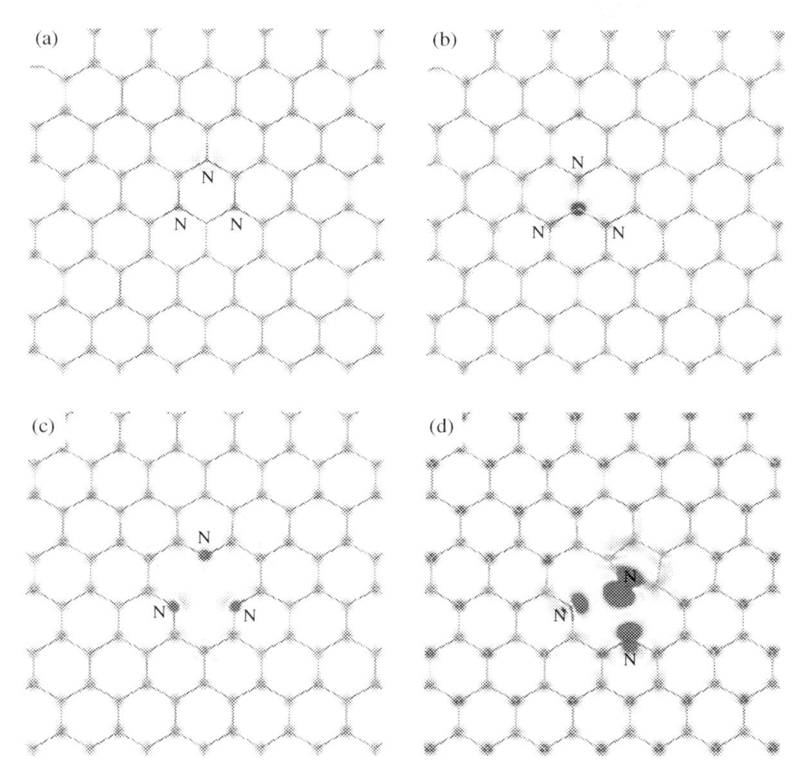

FIG. 9.13 Magnetization density for graphitic and vacancy-containing N complexes: (A) triazine, (B) graphitic-3N, (C) trimerized pyridine, and (D) trimerized pyrrolic defect. *Red* (*light gray* in the print version) and *blue colors* (*dark gray* in the print version) indicate the up and down spin density, respectively. While the magnetism in graphitic N-doped graphene is mostly itinerant in nature, the magnetization density demonstrates that a very significant fraction of the moment is localized at the N sites for the vacancy-containing N-doped graphene. *Reprinted with permission from R. Babar, M. Kabir, Ferromagnetism in nitrogen-doped graphene. Phys. Rev. B 99 (11) (2019) 115442. Copyright 2019 by the American Physical Society.*

the maximum of the C-N peak witnessed for the three N-doped graphene structures can be explained in terms of impossibility to distinguish differently coordinated nitrogen atoms with carbon atoms in graphene with similar binding energy values in the C(1s) domain and the significant overlap of the C-N and C-O spectral components [100]. In the available literature, N(1s) core-level signal was most often interpreted by fitting the experimental spectra of three peaks to graphitic, pyrrolic, and pyridinic nitrogen doped in graphene [127]. The results of Hartree-Fock calculations reveal that graphitic and pyrrolic N(1s) core-level binding energies of XPS are too close to be distinguished. Therefore, the general interpretation of three peaks of the N(1s) XPS in N-doped graphene may be an artifact caused by the fitting procedure. On the basis of a combined computational-experimental study, Figueras et al. urge caution in interpreting XPS spectra of N-doped graphene because graphitic and pyrrolic N types cannot be distinguished with enough accuracy to provide physical meaning [127].

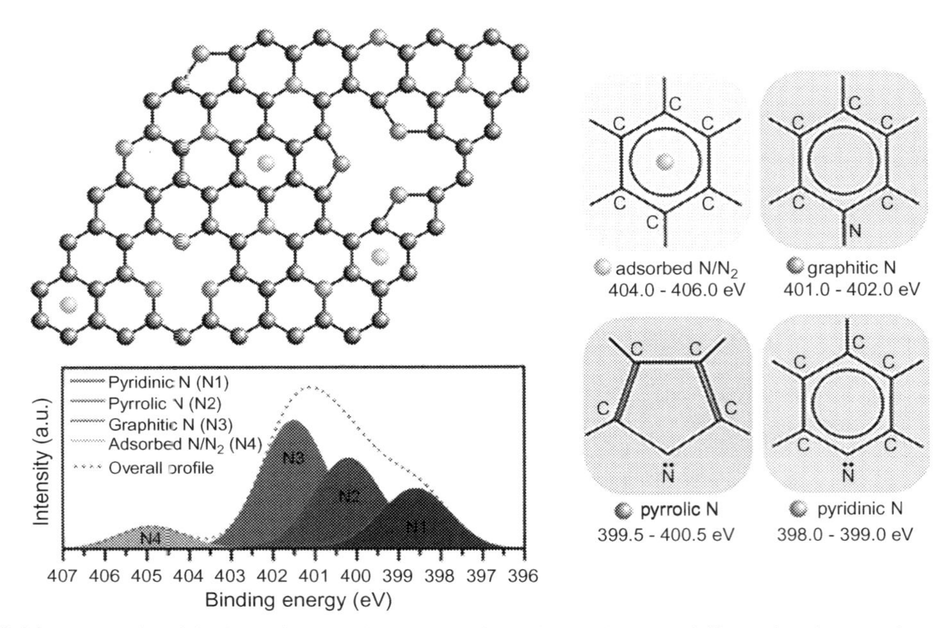

FIG. 9.14 A simulated high-resolution N(1s) XPS with a scheme showing different bonding configurations of nitrogen in N-doped graphene and corresponding peaks. *This was reproduced from P. Błoński, J. Tuček, Z. Sofer, V. Mazánek, M. Petr, M. Pumera, M. Otyepka, R. Zbořil, Doping with graphitic nitrogen triggers ferromagnetism in graphene. J. Am. Chem. Soc. 139 (8) (2017) 3171–3180 with permission (https://pubs.acs.org/doi/10.1021/jacs.6b12934). Copyright 2017 American Chemical Society.*

3.7 N-doped graphene bilayers

Bilayer graphene structures of both AA and AB stacking motifs are known to be separated at a distance corresponding to van der Waals (vdW) interactions. Tian et al. reported unique chemical bonding between N-doped graphene systems using DFT calculations. They also examined the effect of changing the N-doping levels on the interlayer separations. Graphitic N-doping can induce covalent-like π-π bonding (pancake bonding) in N-doped graphene bilayers [128]. The interlayer binding energies for N-doped graphene bilayer were observed to be increased significantly compared to pristine graphene bilayers. In the study by Tian et al. [128], DFT calculations were performed using the projector-augmented wave (PAW) method [129], and the exchange and correlation interactions of valence electrons were described with the Perdew-Burke-Ernzerhof (PBE) [126] functional as implemented in the Vienna Ab Initio Simulation Package (VASP) [103,104]. Grimme's dispersion corrections [130,131] of D2 and D3, and Tkatchenko-Scheffler (TS) methods without and with self-consistent screening (TS+SCS) [132,133] were included with PBE functional to address van der Waals interactions in bilayer N-doped graphene systems. In this study, monolayer was taken with 32 atoms in $4a \times 4b$ supercell. The bilayer supercells with 64 atoms were stacked with AA and AB types as shown in Fig. 9.15. The computational study revealed the existence of a competition between π-π bonding and reduced exchange repulsion favoring AA and AB stacking configurations, respectively.

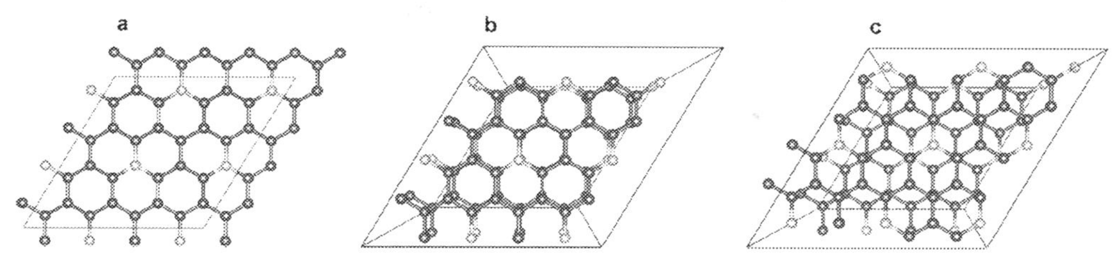

FIG. 9.15 Structural models for N-doped graphene. (A) The monolayer consists of 28 carbon and 4 nitrogen atoms per unit cell. (B) and (C) Perspective views of the AA- and AB-stacked bilayers. *This was reproduced from Y.-H. Tian, J. Huang, X. Sheng, B.G. Sumpter, M. Yoon, M. Kertesz, Nitrogen doping enables covalent-like π–π bonding between graphenes. Nano Lett. 15 (8) (2015) 5482–5491 with permission. Copyright 2015 American Chemical Society.*

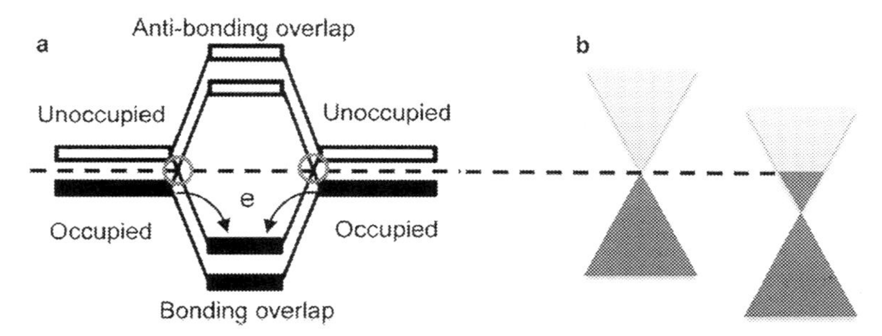

FIG. 9.16 Schematic diagram showing 2D covalent-like π-π bond formation. (A) Energy-level splittings around the Fermi level E_f (*dashed line*) lead to level crossings (*red circles* (*light* in the print version)) as a result of the interlayer π-π orbital overlaps. The occupied and empty levels in the monomers (in this case the monolayer) are artificially separated by a gap to guide the eye. (B) Relative positions of E_f between the pristine and the N-doped graphenes. E_f is shifted from the Dirac point where the density of states (DOS) is low to a place with enhanced DOS, making for a significant π-π bonding effect. *This was reproduced from Y.-H. Tian, J. Huang, X. Sheng, B.G. Sumpter, M. Yoon, M. Kertesz, Nitrogen doping enables covalent-like π–π bonding between graphenes. Nano Lett. 15 (8) (2015) 5482–5491 with permission. Copyright 2015 American Chemical Society.*

The high binding affinity for N-doped graphene bilayers was attributed to the π-π orbital overlaps (Fig. 9.16) across the vdW gap, whereas the individual layers maintain their in-plane π-conjugation. Fig. 9.16 shows a qualitative description of the π-π bonding formation in 2D systems of N-doped graphene bilayers. Electronic structure calculations and crystal orbital overlap population (COOP) analyses confirm the existence of interlayer covalent-like bonding. It has been demonstrated that π-π bonding between AA-stacked bilayers in N-doped graphene can raise the interlayer binding energies by almost 50% in comparison with the pristine graphene bilayers.

3.8 Graphene nanoribbons (GNRs)

DFT calculations and molecular dynamics (MD) simulations were used to investigate nitrogen doping precisely on the edge of graphene nanoribbons (GNRs) [134]. Different types of nitrogen dopants (NDs) were determined using ammonia (NH_3) and acetonitrile (CH_3CN)

as a nitrogen source in MD simulations. In the case of DFT calculations, acetonitrile was used as a nitrogen source for N-doped graphene. The MD simulations were performed using LAMMPS package [135] at a timescale of 40 ns. DFT calculations were performed using soft projector-augmented wave (PAW) pseudopotentials [136] and PBE functional [126] as employed in QUANTUM ESPRESSO software [137,138]. Four types of edges were considered: armchair (A), zigzag (Z), AC, and ZW edges, where the letters "C" and "W" represent the shapes formed by the removal of carbon atoms from the edge [134].

Nitrogen dopants (NDs) can be categorized into graphitic, pyrrolic, pyridinic, pyrazole, amine, and nitrile types as displayed in Fig. 9.17. Among them, only the graphitic NDs are in the plane of the graphene. All of the others are located on the edges. In the case of NH_3 as the nitrogen source, most of the doping configuration belonged to the amine-type doping. On the contrary, pyrrolic and pyridinic dopants were dominant among other doping configurations when CH_3CN was used as the nitrogen source.

MD results revealed that the ammonia produces almost all amine-type dopants, whereas the acetonitrile produces a considerable amount of pyrrolic and pyridinic nitrogen dopants. DFT calculation clarified the reaction mechanism in these different types of the GNRs when using acetonitrile as the nitrogen source. DFT calculations showed the thermodynamic stability of different types of nitrogen-doped graphenes and the results agree with the MD simulations [134]. Fig. 9.18 shows the quantitative analysis of the pyrrolic and pyridinic NDs in these four types of the GNRs. The number of the NDs in all four types of the GNRs rapidly increases at the beginning of 10-ns and then stabilizes after 20-ns simulation, which indicates that reaction is completed in such a timescale. The atomic percentage (at%) of the nitrogen

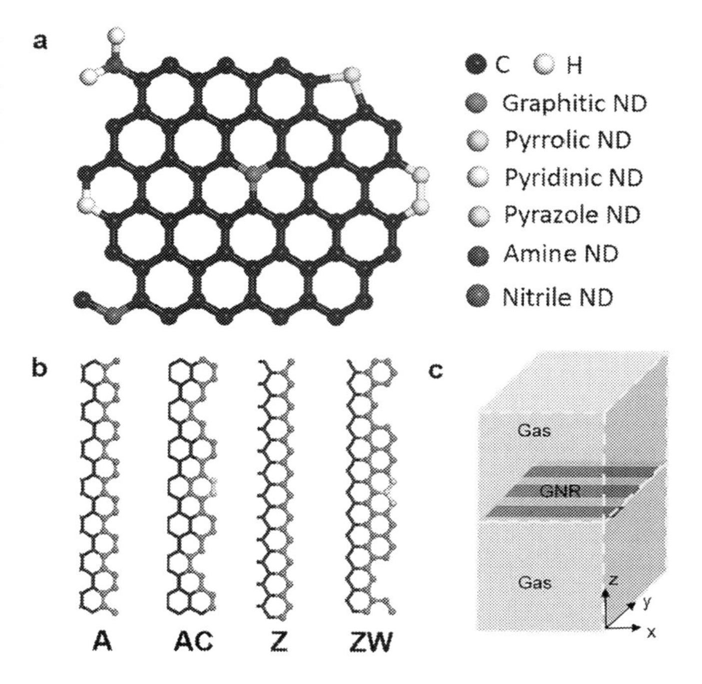

FIG. 9.17 (A) Different types of NDs in graphene. (B) Different types of edges in GNRs. The carbon atoms on the edges are marked with *red* (*light gray* in the print version). The *gray atoms* indicate the carbon atoms to be removed in order to form AC and ZW types of edges. (C) Scheme of a system setup containing the GNRs and gas molecules serving as the nitrogen dopant sources. *Republished with permission of Institute of Physics, IOP Publishing, Ltd., from ref. Y. Dong, M.T. Gahl, C. Zhang, J. Lin, Computational study of precision nitrogen doping on graphene nanoribbon edges. Nanotechnology 28 (50) (2017) 505602; permission conveyed through Copyright Clearance Center, Inc.*

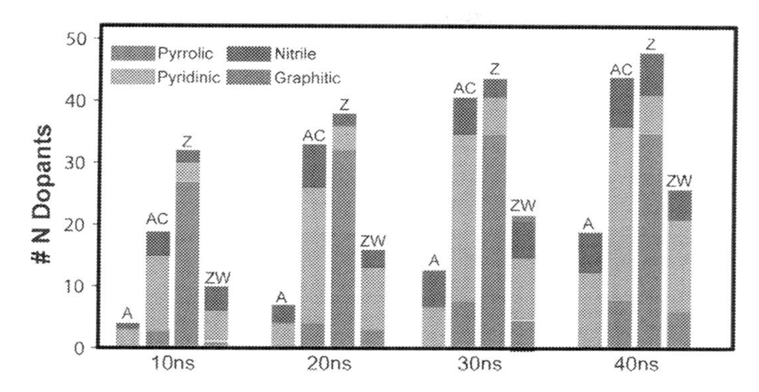

FIG. 9.18 A number of nitrogen dopants for different types of nitrogen doping in A-, AC, Z-, and ZW-GNRs when using CH_3CN as the nitrogen source after 40-ns annealing. *Republished with permission of Institute of Physics, IOP Publishing, Ltd., Y. Dong, M.T. Gahl, C. Zhang, J. Lin, Computational study of precision nitrogen doping on graphene nanoribbon edges. Nanotechnology 28 (50) (2017) 505602; permission conveyed through Copyright Clearance Center, Inc.*

atoms is 5.53 at%, 4.49 at%, 3.70 at%, and 4.61 at% for A-, AC-, Z-, and ZW-GNRs, respectively. It is clearly shown that amine (—NH_2) and imine (=NH) types are the major groups of the nitrogen dopant types. Based on computational modeling, the proposed mechanism of controllable nitrogen doping on the edges of the GNRs was expected to provide the guidance for experimental realization of controlled N-doped graphene nanoribbons.

The importance of heteroatom-doped graphene (e.g., N, B, and S doped) has been increasing in recent years. The computational methods including MD simulations help to understand the stability and reaction mechanism for obtaining doped graphene. DFT calculations were used to trace precisely specific reaction paths, and they were also used to study the synthesis mechanism of graphene-based materials and beyond. Thus, the computational exploration of reaction mechanisms could screen the optimum precursors and thermodynamic reaction conditions. Further, the knowledge from computational results could guide the design of experiments to obtain a high yield of desired material components and structures for a wide range of applications.

4 Applications of N-doped graphene

N-doped carbon materials, such as N-doped graphite (NGt), N-doped graphene (NG), and N-doped carbon nanotube (NCNT), have been successfully applied in supercapacitors [83,139], lithium-ion batteries [51], advanced catalyst support [140], or as active electrocatalysts of oxygen-reduction reactions (ORRs) in fuel cells [49,72,82]. It is important to understand the relationships between the molecular-scale structures and functions of the graphene-based energy materials for realizing efficient, economical, and safe products involving graphene. Application of computational chemistry aids understanding structure-property relationships of graphene-based materials including N-doped graphene for fuel cells, batteries, photovoltaics, and supercapacitors [141]. Nitrogen-doped carbon-based nanomaterials exhibit promising applications in different areas as shown in Fig. 9.19. Nitrogen doping in carbon-based nanomaterials generates extraordinary properties with electronic surfaces or active catalytic sites due to the electronegativity difference between carbon and nitrogen and also alters the

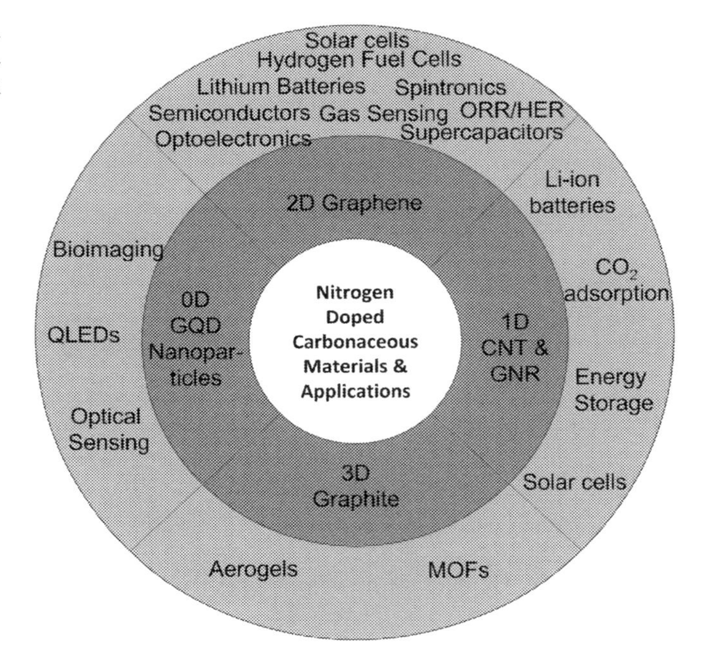

FIG. 9.19 Some of the potential applications or significance of N-doped carbon-based materials of different dimensional systems.

electronic properties, which have a great potential in design and fabrication of tunable devices. Jeong et al. reported histidine-derived nitrogen-doped carbon materials and their electrochemical capacitive performance in energy storage applications. These N-doped materials were identified as rolled planar structures as well as thick 2D-like planar structures with a specific surface area of 455 m^2/g. These materials showed a notable specific capacitance of 58 F/g at current densities of 0.1 A/g and superior stability without deterioration of performance values up to 2000 cycles [142].

Among different low-dimensional materials, N-doped graphene materials show a wide range of applications. Nitrogen-doped graphene systems were reported as active materials for Li-ion batteries [51,143]; fuel cells [26,49]; field-effect transistors [144]; ultra- and supercapacitors [83,145,146]; and in the fields of photocatalysis [147], small molecule/gas sensing [32,36,148], and electrochemical biosensing [82]. In substitutional nitrogen doping of graphene, n-type doping is common [20]. The n-type semiconductor properties are weakened when electron-accepting pyridinic and pyrrolic nitrogen doping coexist with electron-donating substitutional nitrogen doping. Thus, the electron mobility decreased for N-doped graphene compared to pristine graphene. Ultrahigh-electron-donating substitutional N-doping enhances the electrochemical performance of devices [149].

Pyrrolic nitrogen plays an important role in enhancing capacitance [27]. Nitrogen-doped graphene with pyrrolic-type defect was reported to have a high potential for highly efficient next-generation supercapacitors. Nitrogen doping in graphene increased the electrical double-layer supercapacitance to as high as 194 F g^{-1}. Temperature has a crucial impact on the supercapacitor performance of N-doped graphene, and the optimized results were achieved at 130 °C with a high supercapacitance of 194 F g^{-1}. Energy density was estimated to be as high as 6.7 Wh kg^{-1} and the maximum power density reached 3.7 kW kg^{-1} [150]. In the case of N-doped graphene, pyridinic-N and graphitic-N types can increase the total capacitance by increasing quantum capacitance, but the pyrrolic configuration limits the total capacitance due to its much lower quantum capacitance than the other two nitrogen-doping configurations. Thus, it is possible to optimize the capacitance by controlling the type and concentration of N-doping [145].

As demonstrated theoretically by Yu, nitrogen-doped graphene without and with defects could enhance the adsorption of lithium ions for lithium-ion battery applications [99]. Pyridinic N_2V_2 defect in graphene system has the highest lithium adsorption capacity of 249 mA h g^{-1} among the others. The pyridinic N_2V_2 defect in graphene was reported to show a ferromagnetic spin structure with a high magnetic moment and magnetic stabilization energy. All nitrogen-doped defects do not essentially improve the capacity of the lithium-ion batteries [99]. Nitrogen-doped graphene films were reported to enhance Li-battery performance. Reversible discharge capacity of N-doped graphene was noted approximately double compared to pristine graphene due to the large number of surface defects induced by N-doping. The high-performance N-doped graphene electrodes could find vast applications in flexible thin-film batteries [51].

N-doped graphene exhibits much higher electrocatalytic activity toward oxygen reduction and H_2O_2 reduction than graphene [42]. Furthermore, it showed much higher durability and selectivity than the widely used expensive Pt for oxygen reduction. The excellent electrochemical performance of N-doped graphene is attributed to doped nitrogen and the specific properties of graphene. Thus, N-doped graphene has promising applications in electrochemical energy devices (fuel cells, metal-air batteries) [151,152] and biosensors [42]. The oxygen reduction reaction (ORR) is a key reaction in many electrochemical energy devices such as fuel cells and metal-air batteries. H_2O_2 is a general enzymatic product of oxidases and a substrate of peroxidases, which are important in biological processes and biosensors. H_2O_2 is also an essential substance in food, pharmaceutical, clinical, industrial, and environmental analysis. One of the advantages of N-doped graphene over Pt as an ORR electrocatalyst is that oxygen reduction on N-doped graphene is not influenced by fuel molecules (e.g., methanol) unlike Pt. The high selectivity of N-doped graphene toward ORR makes it very promising in direct liquid fuel cells [42]. Nitrogen-doped graphene with sheet-like nanostructures were shown to improve the ORR performance in nonaqueous Li-O_2 batteries and obtained a comparable electrocatalytic ability to Pt/C catalysts [151].

Nitrogen-doped graphene systems were demonstrated to show high catalytic activity. Computational study revealed the selective oxidation of ethylbenzene to acetophenone over nitrogen-doped graphene that serves as a catalyst [98]. This study provides insight into the characteristics of the active sites on the surface of the catalyst and produces information on the mechanistic details [98]. The combined experimental and computational study indicated that the binding of Pt to N-graphene improves the catalytic durability of Pt in ORR [26]. Nitrogen-doped tetragonal-shaped single-crystal graphene was reported to have electromagnetic properties useful toward n-type semiconductors. High electrocatalytic activity and a smooth two-step reaction have suggested their potential as catalysts for hydrogen fuel cells [153]. Geng et al. revealed that nitrogen-doped graphene can be used as a metal-free catalyst for oxygen reduction. The N-doped graphene catalyst had a very high oxygen reduction reaction (ORR) activity through a four-electron transfer process in oxygen-saturated 0.1 M KOH [72]. Nitrogen-doped three-dimensional graphene aerogel was reported as a highly efficient electrocatalyst for methanol oxidation [154]. Nitrogen-doped graphene of graphitic-N-type displayed ferromagnetism. Graphitic nitrogen was found to be a significant source for the magnetic properties. Applications in the field of spintronics and optoelectronics can be realized by ferromagnetic N-doped graphene [100].

Graphene-based materials were reported as electrochemical, ultrasensitive explosive, gas, glucose, and biological sensors for various molecules with greater sensitivity, selectivity, and a low limit of detection [155–157]. Doping graphene with heteroatoms (B, N, P, and S) alters the electronic and chemical properties of graphene. Changes in these properties make graphene-based materials suitable for the construction of economical sensors of practical utility. Heteroatom-doped graphene materials can be used as sensors for the detection of NH_3, NO_2, H_2O_2, heavy metal ions, dopamine, bleomycin sulfate, acetaminophen, caffeic acid, chloramphenicol, and trinitrotoluene. Gas sensors are expected to be small, inexpensive, efficient, and low power-consumption devices. They are classified into different types such as catalytic gas sensors, electrochemical gas sensors, thermal conductivity gas sensors, optical gas sensors, and acoustic gas sensors [155]. Nitrogen-doped graphene is an excellent candidate for selectively sensing CO from air, because only CO can be chemisorbed on the pyridinic-like N-doped graphene with a large concomitant charge transfer. The N-doped graphene was reported to detect CO selectively despite the presence of NO in the environment [156].

Density-functional theory (DFT) study revealed that the structural and electronic properties of the complexes of graphene-gas molecule adsorbed are strongly dependent on the graphene structure and the molecular adsorption configuration. N-doped graphene showed weak interactions with CO, NO, and NH_3, but strong binding with NO_2 with an adsorption energy of -0.98 eV [158]. Experimental evidence was reported for the application of N-doped graphene as NO_2 sensor to validate the computational results predicted almost ten years ago [158,159]. Srivastava et al. synthesized nitrogen-doped graphene using chemical vapor deposition (CVD) method for the fast detection of NO_2 gas at the room temperature. Experimental study demonstrated the improved gas sensitivity of the nitrogen-doped graphene over the pristine graphene nanosheets for NO_2 gas [159].

N-doped graphene is an excellent candidate for the fabrication of selective biosensors for bioelectronics and other biocatalytic applications [148,155]. N-doped graphenes with gold nanoparticles demonstrated for glucose sensor application. Nitrogen-doped GO materials exhibit potential for electrochemical sensing of different biomolecules such as ascorbic acid, uric acid, dopamine, NADH, adenine, and cytosine. Electrochemical sensors based on nitrogen-doped graphene sheets were used as anticancer drug sensors [155]. Metal electrodes coated with nitrogen-doped graphene showed enhanced electron transfer properties, which make them useful in electrochemical sensing [88]. Nitrogen-doped graphene quantum dots

(N-GQDs) exhibit applications in electronics, photonics, and bioimaging [160]. N-GQDs can be used for the detection and estimation of bleomycin sulfate in human serum samples. The carboxyl groups on N-GQDs play an important role in the strong adsorption of ssDNA on its surface, resulting in effective fluorescence quenching of N-GQDs [155].

5 Summary

In this chapter, we reviewed the synthesis methods, characterization techniques, computational approaches employed to study the properties, and applications of N-doped graphene. Characterization and computational techniques are used as tools to elucidate structures and properties. Though computations play more of a role to assist in the understanding, they are demonstrated to be as important as characterization. This chapter highlights key findings from the literature that may generate insight to advance the research and applications involving N-doped graphene. Synthesis conditions such as selection of precursors, ratio of precursors, temperature, flow rate, and carbon source are commonly used toward controlling N content. Selections of metal catalyst substrates may play a role as they varied in literature but are not of primary focus for controlled doping. The type of N-doping can also be controlled by temperature along with other combinations of synthesis conditions previously mentioned.

The most popular characterization techniques are: (1) XPS for determining types of N-doping; (2) Raman spectroscopy for relative determination of defects, thickness of graphene sheets, and N concentration; and (3) STM for confirmation of substitutional N-doping and defects and determining N-doped locations. Theoretical calculations were employed to compliment the experimental results obtained through characterization techniques and revealed some key points: (1) pyridinic configurations prefer to locate in the edge of the graphene systems; (2) in substitutional N-doping, two nitrogen atoms substituted within the six-member ring of graphene are more preferred than nitrogen atoms placed in a defect configuration; two nitrogen atoms generally prefer to be apart from each other; (3) based on MD simulations, as the N concentration increases dramatically, the C—C bonds destabilize, thus breaking C-C bonds which was responsible for the thermal decomposition of N-doped graphene structures; (4) simulated XPS showed that graphitic and pyrrolic N types cannot be accurately distinguished; and (5) the computational study revealed the existence of a competition between π-π bonding and reduced exchange repulsion favoring AA and AB stacking configurations, respectively.

N-doped graphene materials show a wide range of applications including Li-ion batteries, fuel cells, field-effect transistors, ultra- and supercapacitors, photocatalyzers, and photosensors. Nitrogen-doped graphene with pyrrolic-type defect was reported to have a high potential for highly efficient next-generation supercapacitors. It is possible to optimize the capacitance by controlling the type and concentration of N-doping. N-doped graphene exhibits much higher electrocatalytic activity toward oxygen reduction reaction (ORR) and H_2O_2 reduction than graphene. Computational studies provided data and knowledge to understand the reaction mechanisms or binding properties relevant to applications of N-doped graphene as catalysts, batteries, sensors, and capacitors. N-doped graphene is an excellent candidate for the fabrication of selective biosensors for bioelectronics and other biocatalytic applications.

Acknowledgments

T.D. acknowledges the National Science Foundation (NSF) for the grants through HBCU-UP RIA (Grant Number: 1601071) and RISE (Grant Number: 1924204) for financial support. Graduate Studies Office at Clark Atlanta University is thanked for the graduate student support to J.L. and BH through the Title III program. N.A. acknowledges the Saudi Arabian Cultural Mission (SACM) for the scholarship. The Extreme Science and Engineering Discovery Environment (XSEDE) is acknowledged for the computational resources (resource allocation grant DMR 160170).

References

[1] K.E. Kitko, Q. Zhang, Graphene-based nanomaterials: from production to integration with modern tools in neuroscience, Front. Syst. Neurosci. 13 (2019) 26.
[2] V. Barone, O. Hod, G.E. Scuseria, Electronic structure and stability of semiconducting graphene nanoribbons, Nano Lett. 6 (12) (2006) 2748–2754.
[3] D.V. Kosynkin, A.L. Higginbotham, A. Sinitskii, J.R. Lomeda, A. Dimiev, B.K. Price, J.M. Tour, Longitudinal unzipping of carbon nanotubes to form graphene nanoribbons, Nature 458 (7240) (2009) 872–876.
[4] X. Li, X. Wang, L. Zhang, S. Lee, H. Dai, Chemically derived, ultrasmooth graphene nanoribbon semiconductors, Science 319 (5867) (2008) 1229.
[5] L.A. Ponomarenko, F. Schedin, M.I. Katsnelson, R. Yang, E.W. Hill, K.S. Novoselov, A.K. Geim, Chaotic dirac billiard in graphene quantum dots, Science 320 (5874) (2008) 356.
[6] K.A. Ritter, J.W. Lyding, The influence of edge structure on the electronic properties of graphene quantum dots and nanoribbons, Nat. Mater. 8 (3) (2009) 235–242.
[7] L. Jiao, L. Zhang, X. Wang, G. Diankov, H. Dai, Narrow graphene nanoribbons from carbon nanotubes, Nature 458 (7240) (2009) 877–880.
[8] L. Kittiratanawasin, S. Hannongbua, The effect of edges and shapes on band gap energy in graphene quantum dots, Integr. Ferroelectr. 175 (1) (2016) 211–219.
[9] F. Schedin, A.K. Geim, S.V. Morozov, E.W. Hill, P. Blake, M.I. Katsnelson, K.S. Novoselov, Detection of individual gas molecules adsorbed on graphene, Nat. Mater. 6 (9) (2007) 652–655.
[10] T.O. Wehling, K.S. Novoselov, S.V. Morozov, E.E. Vdovin, M.I. Katsnelson, A.K. Geim, A.I. Lichtenstein, Molecular doping of graphene, Nano Lett. 8 (1) (2008) 173–177.
[11] G. Giovannetti, P.A. Khomyakov, G. Brocks, V.M. Karpan, J. van den Brink, P.J. Kelly, Doping graphene with metal contacts, Phys. Rev. Lett. 101 (2) (2008) 026803.
[12] H. Pinto, R. Jones, J.P. Goss, P.R. Briddon, Unexpected change in the electronic properties of the Au-graphene interface caused by toluene, Phys. Rev. B 82 (12) (2010) 125407.
[13] W. Chen, S. Chen, D.C. Qi, X.Y. Gao, A.T.S. Wee, Surface transfer p-type doping of epitaxial graphene, J. Am. Chem. Soc. 129 (34) (2007) 10418–10422.
[14] X. Wang, J.-B. Xu, W. Xie, J. Du, quantitative analysis of graphene doping by organic molecular charge transfer, J. Phys Chem. C 115 (15) (2011) 7596–7602.
[15] H.R. Jiang, T.S. Zhao, L. Shi, P. Tan, L. An, First-principles study of nitrogen-, boron-doped graphene and Co-doped graphene as the potential catalysts in nonaqueous Li–O2 batteries, J. Phys. Chem. C 120 (12) (2016) 6612–6618.
[16] S. Varghese, S. Swaminathan, K. Singh, V. Mittal, Energetic stabilities, structural and electronic properties of monolayer graphene doped with boron and nitrogen atoms, Electronics 5 (2016) 91.
[17] H. Liu, Y. Liu, D. Zhu, Chemical doping of graphene, J. Mater. Chem. 21 (10) (2011) 3335–3345.
[18] X.-F. Li, K.-Y. Lian, L. Liu, Y. Wu, Q. Qiu, J. Jiang, M. Deng, Y. Luo, Unraveling the formation mechanism of graphitic nitrogen-doping in thermally treated graphene with ammonia, Sci. Rep. 6 (1) (2016) 23495.
[19] M. Hu, Q. Lv, R. Lv, Controllable synthesis of nitrogen-doped graphene oxide by tablet-sintering for efficient lithium/sodium-ion storage, ES Energy Environ. 3 (2019) 45–54.

[20] H. Pinto, A. Markevich, Electronic and electrochemical doping of graphene by surface adsorbates, Beilstein J. Nanotechnol. 5 (2014) 1842–1848.
[21] T. Granzier-Nakajima, K. Fujisawa, V. Anil, M. Terrones, Y.-T. Yeh, Controlling nitrogen doping in graphene with atomic precision synthesis and characterization, Nanomaterials 9 (3) (2019).
[22] H. Miao, S. Li, Z. Wang, S. Sun, M. Kuang, Z. Liu, J. Yuan, Enhancing the pyridinic N content of Nitrogen-doped graphene and improving its catalytic activity for oxygen reduction reaction, Int. J. Hydrog. Energy 42 (47) (2017) 28298–28308.
[23] H. Xu, L. Ma, Z. Jin, Nitrogen-doped graphene: synthesis, characterizations and energy applications, J. Energy Chem. 27 (1) (2018) 146–160.
[24] C.P. Ewels, M. Glerup, nitrogen doping in carbon nanotubes, J. Nanosci. Nanotechnol. 5 (9) (2005) 1345–1363.
[25] J. Casanovas, J.M. Ricart, J. Rubio, F. Illas, J.M. Jiménez-Mateos, Origin of the large N 1s binding energy in X-ray photoelectron spectra of calcined carbonaceous materials, J. Am. Chem. Soc. 118 (34) (1996) 8071–8076.
[26] H. Wang, T. Maiyalagan, X. Wang, Review on recent progress in nitrogen-doped graphene: synthesis, characterization, and its potential applications, ACS Catal. 2 (5) (2012) 781–794.
[27] S.N. Faisal, E. Haque, N. Noorbehesht, W. Zhang, A.T. Harris, T.L. Church, A.I. Minett, Pyridinic and graphitic nitrogen-rich graphene for high-performance supercapacitors and metal-free bifunctional electrocatalysts for ORR and OER, RSC Adv. 7 (29) (2017) 17950–17958.
[28] Q. Lv, W. Si, J. He, L. Sun, C. Zhang, N. Wang, Z. Yang, X. Li, X. Wang, W. Deng, Y. Long, C. Huang, Y. Li, Selectively nitrogen-doped carbon materials as superior metal-free catalysts for oxygen reduction, Nat. Commun. 9 (1) (2018) 3376.
[29] H. Huan, Y. Qingwei, X. Shaohua, H. Xiaoyan, L. Qin, L. Kangle, H. Jingping, T. Dingguo, K. Deng, Investigation of edge-selectively nitrogen-doped metal free graphene for oxygen reduction reaction, J. Adv. Nanotechnol. 1 (2) (2020) 5–13.
[30] J. Zhou, Q. Chen, Y. Han, S. Zheng, Enhanced catalytic hydrodechlorination of 2,4-dichlorophenol over Pd catalysts supported on nitrogen-doped graphene, RSC Adv. 5 (111) (2015) 91363–91371.
[31] M. Hou, X. Zhang, S. Yuan, W. Cen, Double graphitic-N doping for enhanced catalytic oxidation activity of carbocatalysts, Phys. Chem. Chem. Phys. 21 (10) (2019) 5481–5488.
[32] S. Feng, M.C. dos Santos, B.R. Carvalho, R. Lv, Q. Li, K. Fujisawa, A.L. Elías, Y. Lei, N. Perea-López, M. Endo, M. Pan, M.A. Pimenta, M. Terrones, Ultrasensitive molecular sensor using N-doped graphene through enhanced Raman scattering, Sci. Adv. 2 (7) (2016) e1600322.
[33] I. Baldea, D. Olteanu, G.A. Filip, F. Pogacean, M. Coros, M. Suciu, S.C. Tripon, M. Cenariu, L. Magerusan, R.-I. Stefan-van Staden, S. Pruneanu, Cytotoxicity mechanisms of nitrogen-doped graphene obtained by electrochemical exfoliation of graphite rods, on human endothelial and colon cancer cells, Carbon 158 (2020) 267–281.
[34] T. Wang, S. Zhu, X. Jiang, Toxicity mechanism of graphene oxide and nitrogen-doped graphene quantum dots in RBCs revealed by surface-enhanced infrared absorption spectroscopy, Toxicol. Res. 4 (4) (2015) 885–894.
[35] T. Wang, L. Wang, D. Wu, W. Xia, H. Zhao, D. Jia, Hydrothermal synthesis of nitrogen-doped graphene hydrogels using amino acids with different acidities as doping agents, J. Mater. Chem. A 2 (22) (2014) 8352–8361.
[36] R. Lv, Q. Li, A.R. Botello-Méndez, T. Hayashi, B. Wang, A. Berkdemir, Q. Hao, A.L. Elías, R. Cruz-Silva, H.R. Gutiérrez, Y.A. Kim, H. Muramatsu, J. Zhu, M. Endo, H. Terrones, J.-C. Charlier, M. Pan, M. Terrones, Nitrogen-doped graphene: beyond single substitution and enhanced molecular sensing, Sci. Rep. 2 (2012) 586.
[37] Mageed, A. K. R., Dayang;; Salmiaton, A. I., Shamsul; Razak, Musab Abdul; Yusoff, H. M. Y., F.M.; Kamarudin, Suryani; Ali, Buthainah, Preparation and characterization of nitrogen doped reduced graphene oxide sheet. Int. J. Appl. Chem. 2016, 12 (1) 104-108.
[38] R. Yadav, C.K. Dixit, Synthesis, characterization and prospective applications of nitrogen-doped graphene: a short review, J. Sci. Adv. Mater. Dev. 2 (2) (2017) 141–149.
[39] D. Wei, Y. Liu, Y. Wang, H. Zhang, L. Huang, G. Yu, Synthesis of N-doped graphene by chemical vapor deposition and its electrical properties, Nano Lett. 9 (5) (2009) 1752–1758.
[40] B. Guo, Q. Liu, E. Chen, H. Zhu, L. Fang, J.R. Gong, Controllable N-doping of graphene, Nano Lett. 10 (12) (2010) 4975–4980.
[41] X. Wang, X. Li, L. Zhang, Y. Yoon, P.K. Weber, H. Wang, J. Guo, H. Dai, N-doping of graphene through electrothermal reactions with ammonia, Science 324 (5928) (2009) 768–771.
[42] Y. Shao, S. Zhang, M.H. Engelhard, G. Li, G. Shao, Y. Wang, J. Liu, I.A. Aksay, Y. Lin, Nitrogen-doped graphene and its electrochemical applications, J. Mater. Chem. 20 (35) (2010) 7491–7496.

[43] M. Telychko, P. Mutombo, M. Ondráček, P. Hapala, F.C. Bocquet, J. Kolorenč, M. Vondráček, P. Jelínek, M. Švec, Achieving high-quality single-atom nitrogen doping of graphene/SiC(0001) by ion implantation and subsequent thermal stabilization, ACS Nano 8 (7) (2014) 7318–7324.
[44] D. Long, W. Li, L. Ling, J. Miyawaki, I. Mochida, S.-H. Yoon, Preparation of nitrogen-doped graphene sheets by a combined chemical and hydrothermal reduction of graphene oxide, Langmuir 26 (20) (2010) 16096–16102.
[45] Z.-H. Sheng, L. Shao, J.-J. Chen, W.-J. Bao, F.-B. Wang, X.-H. Xia, Catalyst-free synthesis of nitrogen-doped graphene via thermal annealing graphite oxide with melamine and its excellent electrocatalysis, ACS Nano 5 (6) (2011) 4350–4358.
[46] Z. Luo, S. Lim, Z. Tian, J. Shang, L. Lai, B. MacDonald, C. Fu, Z. Shen, T. Yu, J. Lin, Pyridinic N doped graphene: synthesis, electronic structure, and electrocatalytic property, J. Mater. Chem. 21 (22) (2011) 8038–8044.
[47] L. Zhao, R. He, K.T. Rim, T. Schiros, K.S. Kim, H. Zhou, C. Gutiérrez, S.P. Chockalingam, C.J. Arguello, L. Pálová, D. Nordlund, M.S. Hybertsen, D.R. Reichman, T.F. Heinz, P. Kim, A. Pinczuk, G.W. Flynn, A.N. Pasupathy, Visualizing individual nitrogen dopants in monolayer graphene, Science 333 (6045) (2011) 999–1003.
[48] D. Geng, S. Yang, Y Zhang, J. Yang, J. Liu, R. Li, T.-K. Sham, X. Sun, S. Ye, S. Knights, Nitrogen doping effects on the structure of graphene, Appl. Surf. Sci. 257 (21) (2011) 9193–9198.
[49] L. Qu, Y. Liu, J.-B. Baek, L. Dai, Nitrogen-doped graphene as efficient metal-free electrocatalyst for oxygen reduction in fuel cells, ACS Nano 4 (3) (2010) 1321–1326.
[50] Di, C.-a.; Wei, D.; Yu, G.; Liu, Y.; Guo, Y.; Zhu, D., Patterned graphene as source/drain electrodes for bottom-contact organic field-effect transistors. Adv. Mater. 2008, 20 (17) 3289-3293.
[51] A.L.M. Reddy, A. Srivastava, S.R. Gowda, H. Gullapalli, M. Dubey, P.M. Ajayan, Synthesis of nitrogen-doped graphene films for lithium battery application, ACS Nano 4 (11) (2010) 6337–6342.
[52] Z. Jin, J. Yao, C. Kittrell, J.M. Tour, Large-scale growth and characterizations of nitrogen-doped monolayer graphene sheets, ACS Nano 5 (5) (2011) 4112–4117.
[53] Y.J. Cho, H.S. Kim, S.Y. Baik, Y. Myung, C.S. Jung, C.H. Kim, J. Park, H.S. Kang, Selective nitrogen-doping structure of nanosize graphitic layers, J. Phys. Chem. C 115 (9) (2011) 3737–3744.
[54] A.G. Kudashov, A.V. Okotrub, L.G. Bulusheva, I.P. Asanov, Y.V. Shubin, N.F. Yudanov, L.I. Yudanova, V.S. Danilovich, O.G. Abrosimov, Influence of Ni−Co catalyst composition on nitrogen content in carbon nanotubes, J. Phys. Chem. B 108 (26) (2004) 9048–9053.
[55] J. Liu, S. Webster, D.L. Carroll, Temperature and flow rate of NH3 effects on nitrogen content and doping environments of carbon nanotubes grown by injection CVD method, J. Phys. Chem. B 109 (33) (2005) 15769–15774.
[56] W.-X. Lv, R. Zhang, T.-L. Xia, H.-M. Bi, K.-Y. Shi, Influence of NH3 flow rate on pyridine-like N content and NO electrocatalytic oxidation of N-doped multiwalled carbon nanotubes, J. Nanopart. Res. 13 (6) (2011) 2351–2360.
[57] Y. Sui, B. Zhu, H. Zhang, H. Shu, Z. Chen, Y. Zhang, Y. Zhang, B. Wang, C. Tang, X. Xie, G. Yu, Z. Jin, X. Liu, Temperature-dependent nitrogen configuration of N-doped graphene by chemical vapor deposition, Carbon 81 (2015) 814–820.
[58] T. Schiros, D. Nordlund, L. Pálová, D. Prezzi, L. Zhao, K.S. Kim, U. Wurstbauer, C. Gutiérrez, D. Delongchamp, C. Jaye, D. Fischer, H. Ogasawara, L.G.M. Pettersson, D.R. Reichman, P. Kim, M.S. Hybertsen, A.N. Pasupathy, Connecting dopant bond type with electronic structure in N-doped graphene, Nano Lett. 12 (8) (2012) 4025–4031.
[59] C. Zhang, L. Fu, N. Liu, M. Liu, Y. Wang, Z. Liu, Synthesis of nitrogen-doped graphene using embedded carbon and nitrogen sources, Adv. Mater. 23 (8) (2011) 1020–1024.
[60] D. Deng, X. Pan, L. Yu, Y. Cui, Y. Jiang, J. Qi, W.-X. Li, Q. Fu, X. Ma, Q. Xue, G. Sun, X. Bao, Toward N-doped graphene via solvothermal synthesis, Chem. Mater. 23 (5) (2011) 1188–1193.
[61] D. O'Hare, Hydrothermal synthesis, in: K.H.J. Buschow, R.W. Cahn, M.C. Flemings, B. Ilschner, E.J. Kramer, S. Mahajan, P. Veyssière (Eds.), Encyclopedia of Materials: Science and Technology, Elsevier, Oxford, 2001, pp. 3989–3992.
[62] Z. Xing, Z. Ju, Y. Zhao, J. Wan, Y. Zhu, Y. Qiang, Y. Qian, One-pot hydrothermal synthesis of nitrogen-doped graphene as high-performance anode materials for lithium ion batteries, Sci. Rep. 6 (1) (2016) 26146.
[63] N.-J. Kuo, Y.-S. Chen, C.-W. Wu, C.-Y. Huang, Y.-H. Chan, I.W.P. Chen, One-pot synthesis of hydrophilic and hydrophobic N-doped graphene quantum dots via exfoliating and disintegrating graphite flakes, Sci. Rep. 6 (1) (2016) 30426.
[64] J. Ju, R. Zhang, S. He, W. Chen, Nitrogen-doped graphene quantum dots-based fluorescent probe for the sensitive turn-on detection of glutathione and its cellular imaging, RSC Adv. 4 (94) (2014) 52583–52589.

[65] Agusu, L.; Ahmad, L. O.; Alimin; Nurdin, M.; Herdianto; Mitsudo, S.; Kikuchi, H., Hydrothermal synthesis of reduced graphene oxide using urea as reduction agent: excellent x-band electromagnetic absorption properties. IOP Conf. Ser. Mater. Sci. Eng. 2018, 367, 012002.

[66] Z. Yang, G. Xing, P. Hou, D. Han, Amino acid-mediated N-doped graphene aerogels and its electrochemical properties, Mater. Sci. Eng. B 228 (2018) 198–205.

[67] M. Khandelwal, A. Kumar, One-pot environmentally friendly amino acid mediated synthesis of N-doped graphene–silver nanocomposites with an enhanced multifunctional behavior, Dalton Trans. 45 (12) (2016) 5180–5195.

[68] R. Droppa, P. Hammer, A.C.M. Carvalho, M.C. dos Santos, F. Alvarez, Incorporation of nitrogen in carbon nanotubes, J. Non-Cryst. Solids 299-302 (2002) 874–879.

[69] C. Journet, W.K. Maser, P. Bernier, A. Loiseau, M.L. de la Chapelle, S. Lefrant, P. Deniard, R. Lee, J.E. Fischer, Large-scale production of single-walled carbon nanotubes by the electric-arc technique, Nature 388 (6644) (1997) 756–758.

[70] L.S. Panchakarla, K.S. Subrahmanyam, S.K. Saha, A. Govindaraj, H.R. Krishnamurthy, U.V. Waghmare, C.N.R. Rao, Synthesis, structure, and properties of boron- and nitrogen-doped graphene, Adv. Mater. 21 (46) (2009) 4726–4730.

[71] A. Ghosh, D.J. Late, L.S. Panchakarla, A. Govindaraj, C.N.R. Rao, NO2 and humidity sensing characteristics of few-layer graphenes, J. Exp. Nanosci. 4 (4) (2009) 313–322.

[72] D. Geng, Y. Chen, Y. Chen, Y. Li, R. Li, X. Sun, S. Ye, S. Knights, High oxygen-reduction activity and durability of nitrogen-doped graphene, Energy Environ. Sci. 4 (3) (2011) 760–764.

[73] K. Kinoshita, Carbon: Electrochemical and Physicochemical Properties, Wiley, New York, 1988.

[74] X. Li, H. Wang, J.T. Robinson, H. Sanchez, G. Diankov, H. Dai, Simultaneous nitrogen doping and reduction of graphene oxide, J. Am. Chem. Soc. 131 (43) (2009) 15939–15944.

[75] D. Gu, Y. Zhou, R. Ma, F. Wang, Q. Liu, J. Wang, Facile synthesis of N-doped graphene-like carbon nanoflakes as efficient and stable electrocatalysts for the oxygen reduction reaction, Nano-Micro Lett. 10 (2) (2017) 29.

[76] S.S. Hassani, L. Samiee, E. Ghasemy, A. Rashidi, M.R. Ganjali, S. Tasharrofi, Porous nitrogen-doped graphene prepared through pyrolysis of ammonium acetate as an efficient ORR nanocatalyst, Int. J. Hydrog. Energy 43 (33) (2018) 15941–15951.

[77] W.-D. Wang, X.-Q. Lin, H.-B. Zhao, Q.-F. Lü, Nitrogen-doped graphene prepared by pyrolysis of graphene oxide/polyaniline composites as supercapacitor electrodes, J. Anal. Appl. Pyrolysis 120 (2016) 27–36.

[78] D. Golberg, Y. Bando, L. Bourgeois, K. Kurashima, T. Sato, Large-scale synthesis and HRTEM analysis of single-walled B- and N-doped carbon nanotube bundles, Carbon 38 (14) (2000) 2017–2027.

[79] C. Morant, J. Andrey, P. Prieto, D. Mendiola, J.M. Sanz, E. Elizalde, XPS characterization of nitrogen-doped carbon nanotubes, Phys. Stat. Solidi (a) 203 (15) (2006) 3893.

[80] K. Suenaga, M.P. Johansson, N. Hellgren, E. Broitman, L.R. Wallenberg, C. Colliex, J.E. Sundgren, L. Hultman, Carbon nitride nanotubulite—densely-packed and well-aligned tubular nanostructures, Chem. Phys. Lett. 300 (5) (1999) 695–700.

[81] R. Imran Jafri, N. Rajalakshmi, S. Ramaprabhu, Nitrogen doped graphene nanoplatelets as catalyst support for oxygen reduction reaction in proton exchange membrane fuel cell, J. Mater. Chem. 20 (34) (2010) 7114–7117.

[82] Y. Wang, Y. Shao, D.W. Matson, J. Li, Y. Lin, Nitrogen-doped graphene and its application in electrochemical biosensing, ACS Nano 4 (4) (2010) 1790–1798.

[83] H.M. Jeong, J.W. Lee, W.H. Shin, Y.J. Choi, H.J. Shin, J.K. Kang, J.W. Choi, Nitrogen-doped graphene for high-performance ultracapacitors and the importance of nitrogen-doped sites at basal planes, Nano Lett. 11 (6) (2011) 2472–2477.

[84] N. Bundaleska, J. Henriques, M. Abrashev, A.M. Botelho do Rego, A.M. Ferraria, A. Almeida, F.M. Dias, E. Valcheva, B. Arnaudov, K.K. Upadhyay, M.F. Montemor, E. Tatarova, Large-scale synthesis of free-standing N-doped graphene using microwave plasma, Sci. Rep. 8 (1) (2018) 12595.

[85] C.R.S.V. Boas, B. Focassio, E. Marinho, D.G. Larrude, M.C. Salvadori, C.R. Leão, D.J. dos Santos, Characterization of nitrogen doped graphene bilayers synthesized by fast, low temperature microwave plasma-enhanced chemical vapour deposition, Sci. Rep. 9 (1) (2019) 13715.

[86] U. Bangert, W. Pierce, D.M. Kepaptsoglou, Q. Ramasse, R. Zan, M.H. Gass, J.A. Van den Berg, C.B. Boothroyd, J. Amani, H. Hofsäss, Ion implantation of graphene—toward IC compatible technologies, Nano Lett. 13 (10) (2013) 4902–4907.

[87] C.D. Cress, S.W. Schmucker, A.L. Friedman, P. Dev, J.C. Culbertson, J.W. Lyding, J.T. Robinson, Nitrogen-doped graphene and twisted bilayer graphene via hyperthermal ion implantation with depth control, ACS Nano 10 (3) (2016) 3714–3722.

[88] D.-W. Wang, I.R. Gentle, G.Q. Lu, Enhanced electrochemical sensitivity of PtRh electrodes coated with nitrogen-doped graphene, Electrochem. Commun. 12 (10) (2010) 1423–1427.
[89] Y.-C. Lin, C.-Y. Lin, P.-W. Chiu, Controllable graphene N-doping with ammonia plasma, Appl. Phys. Lett. 96 (13) (2010) 133110.
[90] Z.H. Ni, H.M. Wang, J. Kasim, H.M. Fan, T. Yu, Y.H. Wu, Y.P. Feng, Z.X. Shen, Graphene thickness determination using reflection and contrast spectroscopy, Nano Lett. 7 (9) (2007) 2758–2763.
[91] D. Graf, F. Molitor, K. Ensslin, C. Stampfer, A. Jungen, C. Hierold, L. Wirtz, Spatially resolved raman spectroscopy of single- and few-layer graphene, Nano Lett. 7 (2) (2007) 238–242.
[92] Y.H. Wu, T. Yu, Z.X. Shen, Two-dimensional carbon nanostructures: Fundamental properties, synthesis, characterization, and potential applications, J. Appl. Phys. 108 (7) (2010) 071301.
[93] C. Casiraghi, S. Pisana, K.S. Novoselov, A.K. Geim, A.C. Ferrari, Raman fingerprint of charged impurities in graphene, Appl. Phys. Lett. 91 (23) (2007) 233108.
[94] A. Gupta, G. Chen, P. Joshi, S. Tadigadapa, Eklund, raman scattering from high-frequency phonons in supported N-graphene layer films, Nano Lett. 6 (12) (2006) 2667–2673.
[95] A.C. Ferrari, Raman spectroscopy of graphene and graphite: disorder, electron–phonon coupling, doping and nonadiabatic effects, Solid State Commun. 143 (1) (2007) 47–57.
[96] M. Herz, F.J. Giessibl, J. Mannhart, Probing the shape of atoms in real space, Phys. Rev. B 68 (4) (2003) 045301.
[97] B. Zheng, P. Hermet, L. Henrard, Scanning tunneling microscopy simulations of nitrogen- and boron-doped graphene and single-walled carbon nanotubes, ACS Nano 4 (7) (2010) 4165–4173.
[98] C. Ricca, F. Labat, N. Russo, C. Adamo, E. Sicilia, Oxidation of ethylbenzene to acetophenone with N-doped graphene: insight from theory, J. Phys. Chem. C 118 (23) (2014) 12275–12284.
[99] Y.-X. Yu, Can all nitrogen-doped defects improve the performance of graphene anode materials for lithium-ion batteries? Phys. Chem. Chem. Phys. 15 (39) (2013) 16819–16827.
[100] P. Błoński, J. Tuček, Z. Sofer, V. Mazánek, M. Petr, M. Pumera, M. Otyepka, R. Zbořil, Doping with graphitic nitrogen triggers ferromagnetism in graphene, J. Am. Chem. Soc. 139 (8) (2017) 3171–3180.
[101] D. Herath, T. Dinadayalane, Computational investigation of double nitrogen doping on graphene, J. Mol. Model. 24 (1) (2018) 26.
[102] R.S. Mulliken, Electronic population analysis on LCAO–MO molecular wave functions. II. Overlap populations, bond orders, and covalent bond energies, J. Chem. Phys. 23 (10) (1955) 1841–1846.
[103] G. Kresse, J. Furthmüller, Efficient iterative schemes for ab initio total-energy calculations using a plane-wave basis set, Phys. Rev. B 54 (16) (1996) 11169–11186.
[104] G. Kresse, J. Furthmüller, Efficiency of ab-initio total energy calculations for metals and semiconductors using a plane-wave basis set, Comput. Mater. Sci. 6 (1) (1996) 15–50.
[105] W.A. Hofer, Challenges and errors: interpreting high resolution images in scanning tunneling microscopy, Prog. Surf. Sci. 71 (5) (2003) 147–183.
[106] K. Palotás, W. Hofer, Multiple scattering in a vacuum barrier obtained from real-space wavefunctions, J. Phys. Condens. Matter 17 (2005) 2705–2713.
[107] J. Neilson, H. Chinkezian, H. Phirke, A. Osei-Twumasi, Y. Li, C. Chichiri, J. Cho, K. Palotás, L. Gan, S.J. Garrett, K.C. Lau, L. Gao, Nitrogen-doped graphene on copper: edge-guided doping process and doping-induced variation of local work function, J. Phys. Chem. C 123 (14) (2019) 8802–8812.
[108] J.P. Perdew, Y. Wang, Accurate and simple analytic representation of the electron-gas correlation energy, Phys. Rev. B 45 (23) (1992) 13244–13249.
[109] Y. Inada, H. Orita, Efficiency of numerical basis sets for predicting the binding energies of hydrogen bonded complexes: evidence of small basis set superposition error compared to Gaussian basis sets, J. Comput. Chem. 29 (2) (2008) 225–232.
[110] B. Delley, An all-electron numerical method for solving the local density functional for polyatomic molecules, J. Chem. Phys. 92 (1) (1990) 508–517.
[111] A.V. Vorontsov, E.V. Tretyakov, Determination of graphene's edge energy using hexagonal graphene quantum dots and PM7 method, Phys. Chem. Chem. Phys. 20 (21) (2018) 14740–14752.
[112] W. Wang, Y. Zhang, C. Shen, Y. Chai, Adsorption of CO molecules on doped graphene: a first-principles study, AIP Adv. 6 (2) (2016) 025317.
[113] M. Yang, L. Wang, M. Li, T. Hou, Y. Li, Structural stability and O2 dissociation on nitrogen-doped graphene with transition metal atoms embedded: a first-principles study, AIP Adv. 5 (6) (2015) 067136.
[114] Y. Fujimoto, Formation, energetics, and electronic properties of graphene monolayer and bilayer doped with heteroatoms, Adv. Condens. Matter Phys. 2015 (2015) 1–14.

[115] V.V. Chaban, O.V. Prezhdo, Nitrogen–nitrogen bonds undermine stability of N-doped graphene, J. Am. Chem. Soc. 137 (36) (2015) 11688–11694.
[116] P. Rani, V.K. Jindal, Designing band gap of graphene by B and N dopant atoms, RSC Adv. 3 (3) (2013) 802–812.
[117] X. Chen, A.R. McDonald, Functionalization of two-dimensional transition-metal dichalcogenides, Adv. Mater. 28 (27) (2016) 5738–5746.
[118] J. Tao, J.P. Perdew, V.N. Staroverov, G.E. Scuseria, Climbing the density functional ladder: nonempirical meta-generalized gradient approximation designed for molecules and solids, Phys. Rev. Lett. 91 (14) (2003) 146401.
[119] V.N. Staroverov, G.E. Scuseria, J. Tao, J.P. Perdew, Comparative assessment of a new nonempirical density functional: molecules and hydrogen-bonded complexes, J. Chem. Phys. 119 (23) (2003) 12129–12137.
[120] J. Heyd, G.E. Scuseria, M. Ernzerhof, Hybrid functionals based on a screened Coulomb potential, J. Chem. Phys. 118 (18) (2003) 8207–8215.
[121] J. Heyd, G.E. Scuseria, M. Ernzerhof, Erratum: "Hybrid functionals based on a screened Coulomb potential" [J. Chem. Phys. 118, 8207 (2003)], J. Chem. Phys. 124 (21) (2006) 219906.
[122] M. Valiev, E.J. Bylaska, N. Govind, K. Kowalski, T.P. Straatsma, H.J.J. Van Dam, D. Wang, J. Nieplocha, E. Apra, T.L. Windus, W.A. de Jong, NWChem: a comprehensive and scalable open-source solution for large scale molecular simulations, Comput. Phys. Commun. 181 (9) (2010) 1477–1489.
[123] Z. Shi, A. Kutana, B.I. Yakobson, How much N-doping can graphene sustain? J. Phys. Chem. Lett. 6 (1) (2015) 106–112.
[124] R. Babar, M. Kabir, Ferromagnetism in nitrogen-doped graphene, Phys. Rev. B 99 (11) (2019) 115442.
[125] Z. Wang, S. Qin, C. Wang, Electronic and magnetic properties of single-layer graphene doped by nitrogen atoms, Eur. Phys. J. B 87 (4) (2014) 88.
[126] J.P. Perdew, K. Burke, M. Ernzerhof, Generalized gradient approximation made simple, Phys. Rev. Lett. 77 (18) (1996) 3865–3868.
[127] M. Figueras, I.J. Villar-Garcia, F. Viñes, C. Sousa, V.A. de la Peña O'Shea, F. Illas, Correcting flaws in the assignment of nitrogen chemical environments in N-doped graphene, J. Phys. Chem. C 123 (17) (2019) 11319–11327.
[128] Y.-H. Tian, J. Huang, X. Sheng, B.G. Sumpter, M. Yoon, M. Kertesz, Nitrogen doping enables covalent-like π–π bonding between graphenes, Nano Lett. 15 (8) (2015) 5482–5491.
[129] P.E. Blöchl, Projector augmented-wave method, Phys. Rev. B 50 (24) (1994) 17953–17979.
[130] S. Grimme, Semiempirical GGA-type density functional constructed with a long-range dispersion correction, J. Comput. Chem. 27 (15) (2006) 1787–1799.
[131] S. Grimme, J. Antony, S. Ehrlich, H. Krieg, A consistent and accurate ab initio parametrization of density functional dispersion correction (DFT-D) for the 94 elements H-Pu, J. Chem. Phys. 132 (15) (2010) 154104.
[132] A. Tkatchenko, M. Scheffler, Accurate molecular Van Der waals interactions from ground-state electron density and free-atom reference data, Phys. Rev. Lett. 102 (7) (2009) 073005.
[133] A. Tkatchenko, R.A. DiStasio, R. Car, M. Scheffler, Accurate and efficient method for many-body van der Waals interactions, Phys. Rev. Lett. 108 (23) (2012) 236402.
[134] Y. Dong, M.T. Gahl, C. Zhang, J. Lin, Computational study of precision nitrogen doping on graphene nanoribbon edges, Nanotechnology 28 (50) (2017) 505602.
[135] S. Plimpton, Fast parallel algorithms for short-range molecular dynamics, J. Comput. Phys. 117 (1) (1995) 1–19.
[136] G. Kresse, D. Joubert, From ultrasoft pseudopotentials to the projector augmented-wave method, Phys. Rev. B 59 (3) (1999) 1758–1775.
[137] P. Giannozzi, S. Baroni, N. Bonini, M. Calandra, R. Car, C. Cavazzoni, D. Ceresoli, G.L. Chiarotti, M. Cococcioni, I. Dabo, A. Dal Corso, S. de Gironcoli, S. Fabris, G. Fratesi, R. Gebauer, U. Gerstmann, C. Gougoussis, A. Kokalj, M. Lazzeri, L. Martin-Samos, N. Marzari, F. Mauri, R. Mazzarello, S. Paolini, A. Pasquarello, L. Paulatto, C. Sbraccia, S. Scandolo, G. Sclauzero, A.P. Seitsonen, A. Smogunov, P. Umari, R.M. Wentzcovitch, QUANTUM ESPRESSO: a modular and open-source software project for quantum simulations of materials, J. Phys. Condens. Matter 21 (39) (2009) 395502.
[138] Giannozzi, P.; Baseggio, O.; Bonfà, P.; Brunato, D.; Car, R.; Carnimeo, I.; Cavazzoni, C.; Gironcoli, S.D.; Delugas, P.; Ruffino, F. F.; Ferretti, A.; Marzari, N.; Timrov, I.; Urru, A.; Baroni, S., Quantum ESPRESSO toward the exascale. J. Chem. Phys. 2020, 152 (15) 154105.
[139] B. Zhao, C. Song, F. Wang, W. Zi, H. Du, Facile synthesis of microporous N-doped carbon material and its application in supercapacitor, Microporous Mesoporous Mater. 306 (2020) 110483.

[140] L.-S. Zhang, X.-Q. Liang, W.-G. Song, Z.-Y. Wu, Identification of the nitrogen species on N-doped graphene layers and Pt/NG composite catalyst for direct methanol fuel cell, Phys. Chem. Chem. Phys. 12 (38) (2010) 12055–12059.
[141] Z.E. Hughes, T.R. Walsh, Computational chemistry for graphene-based energy applications: progress and challenges, Nanoscale 7 (16) (2015) 6883–6908.
[142] H. Jeong, H.J. Kim, Y.J. Lee, J.Y. Hwang, O.-K. Park, J.-H. Wee, C.-M. Yang, B.-C. Ku, J.K. Lee, Amino acids derived nitrogen-doped carbon materials for electrochemical capacitive energy storage, Mater. Lett. 145 (2015) 273–278.
[143] Y. Zhao, X. Li, B. Yan, D. Li, S. Lawes, X. Sun, Significant impact of 2D graphene nanosheets on large volume change tin-based anodes in lithium-ion batteries: a review, J. Power Sources 274 (2015) 869–884.
[144] O.S. Kwon, S.J. Park, J.-Y. Hong, A.R. Han, J.S. Lee, J.S. Lee, J.H. Oh, J. Jang, Flexible FET-Type VEGF aptasensor based on nitrogen-doped graphene converted from conducting polymer, ACS Nano 6 (2) (2012) 1486–1493.
[145] Zhan, C.; Zhang, Y.; Cummings, P. T.; Jiang, D.-e., Enhancing graphene capacitance by nitrogen: effects of doping configuration and concentration. Phys. Chem. Chem. Phys. 2016, 18 (6) 4668-4674.
[146] L.-F. Chen, X.-D. Zhang, H.-W. Liang, M. Kong, Q.-F. Guan, P. Chen, Z.-Y. Wu, S.-H. Yu, Synthesis of nitrogen-doped porous carbon nanofibers as an efficient electrode material for supercapacitors, ACS Nano 6 (8) (2012) 7092–7102.
[147] X. Huang, X. Qi, F. Boey, H. Zhang, Graphene-based composites, Chem. Soc. Rev. 41 (2) (2012) 666–686.
[148] Q. Min, X. Zhang, X. Chen, S. Li, J.-J. Zhu, N-doped graphene: an alternative carbon-based matrix for highly efficient detection of small molecules by negative ion MALDI-TOF MS, Anal. Chem. 86 (18) (2014) 9122–9130.
[149] G. Li, Z. Li, X. Xiao, Y. An, W. Wang, Z. Hu, An ultrahigh electron-donating quaternary-N-doped reduced graphene oxide@carbon nanotube framework: a covalently coupled catalyst support for enzymatic bioelectrodes, J. Mater. Chem. A 7 (18) (2019) 11077–11085.
[150] F.M. Hassan, V. Chabot, J. Li, B.K. Kim, L. Ricardez-Sandoval, A. Yu, Pyrrolic-structure enriched nitrogen doped graphene for highly efficient next generation supercapacitors, J. Mater. Chem. A 1 (8) (2013) 2904–2912.
[151] G. Wu, N.H. Mack, W. Gao, S. Ma, R. Zhong, J. Han, J.K. Baldwin, P. Zelenay, Nitrogen-doped graphene-rich catalysts derived from heteroatom polymers for oxygen reduction in nonaqueous lithium–O2 battery cathodes, ACS Nano 6 (11) (2012) 9764–9776.
[152] C. Zhao, C. Yu, S. Liu, J. Yang, X. Fan, H. Huang, J. Qiu, 3d porous N-doped graphene frameworks made of interconnected nanocages for ultrahigh-rate and long-life Li–O2 batteries, Adv. Funct. Mater. 25 (44) (2015) 6913–6920.
[153] Zheng, H.; Zheng, J.-j.; He, L.; Zhao, X., Unique configuration of a nitrogen-doped graphene nanoribbon: potential applications to semiconductor and hydrogen fuel cell. J. Phys. Chem. C 2014, 118 (42) 24723-24729.
[154] X. Zhang, N. Hao, X. Dong, S. Chen, Z. Zhou, Y. Zhang, K. Wang, One-pot hydrothermal synthesis of platinum nanoparticle-decorated three-dimensional nitrogen-doped graphene aerogel as a highly efficient electrocatalyst for methanol oxidation, RSC Adv. 6 (74) (2016) 69973–69976.
[155] S. Kaushal, M. Kaur, N. Kaur, V. Kumari, P. Singh, Heteroatom-doped graphene as sensing materials: a mini review, RSC Adv. 10 (2020) 28608–28629.
[156] C. Ma, X. Shao, D. Cao, Nitrogen-doped graphene as an excellent candidate for selective gas sensing, Sci. China Chem. 57 (6) (2014) 911–917.
[157] M. Kaur, M. Kaur, V.K. Sharma, Nitrogen-doped graphene and graphene quantum dots: a review onsynthesis and applications in energy, sensors and environment, Adv. Colloid Interf. Sci. 259 (2018) 44–64.
[158] Y.-H. Zhang, Y.-B. Chen, K.-G. Zhou, C.-H. Liu, J. Zeng, H.-L. Zhang, Y. Peng, Improving gas sensing properties of graphene by introducing dopants and defects: a first-principles study, Nanotechnology 20 (18) (2009) 185504.
[159] S. Srivastava, P.K. Kashyap, V. Singh, T.D. Senguttuvan, B.K. Gupta, Nitrogen doped high quality CVD grown graphene as a fast responding NO2 gas sensor, New J. Chem. 42 (12) (2018) 9550–9556.
[160] G. Yang, C. Wu, X. Luo, X. Liu, Y. Gao, P. Wu, C. Cai, S.S. Saavedra, Exploring the emissive states of heteroatom-doped graphene quantum dots, J. Phys. Chem. C 122 (11) (2018) 6483–6492.

CHAPTER

10

Toward graphene-based devices for nanospintronics

Macon Magno and Frank Hagelberg

Department of Physics and Astronomy, East Tennessee State University, Johnson City, TN, United States

1 Introduction

Spin-filtering devices, used to generate and control currents with net spin populations, are of major interest for spintronics. Here, the spin is employed to define and manipulate the state of logic units, and thus, it takes the role of the charge in electronic networks. Recent examples for the importance of spin-polarized current are provided by nanoscale realizations of magnetic data storage, in particular of magnetoresistive random access memory (MRAM) devices. These combine non-volatility with high-speed operation, low power consumption, and an unsurpassed read-and-write endurance [1,2]. Recent advances in this field include the use of a *spin-transfer torque* (STT) to change the state of a target magnetic moment in a controlled way [3–5]. This mechanism involves a spin exchange between the ferromagnetic storage medium and a spin-polarized electron current, creating a strong incentive to identify nanomaterials that act as spin filters, since they endow previously unpolarized currents with a distinct and tunable degree of spin polarization.

Half metals—where only one spin channel is conductive, while the other is insulating [2]—achieve, in principle, perfect spin polarization for currents traversing them. Half-metallic MRAM elements currently contain mostly transition metals (TMs), such as Fe, Co, or Ni [3]. These have high magnetic moments but come with drawbacks such as high fabrication cost, relatively high weight, and low efficiency when it comes to power consumption.

Graphene has been proposed as a lighter, less costly, and more energy efficient alternative to TM-based MRAM elements. In addition to these advantages, graphene is known to be an excellent medium for spin transport [6–10], combining low spin-orbit coupling [6–8] with the small hyperfine interaction characteristic for C atoms [9]. Various two-dimensional

Theoretical and Computational Chemistry, Volume 21
https://doi.org/10.1016/B978-0-12-819514-7.00009-9

carbon-based materials with half-metallicity have been proposed, including graphene nanoribbons with applied in-plane electric field [10,11] or different chemical modifications of the edge [12], semi-hydrogenations of graphene [13,14], etc. Experimental confirmation of half-metallicity in graphene-based transmission elements is at the frontier of current research on the fundamentals of spintronics [14]. Pure graphene, on the other hand, is not suitable as a spin-filtering device since its electronic ground state is non-magnetic. Nevertheless, graphene fragments may carry spin-polarized edge states, associated with ground-state magnetism [15,16].

Thus, graphene nanoribbons (GNRs), as obtained by reducing the graphene sheet to strips of finite width with one periodic dimension, have been discussed as potential realizations of graphene-based transmission elements in spintronics circuits [6–15]. In particular, graphene nanoribbons of the zigzag type (zGNRs) are inherently magnetic, as they exhibit ferromagnetic (FM) coordination along their edges. Across the two edges, antiferromagnetic (AFM) order is realized, implying symmetry between up and down spin orientations. Due to this feature, pure zGNRs in their electronic ground states do not act as spin filters [12,17]. However, preparing them in an excited FM state, with parallel magnetic moments along both edges, will, in principle, give rise to the spin-filtering property. In practice, this approach requires applying a locally well-controlled magnetic field, which may be difficult to realize in MRAM networks at the nanoscale. It is further compromised by the low chemical stability of the zigzag edges, which are susceptible to oxidation [18,19].

The edges of armchair graphene nanoribbons (aGNRs) exhibit much higher chemical stability but are non-magnetic, which excludes pure aGNRs as spin-filtering media. On the other hand, aGNRs might act as host materials for impurities with high magnetic moments, such as magnetic metal atoms, and thus yield devices that combine the excellent transport properties of graphene with a low degree of edge reactivity. While this solution reverts to metallic components, it involves metals in small quantities, i.e., as trace elements in extended carbon environments.

Here, we present a comparative investigation of both prototypes: zGNRs that deviate from their antiferromagnetic ground-state configuration and aGNRs with substitutional transition metal atoms, in terms of their spin-filter capabilities.

Most previous studies on the spin-filtering features of GNRs have focused on the former group, zigzag graphene nanoribbons (zGNRs) and systems derived from them [20–22]. While imposing FM order on the unit as a whole generates spin-filter activity, this effect may also be induced by structural modifications of the zGNR lattice [23,24]. Specifically, zGNRs act as spin filters when the ferromagnetic (FM) order of one of the edges is disrupted, such that the zGNR acquires finite spin polarization [24]. This may be realized by carving notches into the zGNR lattice, rectangular, or V-shaped incisions cut into one of the edges, while the other is left intact [25]. As has been shown by tight-binding simulation, applying strain on zGNRs modified by notches in one of the edges can give rise to half-metallicity in these structures [25]. Similarly, subjecting the two zGNR edges to different degrees of strain by in-plane bending of the ribbon has been shown to cause half-metallicity [23].

Turning to the second group of GNR candidates for spin-selective transmission, we comment in the following on aGNRs with atomic TM impurities. Several recent density functional theory (DFT) studies have dealt with TM atoms adsorbing to graphene or to graphene-derived species [26], or substituting for C atoms in graphene networks [20,27,28].

Crook et al. explored ensembles of TM impurities (TM = Mn, Cr, V) in the graphene fabric and characterized superexchange interaction between the magnetic centers [29,30]. Longo et al. [31] analyzed the magnetism of a substitutional Fe atom in various GNRs and reported a magnetic ground state for these systems, in contrast to Fe-substituted graphene, where a non-magnetic ground state was found [27,28]. Jaiswal and Srivastava studied the transport properties of aGNRs with terminating Ni atoms [32]. Further, they considered Fe atoms in terminating or substitutional positions [33], forming monatomic Fe wires that connect the electrodes. Inspecting partial density-of-states (PDOS) distributions for both spin orientations, the authors observed degrees of spin polarization of up to 60% close to the Fermi level (E_F). The magnitude of the spin polarization was shown to depend critically on the site of the Fe line defect, being maximal at the aGNR edge, and vanishing at the center.

A recent estimate, based on hypothetical planar aGNR devices with one or two substitutional Fe atoms, arrived at high degrees of spin current polarization, reaching into the 90% range for certain choices of the bias [17].

In the present work, we focus on representative species of both types, spin filters based on zGNR and on aGNR units. In both cases, emphasis is on the elementary mechanism that turns these systems into spin-filter active devices. With respect to zGNRs, we explore the impact of simple vacancies, obtained by removing selected C atoms from the graphene network, on their spin-filter capacity. Longitudinal strain is applied on the device, and the strain dependence of the filtering efficiency is monitored for different vacancy configurations.

Commenting on aGNRs, we treat units with one or two Fe atoms. Specifically, we define and examine the parameters that affect the performance of substituted Fe atoms as spin polarization-inducing agents in an aGNR. As a pure aGNR provides a non-magnetic medium, all spin-filter activity of an aGNR with substituted TM atoms must be ascribed to the presence of the magnetic impurities. We note that Fe admixtures are likely to occur in graphene as well as in graphene-derived structures. Iron is prevalent as catalyst in graphite, making it probable to find Fe atoms in graphene sheets manufactured from graphite [34].

For both classes of systems, the questions guiding this work are the following:

Which configurations—of vacancies in zGNRs or impurity atoms in aGNRs—give rise to the maximum degree of spin polarization in an initially unpolarized current traversing the GNR? How does the spin-filter effect of the GNR transmission element depend on the bias across the unit? Further, to what extent does a combination of the two defect types considered here, vacancies and impurity atoms, affect the resulting degrees of current spin polarization? Other factors, among them the width of the nanoribbon, or the symmetry of the defect configuration, may be of impact on the current spin polarization but are not covered in the present work. Rather than modeling functional devices of nanospintronics, this contribution explores elementary mechanisms of generating spin-polarized currents by the use of modified aGNRs.

2 Methods

All systems studied in this work were analyzed by electronic structure and transport calculations. In the following, we provide details on both types of computation.

2.1 Electronic structure

To describe the electronic structure, we adopted the spin-polarized generalized-gradient approximation (SGGA) and used Perdew-Burke-Ernzerhof (PBE) functionals for all species considered here. Norm-conserving pseudopotentials of the OpenMX (OMX, [35]) type were used, along with associated localized numerical basis sets [36], allowing for a viable compromise between accuracy and tractability in large-scale DFT computations. Specifically, the Fe (C) atoms were described by treating 14 (four) valence electrons explicitly. This includes the $3p^6$, $3d^6$, and $4s^2$ subshells in the representation of the Fe atom. The average accuracy of OMX pseudopotential calculations in comparison with all-electron approaches has been determined as $\Delta = 2.0$ meV/atom, where Δ denotes the root-mean-square energy difference between the equations of state obtained by the two methods, averaged over a large set of elemental crystals [37].

Representative geometries for the two prototypes considered in this work are shown in Fig. 10.1. All structures were subjected to geometry optimization under periodic boundary conditions in three dimensions. Specifically, we used a supercell where the transport direction was treated as continuous and vacuum layers were applied in the two directions orthogonal to it. These layers were chosen sufficiently spacious to keep the interaction between adjacent images negligibly small. Thus, in the vertical direction (A, see Fig. 10.1), the separation between neighbor GNRs was set to 10 Å, while their horizontal separation (B in Fig. 10.1) was larger or equal to 8 Å.

Geometry optimization was performed with an atomic force threshold of 0.05 eV/Å as nuclear convergence criterion, while the electronic steps were constrained to converge with an accuracy of 1.0×10^{-6} eV. Wherever periodic boundary conditions were imposed on the system, the lattice parameter was included in the optimization procedure. The basis set size was restricted by a cutoff energy of 120 Hartree, and the Monkhorst-Pack method [38] was applied, with a wave number mesh of dimension $1 \times 1 \times 100$.

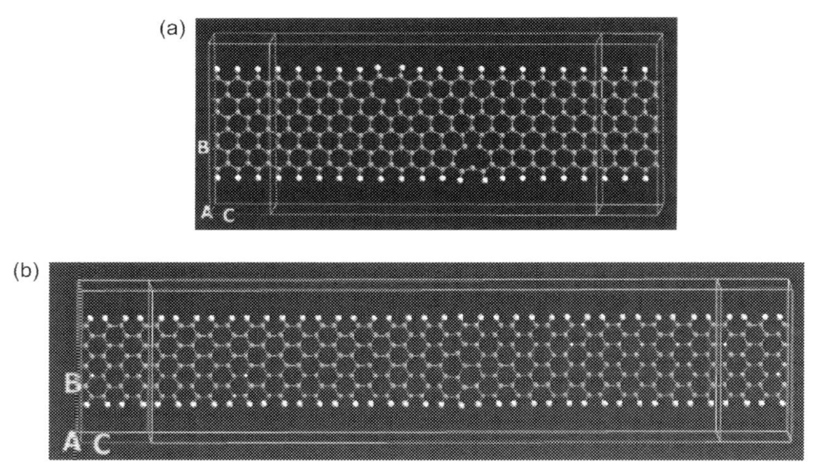

FIG. 10.1 Top views of representative zGNR (A) and aGNR (B) devices, consisting of a transmission element in contact with semi-infinite electrodes. A zGNR (aGNR) of the (12,0) ((0,8)) type is chosen. Both models contain 16 unit cells.

2.2 Transport

The transport geometry for both the zGNR and the aGNR device is presented in Fig. 10.1. In both cases, the transport device consists of two semi-infinite electrodes, enclosing a transmission element derived from a zGNR or an aGNR of the same type.

All transport computations were performed in the framework of the NEGF procedure as implemented in the code Atomistix ToolKit (ATK) [39–41]. The Landauer–Büttiker formula [42] was used to model the spin-dependent current, according to:

$$I_\sigma\left(V_{\text{bias}}\right)=e/h\int T_\sigma(E,V_{\text{bias}})\left[f_{\text{L}}(E-\mu_{\text{L}})-f_{\text{R}}(E-\mu_{\text{R}})\right]\text{d}E. \tag{10.1}$$

In this expression, $f_i(E-\mu_i)$ $(i=\text{L, R})$ denotes Fermi-Dirac distribution functions. The quantities μ_i $(i=\text{L, R})$ are defined as the electrochemical potentials of the left and the right electrode, respectively. They depend on the voltage V_{bias}. Specifically:

$$\mu_{\text{L/R}}\left(V_{\text{bias}}\right)=\mu_i\left(0\right)\pm eV_{\text{bias}}/2. \tag{10.2}$$

The symbol T_σ stands for the spin-dependent transmission:

$$T_\sigma(E,V_{\text{bias}})=Tr\left\{\boldsymbol{\Gamma}_{\mathbf{L}}\mathbf{G}\boldsymbol{\Gamma}_{\mathbf{R}}\mathbf{G}^\dagger\right\}, \tag{10.3}$$

here expressed as function of the energy and the bias. The transmission is defined as the trace over the matrix $\boldsymbol{\Gamma}_{\mathbf{L}}\mathbf{G}\boldsymbol{\Gamma}_{\mathbf{R}}\mathbf{G}^\dagger=\mathbf{t}\,\mathbf{t}^\dagger$, where the symbols $\boldsymbol{\Gamma}_i$ $(i=\text{L, R})$ denote the anti-Hermitian components of the self-energy for the left (L) and the right (R) contact, while **G** stands for the energy-dependent matrix of the Green's function for the transmission element. Further, **t** refers to the *transmission matrix* whose elements t_{nm} are the amplitudes for the transition of an electron from state n of one of the two electrodes into state m of the other [43]. To obtain the transmission at $V_{\text{bias}}\neq 0$, the effective electrostatic potential induced by the electrodes across the transmission element was calculated by use of the Poisson equation.

The spin polarization accumulated by the current as it traverses the device is measured by the magnetocurrent ratio MCR, defined as

$$\text{MCR}=(\text{I}\uparrow-\text{I}\downarrow)/(\text{I}\uparrow+\text{I}\downarrow), \tag{10.4}$$

where $\text{I}\uparrow(\text{I}\downarrow)$ stands for the current with up (down) spin orientation. MCR convergence was used as selection criterion for the transmission element of minimally admissible length. MCR convergence tests were used to determine the minimum GNR length, i.e., 16 unit cells for both types (see Fig. 10.1).

Similar observations were made with respect to the electrodes where a length of 2 aGNR unit cells was found to be sufficient. Thus, for the calculations reported here, a model involving a transmission element (electrodes) with 16 (2) aGNR unit cells was adopted.

The ribbon widths were chosen as a compromise between the extension of the systems, the numbers of their constituents, and their energy gaps. The latter tend to increase with diminishing width [44]. A large GNR energy gap, however, implies that current will flow through the device only beyond a substantial bias threshold. Large bias, on the other hand, may necessitate a transmission element of prohibitive length, to guarantee sufficient electric screening of the electrode regimes.

The zGNRs were modeled under conditions of longitudinal strain. To define the strain, we extended the mid-section of the transmission element, consisting of 10 zGNR unit cells, by fixed margins, stretching it by factors of 1.01, 1.03, and 1.05. Introducing the strain parameter ε by the relation.

$$|\mathbf{T}| = (1+\varepsilon)\,|\mathbf{T}_0|, \tag{10.5}$$

we label the three strain levels by ε_i ($i=0$–3), where $\varepsilon_0=0$, $\varepsilon_1=0.01$, $\varepsilon_2=0.03$, and $\varepsilon_3=0.05$. In relation (*V*), $\mathbf{T}$ is the translational vector of the zGNR, where $\mathbf{T}_0$ refers to the unstrained ribbon. Within the length constraint defined by $|\mathbf{T}|$, full geometry optimizations of the mid-section were carried out. The two bounding regions of the transmission element, each one 3 zGNR unit cells long, were defined as replica of the electrode unit cells.

3 Results and discussion

In this section, we present and discuss our findings on the spin-filtering effects associated with the two prototypes examined in this work, based on zGNR and aGNR units. We first consider current spin polarization caused by vacancies in the GNR fabric (Section 3.1). This is followed by an assessment of the spin-filtering capability of aGNRs with embedded transition metal atoms (Section 3.2).

3.1 Graphene nanoribbons with vacant sites

3.1.1 *Zigzag nanoribbons*

In this section, we present our results on zGNR networks with vacancies in terms of their equilibrium structures, their transmission behavior, and their spin-filtering characteristics as a function of the strain and the bias.

3.1.1.1 EQUILIBRIUM STRUCTURES

We derive the spin-filtering systems studied in this work from a zigzag graphene nanoribbon (ZGNR) of the type (12,0) (see Fig. 10.2A). In particular, they are obtained by excising two carbon atoms from this pattern. We distinguish between two prototypes: *VAC-I*, where the carbon atoms are removed from the polyene chains adjacent to the ribbon edges (Fig. 10.2B), and *VAC-II*, where the atoms are eliminated from the bulk of the ribbon, namely, the third and the fourth polyene chain from either edge (Fig. 10.2D). Upon lattice reconstruction, these geometric defects approach the structure of a five-fold ring (pentagon) merged with a nine-fold ring (nonagon). For system 2(b) [2(d)], the longest distance between two neighboring carbon atoms within the pentagon substructure is found to be 1.75 Å [1.76 Å] and thus substantially elongated as compared with the equilibrium C—C bond distance in graphene, 1.42 Å [40].

Longitudinal strain, applied to the nanoribbons (see Section 2), modifies the structural defects of the prototypes 1(b) and 1(d). Specifically, the orientation of the defects changes, as is seen by comparing 1(b) [1(d)] with the strained structure 1(c) [1(e)]. With reference to the

(a)

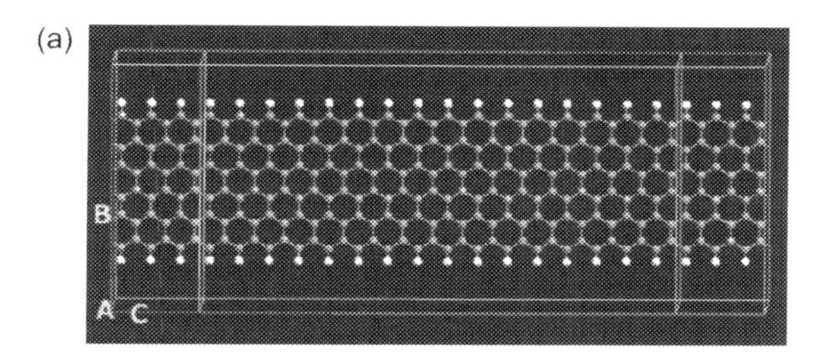

(b)

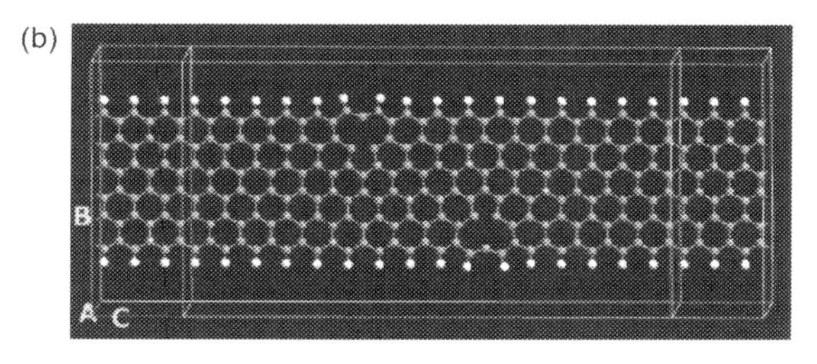

(c)

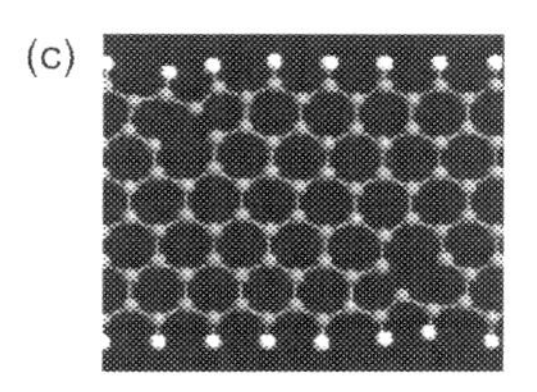

(d)

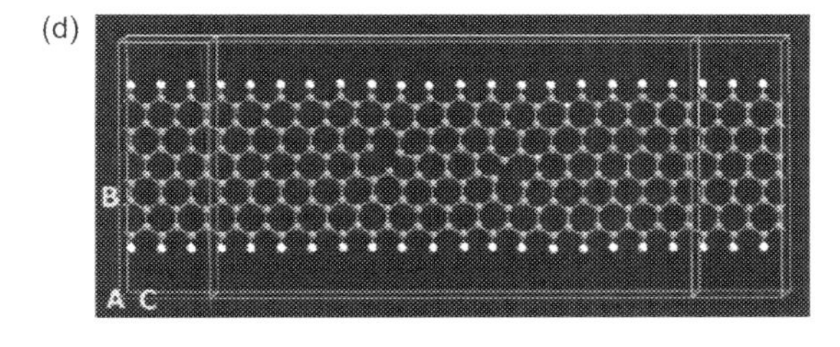

(e)

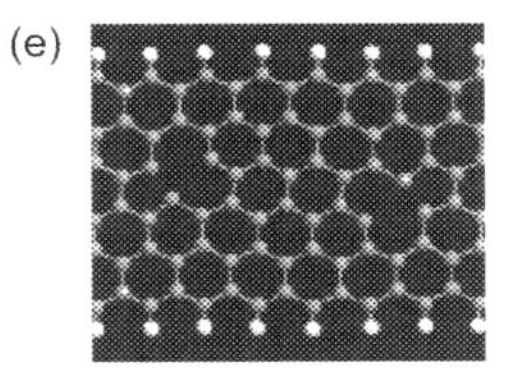

FIG. 10.2 The three zGNR-based device types examined in this work. All are derived from a segment of the hydrogenated (12,0) zGNR, consisting of 16 unit cells. The electrode unit cells are indicated on the left and the right of the transmission element. Panels (A)–(E) show: (A) The intact device, (B) the structure *VAC-I*, involving two symmetrically placed C atom vacancies in the edge regime of the device, (C) deformation of the vacancy substructures of *VAC-I* in response to longitudinal strain at the level ε_3 (= 0.05, as defined by relation (*V*)), (D) the structure *VAC-II*, involving two symmetrically placed C atom vacancies in the bulk regime of the device, and (E) deformation of the vacancy substructures of *VAC-II* in response to longitudinal strain at the level ε_3.

former, the bond shared by the pentagon and the nonagon motives in the unstrained system is cleaved as one applies a sufficient amount of strain. Simultaneously, a C—C distance that connects the edge chain and the chain next to it contracts, giving rise to an effective rotation of the defect by 120°. This behavior is observed for strain levels exceeding ε_1.

3.1.1.2 TRANSMISSION ANALYSIS

Fig. 10.3A–D shows the transmission spectra for the three zGNR prototypes considered here, subdivided according to alpha and beta spin contributions. The results for the intact zGNR (12,0) device, shown in Fig. 10.3A and B, provide a reference for the two alternative systems. To a good approximation, the transmission values for this reference structure are quantized, as expected for transport through zGNRs [45]. We comment first on the case of FM coordination between the ribbon edges (see Fig. 10.3B). For alpha spin, one finds a marked transmission maximum at the energy $E = -0.44$ eV. The transmission at this energy exceeds that of the beta spin moiety by about a factor of three.

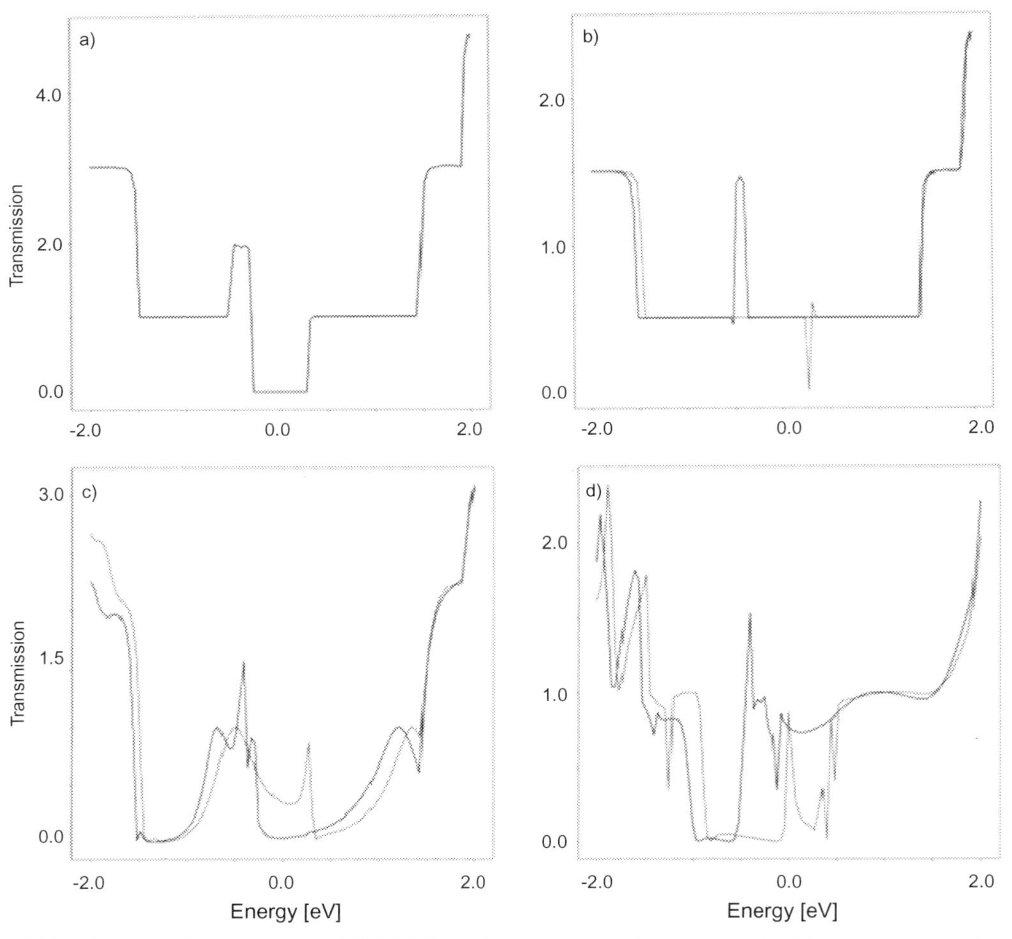

FIG. 10.3 Transmission spectra for zGNR (12,0) devices with and without vacancies. The assignments of panels are as follows: (A) the intact zGNR (12,0) device, with AFM coordination between the ribbon edges, (B) the intact zGNR (12,0) device, with FM coordination between the ribbon edges, (C, D) the structures *VAC-I* (C) and *VAC-II* (D) with FM coordination between the ribbon edges. The *black* (*red, dark gray* in the print version) *line* denotes the spin alpha (beta) component.

TABLE 10.1 Transmission eigenvalues (EVs) for the systems zGNR-(12,0), *VAC-I*, and *VAC-II* at the energy E ($E < E_F$) of the majority spin transmission maximum closest to the Fermi energy

System	Energy (eV)	Alpha EV	Beta EV
zGNR-(12,0)	−0.440	0.997	1.000
		0.982	
		0.962	
VAC-I	−0.400	0.984	0.890
		0.523	
		5.240×10^{-3}	
VAC-II	−0.400	0.912	2.720×10^{-2}
		0.600	
		9.896×10^{-3}	

FM spin configurations are imposed on all three structures. The three entries under Alpha EV *refer to three transmission channels, corresponding to alpha spin transport along the two ribbon edges (the second and third entries) and through the central section of the nanoribbon (the first entry).*

This different transmission behavior of the two spin moieties is reflected by the transmission eigenvalues at $E = -0.44$ eV. The eigenvalues for the spin alpha (beta) system, corresponding to three (one) transport channels, are listed in Table 10.1. All of them are close to unity. Analysis of the corresponding transmission eigenstates reveals that edge current prevails in the alpha channels listed second and third [46].

Comparing the intact system with the *VAC-I* structure, one finds that the transmission maximum decreases and broadens, splitting into three lines, as shown in Fig. 10.3C. At higher energy, a wide alpha-spin gap opens. The transmission eigenvalues of the structure *VAC-I* differ markedly from those of the intact zGNR (see Table 10.1). We note that the appearance of three transmission eigenvalues for the alpha moiety is not directly related to the three peaks shown in Fig. 10.3C. Rather, the eigenvalues are obtained from analyzing the highest spin-alpha transmission peak in the vicinity of the Fermi energy. The maximum of the *VAC-I* majority (alpha) spin transmission spectrum is reached at $E = -0.40$ eV. At this energy, the transmission eigenvalues of the alpha system are significantly lower than those of zGNR-(12,0), indicating an overall decrease of the transmission by about a factor of 2. This trend reflects a very substantial loss of the edge channel transport efficiency. While the vacancy-induced perturbations leave the dominant transport channel largely intact, the two remaining channels are strongly affected by the presence of the vacancies. Thus, the vacancies cause partial blockage of the current through the nanoribbon device, significantly diminishing the current through its edges. At the Fermi energy, one transmission channel is identified. The corresponding eigenvalue for the majority (minority) spin orientation is 0.037 (0.344). Alpha spin transport is thus suppressed with respect to beta spin transport by almost a factor of 10.

Turning to the structure *VAC-II*, we detect reduced transport efficiency for the first of the three transmission eigenvalues when comparing with the corresponding eigenvalues of

VAC-I (see Table 10.1). In the remaining two channels, however, we find an increase of this efficiency. The latter observation appears natural since both the second and the third channel involve edge transport, which is obstructed more in *VAC-I* than in *VAC-II*. The decrease in transmission through channel 1, in contrast, is ascribed to the vacancies in the mid-section of the ribbon which hinder transport through the bulk of the structure. Likewise, the reduced transmission in the *VAC-II* beta channel is associated with preferential transport through the edges where alpha spin polarization prevails.

To summarize, the systems *VAC-I* and *VAC-II* are complementary to each other not only with respect to their geometric structures, but also in terms of their transmission behavior. As established by the preceding analysis, the transmission spectra of these units are determined by marked alpha-spin (*VAC-I*) and beta-spin (*VAC-II*) gaps, respectively, in the vicinity of the Fermi energy.

3.1.1.3 DEPENDENCE OF MAGNETOCURRENT RATIOS ON THE STRAIN AND THE BIAS

Both *VAC-I* and *VAC-II* are metallic and thus do not have energy gaps. However, both show sizable transmission gaps for one of the two spin moieties in the vicinity of the Fermi energy. The size and the spectral position of these gaps can be tuned by applying longitudinal strain. This is confirmed by the entries in Table 10.2, which lists the widths and the upper edges of the beta (alpha) spin gaps identified for *VAC-I* (*VAC-II*) for the transmission spectra of *VAC-I* and *VAC-II*.

Due to the structural changes undergone by both *VAC-I* and *VAC-II* as a function of the strain level (see Fig. 10.2), the change of the gap parameters in response to increasing strain turns out to be somewhat irregular. However, a steady rise of the upper edge of the beta spin gap from −0.04 to 0.20 eV is recorded as the strain level changes from ε_0 to ε_3. Simultaneously, the gap widens.

In what follows, we discuss our MCR results for both devices, *VAC-I* and *VAC-II*. Commenting first on *VAC-II*, we conclude from the existence of a pronounced spin beta gap in the proximity of E_F (Fig. 10.3D) in conjunction with the broadening of this gap with increasing strain level that the MCR will be positive and also increases with the amount of

TABLE 10.2 Spin transmission gaps of the structures *VAC-I* and *VAC-II* in the vicinity of the Fermi energy as a function of longitudinal strain

System	Spin	Strain parameter	Upper edge[a] (eV)	Gap size (eV)
VAC-I	Alpha	ε_1	0.28	0.72
		ε_2	0.20	0.68
		ε_3	0.24	0.64
VAC-II	Beta	ε_0	−0.04	1.00
		ε_1	0.12	0.84
		ε_2	0.16	0.84
		ε_3	0.20	1.32

[a] *Position of the maximum that marks the upper edge of the spin transmission gap.*
No definite gap size could be assigned to VAC-I *in the unstrained state* (ε_0, *see Fig. 10.3C).*

TABLE 10.3 Magnetocurrent ratios MCR for the structures *VAC-I* and *VAC-II* in the voltage regime $V_{\text{bias}} \leq 200$ mV

V_{bias} (mV)	MCR (ε_0)	MCR (ε_1)	MCR (ε_2)	MCR (ε_3)
(i) MCR values for *VAC-I*				
10.0	−0.801	−1.000	−0.777	−0.385
20.0	−0.801	−1.000	−0.773	−0.379
50.0	−0.800	−0.999	−0.770	−0.364
100	−0.799	−0.999	−0.763	−0.340
200	−0.796	−0.998	−0.752	−0.332
(ii) MCR values for *VAC-II*				
10.0	0.217	0.543	0.778	0.890
20.0	0.221	0.538	0.777	0.887
50.0	0.232	0.535	0.769	0.882
100	0.260	0.526	0.752	0.869
200	0.392	0.473	0.687	0.813

longitudinal strain. These conclusions are borne out by the respective results in Table 10.3. As a consequence, the alpha spin transmission at the Fermi energy grows from 47% to 94% as one goes from ε_0 to ε_3. In all strained structures, the alpha spin transmission diminishes as a bias window is opened. Evaluating the net transmission, one still finds the alpha component strongly dominant. However, as the width of the bias window increases, contributions due to the peak spin beta transmission peak at $E = E_F$ (see Fig. 10.3D) enter with increasing weight, lowering the degree of alpha polarization. For the unstrained structure, one encounters the opposite trend, reflecting the near-coincidence of the upper edge of the beta transmission gap with the Fermi energy in this case.

Analogous arguments explain the trends displayed by the MCR values for *VAC-I*. In this case, spin polarization is due to an alpha spin transmission gap, rationalizing the negative sign of all MCRs. Under conditions of zero or low strain, the beta spin transmission strongly exceeds alpha transmission for all considered bias windows, resulting in MCRs of high magnitude (Fig. 10.4).

As one goes to higher strain levels, the alpha spin transmission gap narrows and shifts downward. At the highest level, ε_3, and not too far from the Fermi energy, alpha (beta) spin polarization prevails in the $E > E_F$ ($E < E_F$) regime. In consequence, the MCRs in this region are of sizeably lower magnitude than for unstrained or slightly strained devices. We point out that a current spin polarization of 100% is attained at the strain level ε_1. Exposing *VAC-I* to this amount of strain thus gives rise to perfect spin-filter activity. This finding remains largely uncompromised in the entire bias interval 10.0 mV $\leq V_{\text{bias}} \leq$ 200 mV.

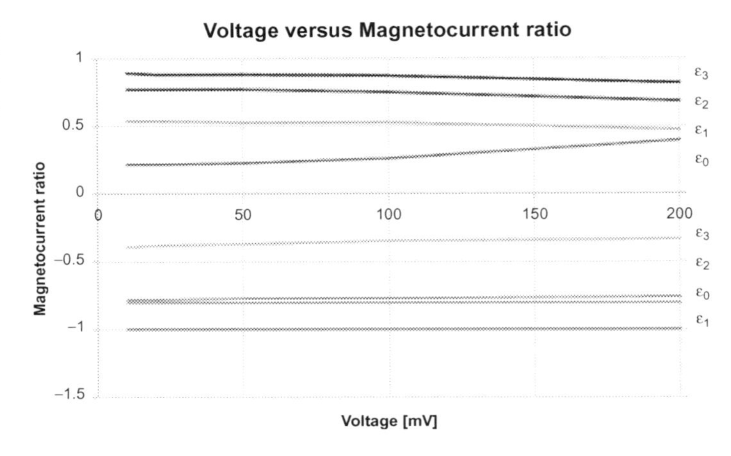

FIG. 10.4 Magnetocurrent ratios MCR for the structures *VAC-II* (*upper panel*) and *VAC-I* (*lower panel*) in the voltage regime $V_{\mathrm{bias}} \leq 200$ mV (for the numerical values, see Table 10.3).

3.1.2 *Armchair nanoribbons*

While voids in zGNRs give rise to marked spin polarization effects, as described above, eliminating two C atoms from selected sites in aGNRs also creates unpaired spins in the graphene network and is thus expected to induce magnetism. In order to explore the spin-filtering features of an (8,0) aGNR device with vacancies, we optimized a geometry comparable with a zGNR model shown in Fig. 10.2D. Once more, the geometric motif that arises from the reconstruction of the hole site may be characterized as a nonagon adjacent to a pentagon. The bond distance shared between the two substructures is markedly cleaved, measuring 1.72 Å. This strongly exceeds the remaining four pentagon (eight nonagon) side lengths, which vary between 1.40 and 1.44 Å (1.36 and 1.52 Å).

This two-hole system was examined by imposing conditions for initial spin orientations on the C atoms surrounding the vacant sites. In contrast to the previously discussed systems, zGNR units under FM conditions, aGNRs with vacant sites in their bulk do not exhibit a favored mode of spin ordering. Thus, *2vac-aGNR* was optimized under spin-parallel or spin-antiparallel conditions, involving identical or opposite orientations of the unpaired spins surrounding the two vacancies, respectively. For the system displayed in Fig. 10.5A, the former alternative turned out to be preferred over the latter by a minute energy margin of 1.4 meV. We note that this figure is so small that the electronic structure method applied here does not allow for an unambiguous comparison between the two spin structures in terms of stability. Fig. 10.5B shows the spin density distribution for the spin-parallel case. Net spins are localized at the sites of dangling bonds that emerge as two C atoms are excised from the aGNR network.

Evaluating the MCR in the bias interval [0 V, 0.5 V], we find that *2vac-aGNR* has a spin-polarizing effect, in analogy to the zGNR devices previously discussed. The spin polarization achieved for aGNR with two vacancies, however, is distinctly lower than in the case of zGNR hosts. This is documented in Fig. 10.6A, which shows the MCR as a function of the bias for both cases considered here, i.e., the spin-parallel and the spin-antiparallel configuration. We point out that higher degrees of current spin polarization have been established for zGNRs with vacant sites [47].

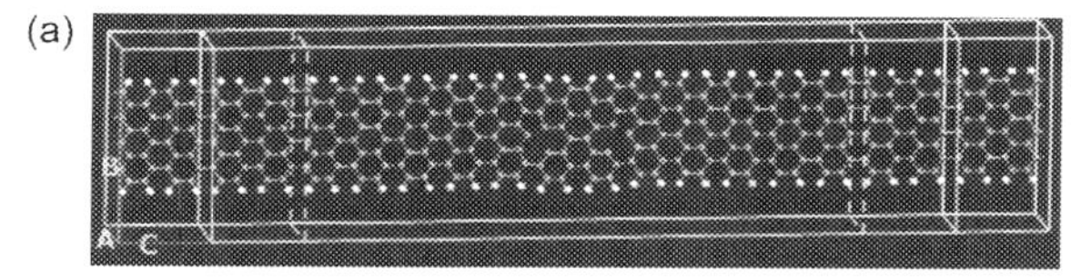

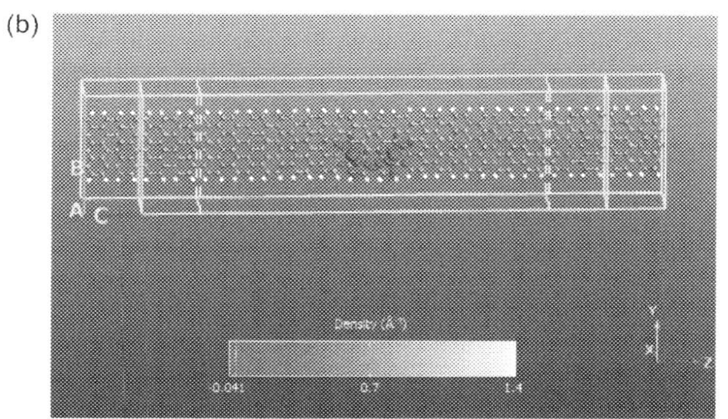

FIG. 10.5 The (8,0) GNR device with vacancies at the sites previously occupied by substitutional Fe atoms (*2vac-aGNR*). (A) Geometric structure of *2vac-aGNR* and (B) spin density distribution of 2 holes-aGNR in spin-parallel conditions. The isovalue for this representation is 0.01.

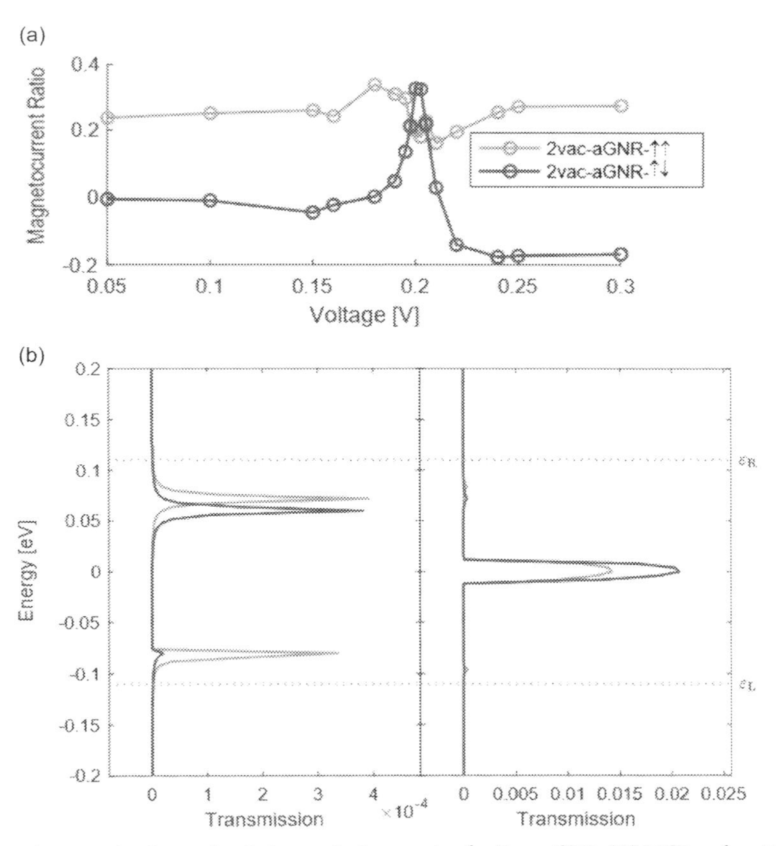

FIG. 10.6 Magnetocurrent ratios and spin transmission spectra for *2vac-aGNR*. (A) MCR as function of the bias for *2vac-aGNR* in the spin-parallel and spin-antiparallel case. (B) Transmission spectra for *2vac-aGNR*↑↓ at two bias ranges: 0.20 V (*left panel*) and 0.24 V (*right panel*). The *blue* [*light gray* in the print version] (*orange* [*dark gray* in the print version]) curve refers to spin-parallel (spin-antiparallel) conditions.

In the spin-parallel case, the MCR for the spin-parallel configuration turns out to be uniformly positive, reflecting the spin polarization of unpaired π electrons at the edges of the vacant sites, which is positive for both vacancies.

A more complex behavior is displayed by the spin-antiparallel configuration (*2vac-aGNR*-$\uparrow\downarrow$). As expected, the MCR is here close to zero in the low-bias regime. After rising to a maximum, it decreases for bias values $V > 0.2$ V. This trend is reflected in Fig. 10.6B, which shows *2vac-aGNR*$\uparrow\downarrow$ for two choices of the bias; 0.20 V (left) and 0.24 V (right). For the latter choice, $V > V_{gap}$, with V_{gap} as the threshold bias that bridges the electrode energy gap. This threshold was found to be 2.02 V. The two vacant sites are here on markedly different electric potentials and thus may have different impact on the current. At $V = 0.20$ V, spin majority outweighs spin minority transmission. The spectrum at $V = 0.24$ V exhibits two maxima, corresponding to opposite spin orientations, around $E = E_F$. This profile is typically found at voltages exceeding the threshold voltage, which matches the aGNR energy gap, by a small amount (compare with Fig. 10.6B). Here, spin minority surpasses spin majority transmission by a small but distinct margin.

3.2 Spin transport properties of armchair graphene nanoribbons with substitutional transition metal atoms

The results presented and discussed in the previous section suggest that GNRs with magnetized edges or vacant sites could be operative as spin-filtering devices, and thus be of interest for nanospintronics. For both GNR families, aGNRs and zGNRs, we identified structures with unpaired spins at the ribbon edges or vacancy sites that turned out to transfer net spin polarization to previously unpolarized current. In terms of the actual applications of these systems as spin filters, the limiting factor is their inherent instability rather than their maximally achievable current spin polarization. Thus, the magnetically active zGNR ribbon edges are chemically fragile, being prone to oxidation. Further, sensitive degrading of the edge spin magnetism in current-loaded zGNRs beyond a voltage threshold of about 0.4 V was established in [48,49]. In view of these concerns, the inherently non-magnetic aGNRs may present a promising alternative to the zGNR type, provided they incorporate magnetic metal atoms as guest species.

In the following, we present our findings on spin transport through various systems based on the (0, 8) system, as considered in Section 3.1.2. In particular, we analyze an aGNR with a single Fe impurity atom (*1Fe-aGNR*), two Fe atoms (*2Fe-aGNR*), and a vacant site enclosed by two Fe atoms (*Fe-vac-Fe-aGNR*). Before turning to transport calculations, we will compare band structure results on the pure aGNR host and on 1Fe-aGNR, as the most elementary of the devices considered here.

3.2.1 A single substitutional Fe atom: Band structure calculations

The unit cell geometry adopted for the case of a single Fe atom substitution for a C atom in the aGNR (*1Fe-aGNR*) is shown in Fig. 10.7A. We choose a substitution site in the central section of the aGNR. Relaxing the geometry causes the Fe atom to move away from the aGNR plane and to locate at a vertical distance of 1.302 Å above this plane. By Mulliken population analysis, the expectation value of the difference between the majority and minority

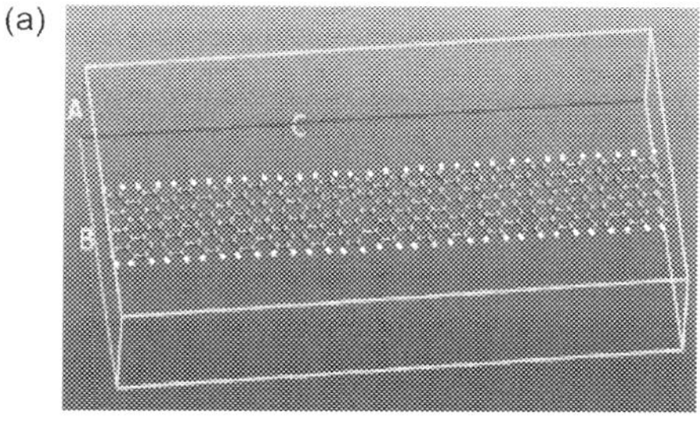

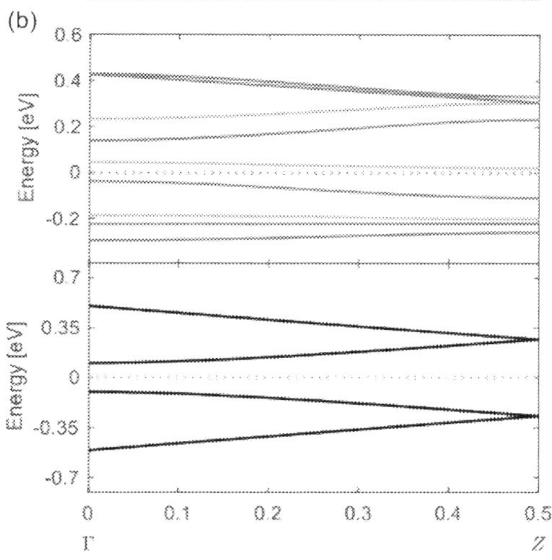

FIG. 10.7 Band structure of aGNR and *1Fe-aGNR*. (A) Scheme of *1Fe-aGNR* at equilibrium geometry. (B) *Upper panel*: band structure of *1Fe-aGNR*. Shown are the bands closest to the Fermi energy E_F, set equal to 0. Majority bands are shown in *orange* (*dark gray* in the print version), while minority bands in *green* (*light gray* in the print version) *color*. *Lower panel*: the same but for the pure aGNR host. The units of the horizontal axis are in $1/c$, where c is the lattice constant of the GNR, measured in Å.

populations of the Fe atom amounts to 1.11 a.u. This margin is largely due to the spin-polarized $3d^6$ and $4s^2$ subshells of the Fe atom, with the former (latter) contributing 0.96 (0.15) a.u to the overall difference.

Fig. 10.7B shows the *1Fe-aGNR* band structure in the vicinity of the Fermi energy (E_F). In this regime, the energy bands are substantially reorganized by the presence of the Fe impurity, as demonstrated by comparison with the band structure of the reference system, the underlying armchair ribbon (see Fig. 10.7B, *lower panel*). As we substitute an Fe atom, the energy gap of pure aGNR, which amounts to 2.02 eV at the Γ point, narrows markedly. Further, the highest occupied and lowest unoccupied bands closest to E_F in aGNR, both two-fold degenerate, each split into two separate bands with spin-up or spin-down character. Specifically, the highest occupied (lowest unoccupied) band of *1Fe-aGNR* has minority (majority) character. Population analysis of the occupied band closest to E_F reveals that it is composed mostly of C(π) orbitals, with a significant admixture due to the Fe(3d) shell, where the latter component varies between 13% ($k=0$) and 22% ($k=\pi/L$, with L as the length of the unit cell). The four occupied bands in the energy interval [−0.3, 0.0] eV are composed mostly of C(π) orbitals, excepting the third band where Fe(3d) orbitals dominate, with contributions that vary from 64% to 69%. As indicated in Fig. 10.7B, three of the four bands display spin-down orientation.

3.2.2 A single substitutional Fe atom: Transport calculations

The device geometry adopted for *1Fe-aGNR*, consisting of the *1Fe-aGNR* transmission element surrounded by aGNR electrodes, is shown in Fig. 10.1. We computed the device density-of-states (DDOS) distribution in a range of [−4.0 eV, +4.0 eV]. The Fermi energy is set equal to zero. A small but distinct energy gap of 0.21 eV separates the occupied and the unoccupied regime (see Fig. 10.8A, *left panel*). While the structure of the DDOS distribution is governed by marked Van-Hove singularities, as expected for a graphene-based system, their clearly noticeable asymmetry is ascribed to the presence of the Fe atom in the aGNR fabric. Close to the Fermi energy, the minority spin orientation is seen to prevail over the majority orientation. This asymmetry between the spin moieties becomes still more pronounced when the device density of states is projected on the Fe site, as shown in the *right panel* of Fig. 10.8A. Similar conclusions are drawn from a bulk calculation with the unit cell defined by the central region shown in Fig. 10.2. Here, analyzing the spectrum close to the Fermi

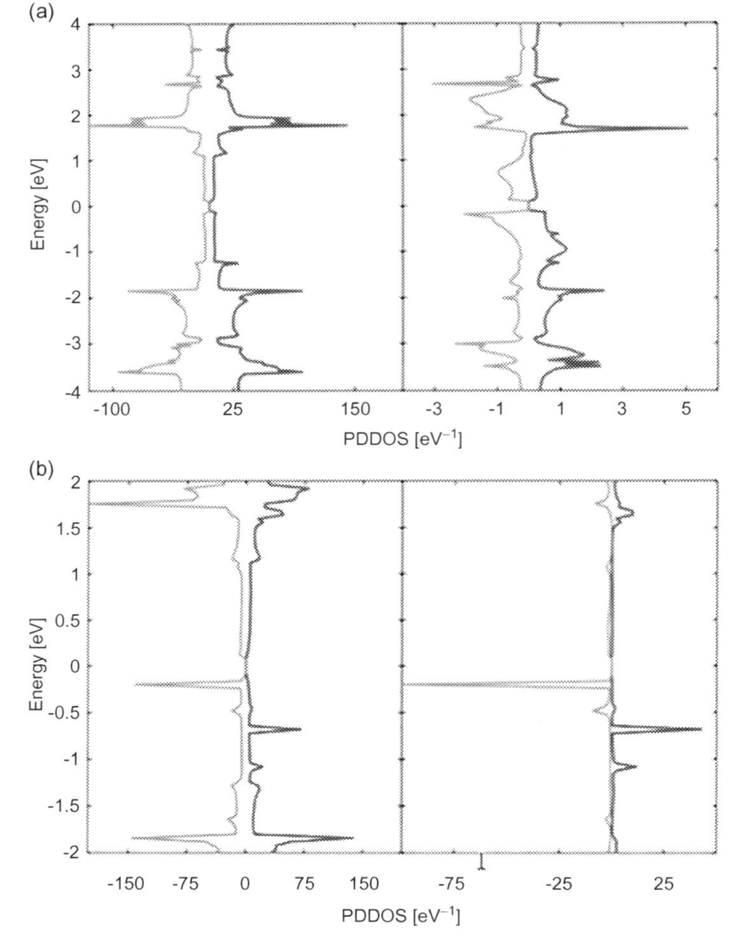

FIG. 10.8 Device density-of-states (DDOS) distributions for *n*Fe-aGNR ($n = 1$, 2). (A) DDOS distributions for *1Fe-aGNR* in the energy regime [−4.0 eV, +4.0 eV]. *Left panel*: total DDOS, *right panel*: partial DDOS (PDDOS), projected on the Fe atom. (B) DDOS distributions for *2Fe-aGNR* in the energy regime [−2.0 eV, +2.0 eV]. *Left panel*: total DDOS, *right panel*: partial DDOS (PDDOS), projected on the two Fe atoms.

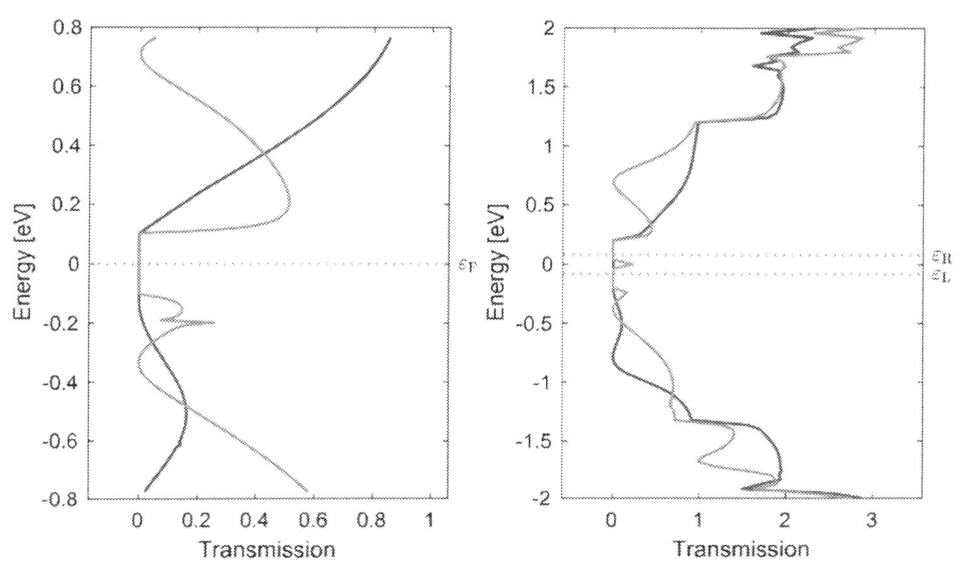

FIG. 10.9 Transmission spectrum of *1Fe-aGNR* in the energy regime [−0.8 eV, +0.8 eV]. The *light-blue* [*light gray* in the print version] (*orange* [*dark gray* in the print version]) *line* refers to majority (minority) spin. *Left panel*: Transmission at 0 V. *Right panel*: Transmission at $V = 0.208$ V. The labels ε_L and ε_R refer to the Fermi energies of the left and the right electrode, respectively.

energy yields -1.49×10^{-2} eV (-1.62×10^{-1} eV) as the energy eigenvalues for the HOMO (HOMO-1) of the minority versus -2.46×10^{-2} eV (-4.27×10^{-1} eV) for the majority subsystem.

This feature is reflected by the transmission spectrum at zero bias and close to the Fermi energy (see Fig. 10.9, *left panel*). As expected, transmission almost vanishes in the gap region. In the regimes adjacent to the gap, minority transport clearly dominates majority transport.

To make a more explicit statement about the transport properties of the *1Fe-aGNR* device, we examined it under finite electric bias, exploring bias windows ΔV in the range [0.1 V, 0.4 V]. The voltage threshold for the onset of substantial transmission within the bias window is given by the electrode (i.e., the aGNR) energy gap. Fig. 10.9, *right panel*, shows the *1Fe-aGNR* transmission spectrum at a bias slightly exceeding this margin, namely, 0.208 V. A distinct minority spin transmission peak appears at $E = E_F$, reflecting the influence of the highest occupied *1Fe-aGNR* band (Fig. 10.7B), which turned out to have spin minority character.

As expected, the current increases very sizably beyond the voltage threshold (see Fig. 10.10, *upper panel*). Further, the pronounced spin asymmetry in the neighborhood of the energy gap is documented by a strong spin polarization of the transmitted current. The respective magnetocurrent ratios are shown in Fig. 10.10, *lower panel*. They are consistently negative in the explored bias regime, corresponding to minority spin prevalence, and assume their maximum magnitude close to the threshold voltage. We emphasize that the spin polarization of the current turns out to be very substantial for bias values that slightly exceed the energy gap, reaching into the 90% range.

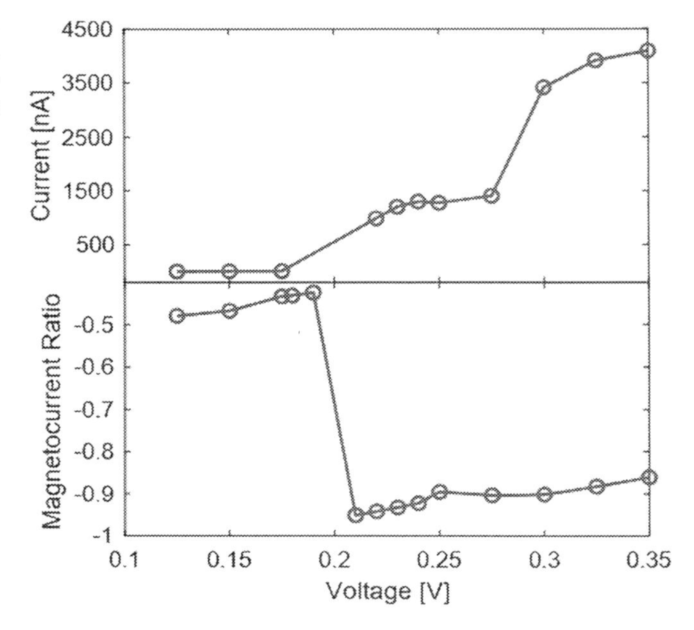

FIG. 10.10 *Upper panel*: Current for 1Fe-aGNR as a function of the bias across the device in the interval [0.125 V, 0.350 V]. *Lower panel*: The corresponding magnetocurrent ratio (MCR).

3.2.3 *Two substitutional Fe atoms*

We discuss first the case of two Fe atoms replacing two C atoms at symmetric sites with respect to the vertical bisecting plane of the chosen aGNR device (*2Fe-aGNR*, see Fig. 10.11). The system exhibits two geometric and two spin distribution prototypes. Excluding the planar structure, the two substitutional atoms may move out of the aGNR plane in the same direction (*2Fe-cis-aGNR* Fig. 10.11A), or in opposite directions (*2Fe-trans-aGNR* Fig. 10.11B). Likewise, the spin distribution of *2Fe-aGNR* may be determined by parallel or antiparallel orientation of the spins centered on the two substitutional atoms. Defining the antiparallel configuration, we adopt the convention that the left (right) Fe center carries alpha (beta) spin polarization. In summary, four principal configurations need to be included in the computational treatment of this system.

In terms of their relative stabilities, geometry optimization yields the hierarchy indicated in Table 10.4. The lowest total energy is obtained for the maximally symmetric device, involving parallel displacement of the two Fe atoms from the aGNR plane as well as parallel spin orientations. Correspondingly, the least symmetric device with respect to both geometry and spin turned out to be of lowest stability. We note that the same sequence of stabilities is found when the FHI [50] rather than the OMX pseudopotential is used.

The size of the magnetic moments, μ(Fe), localized at the Fe centers, is found to be equal at both Fe atoms, as expected. From Table 10.4, the systems with asymmetric spin distribution exhibit smaller local magnetic moments than those with symmetric spins.

Proceeding to the transport properties of *2Fe-aGNR*, and focusing on the most stable one among the four compared species (*2Fe-cis-aGNR-*$\uparrow\uparrow$), we find once more a strongly spin-polarized current. In accordance with our findings for *1Fe-aGNR*, the system displays an

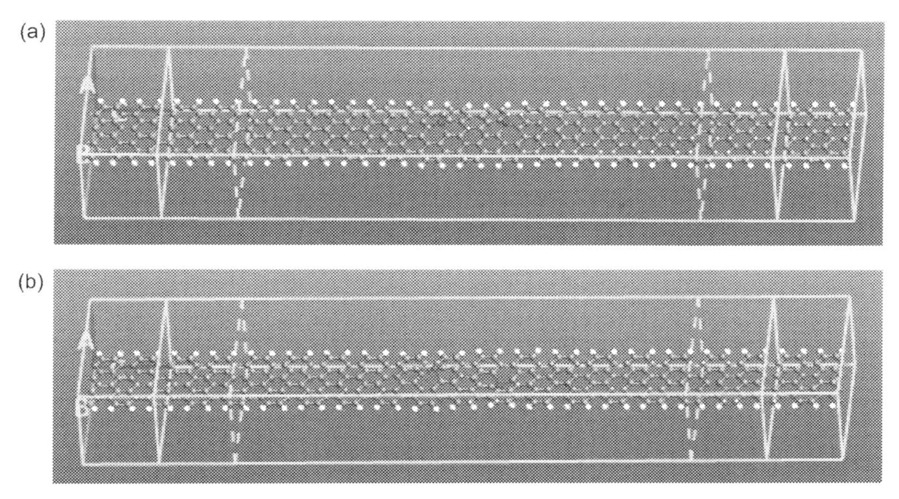

FIG. 10.11 Equilibrium structures of 2Fe-aGNR. Shown are the two geometric prototypes: *2Fe-cis-aGNR* (both Fe atoms above the aGNR plane (A)), *2Fe-trans-aGNR* (one Fe atom above and one below the aGNR plane (B)).

TABLE 10.4 Total energies of the aGNR devices with two substitutional Fe atoms, and magnetic moments localized at either Fe site

System	E^{a} (eV)	$\lvert\mu(\mathrm{Fe})\rvert^{b}$ (a.u.)
2Fe-cis-aGNR-↑↑	0^{c}	0.942
2Fe-cis-aGNR-↑↓	0.141	0.770
2Fe-trans-aGNR-↑↑	0.783	0.940
2Fe-trans-aGNR-↑↓	0.929	0.762

[a] *Total energy.*
[b] *Magnitude of magnetic moment expectation value at the Fe centers.*
[c] E(2Fe-cis-aGNR-↑↑) = −*47,903.21309 eV.*
The energy is defined with reference to the system found most stable, 2Fe-cis-aGNR-↑↑.

energy gap of 0.2 eV, as documented by the DDOS distributions shown in Fig. 10.8B, and the energy regimes surrounding the gap are dominated by spin minority components. In consequence, spin minority outweighs spin majority transport close to the Fermi energy, as confirmed by the transmission spectrum in Fig. 10.12A.

At $V=0$, the transmission spectra for the two spin orientations in *2Fe-cis-aGNR* ↑↑ deviate substantially from each other, while they are identical for *2Fe-cis-aGNR* ↑↓. This is plausible by symmetry, as the two Fe impurities are placed at equivalent sites and exposed to the same (i.e., vanishing) electrostatic potential. This changes as admission is made for a finite voltage. In this case, the two Fe centers of *2Fe-cis-aGNR* ↑↓ are at different electrostatic potential. This is reflected by a noticeable departure of the spin-resolved transmission spectra as $V>0$ (see Fig. 10.12B).

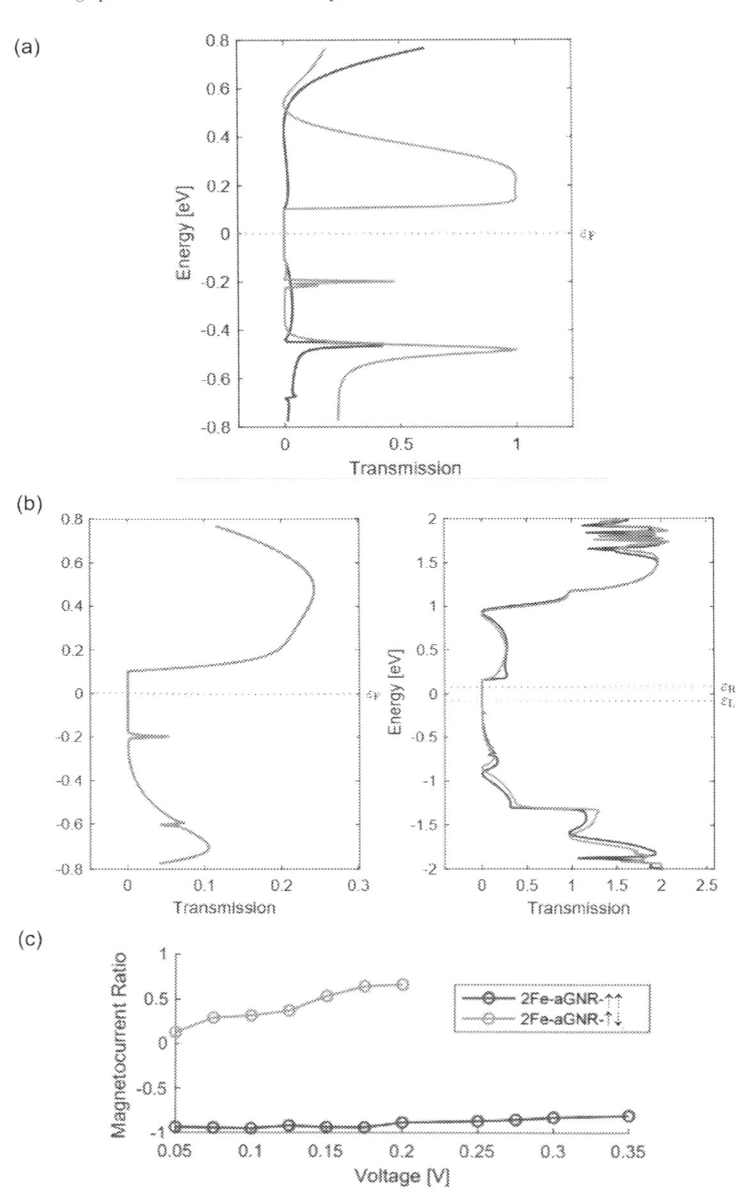

FIG. 10.12 Transport properties of *2Fe-cis-aGNR* with the two Fe atoms on the same ribbon side (see Fig. 10.11B). (A) *2Fe-cis-aGNR*↑↑ transmission spectrum at zero bias in the vicinity of the Fermi energy. (B) *2Fe-cis-aGNR*↑↓ transmission spectra at zero (*left*) and nonzero (*right*, $V=0.125$ V) voltage in the vicinity of the Fermi energy. (C) MCR as function of the bias for *2Fe-cis-aGNR*↑↑ and *2Fe-cis-aGNR*↑↓. The *color* (*different shades of gray* in the print version) *coding* is as in Fig. 10.9.

In consequence, the MCR values found for *2Fe-cis-aGNR*↑↑ and *2Fe-cis-aGNR*↑↓ as a function of the device voltage differ markedly from each other. The comparison between both ratios in Fig. 10.12C is not extended beyond the voltage regime $V \leq 0.2$ V since the antiparallel spin configuration, *2Fe-cis-aGNR*↑↓, turned out to be unstable at $V > 0.2$ V, changing into the spin-parallel ground state in the course of the SCF procedure (see Table 10.1). As implied by the coinciding transmission spectra for the two spin orientations at $V=0$, as well as their marginal deviation at low voltage, the MCR of the system *2Fe-cis-aGNR*↑↓ is seen to be consistently lower than that of *2Fe-cis-aGNR*↑↑ in the considered voltage range. The latter is negative, as in the case of *1Fe-aGNR*, and likewise approaches the limit of half-metallicity, i.e., perfect spin-filtering.

3.2.4 A vacant site between two substitutional Fe atoms

The final model to be discussed in this work was developed from the *2Fe-cis-aGNR* geometry as shown in Fig. 10.13A. Eliminating the two central C atoms between the two Fe centers and repeating the geometry optimization yield the structure displayed in Fig. 10.11A. As expected, the width of the ribbon narrows in the mid-section from which the C atoms were removed. An eight-fold ring emerges, surrounded by two pentagons, each of which contains one Fe impurity.

We studied this system in various magnetic configurations. Specifically, we considered parallel (antiparallel) spins at the Fe sites and combined both choices with initial net spin polarization on all atoms in the eight-fold ring. In keeping with our observations for *2Fe-cis-aGNR*, we find the maximum stability for the model with parallel Fe spins. While the energy difference between the spin-parallel and spin-antiparallel alternatives is in the 100-meV regime for *2Fe-cis-aGNR*, it decreases to $E=5.34$ meV in the present case. The initial spin orientation of the eight-fold ring turned out to be without influence on the total energy of the system. From the magnetic profile shown in Fig. 10.13B for *Fe-vac-Fe-cis-aGNR* in the spin-parallel configuration (*Fe-vac-Fe-cis-aGNR*↑↑), the spin density of the system is predominantly localized on the two Fe centers, with minor admixtures on the eight-fold ring.

The MCR values of *Fe-vac-Fe-cis-aGNR*↑↑ as compared with *2Fe-cis-aGNR*↑↑ are similar at small bias. Beyond $V=0.2$ V, however, the MCR of *Fe-vac-Fe-cis-aGNR*↑↑ decreases rapidly (see Fig. 10.14), in contrast to the trend exhibited by *2Fe-cis-aGNR*↑↑ where the MCR falls

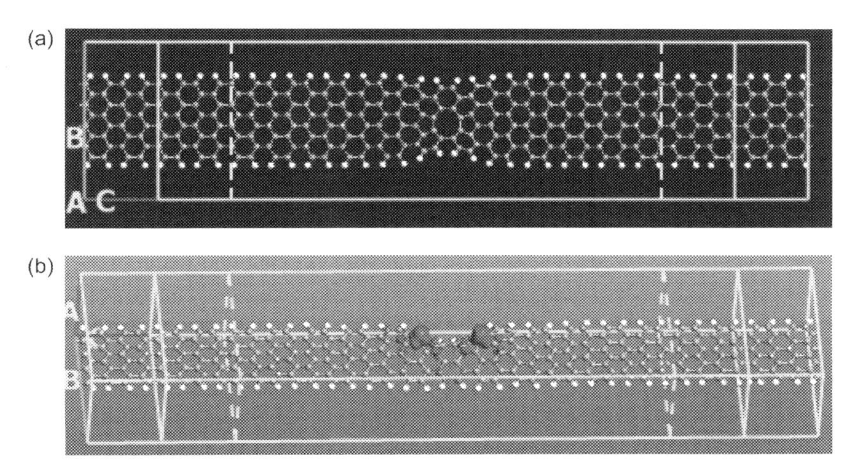

FIG. 10.13 Geometric and magnetic structure of *Fe-vac-Fe-cis-aGNR*. (A) Top view of the device, optimized in spin-parallel conditions. (B) Spin density distribution for *Fe-vac-Fe-cis-aGNR* in spin-parallel conditions. The isosurface was set to 5.0×10^{-3}.

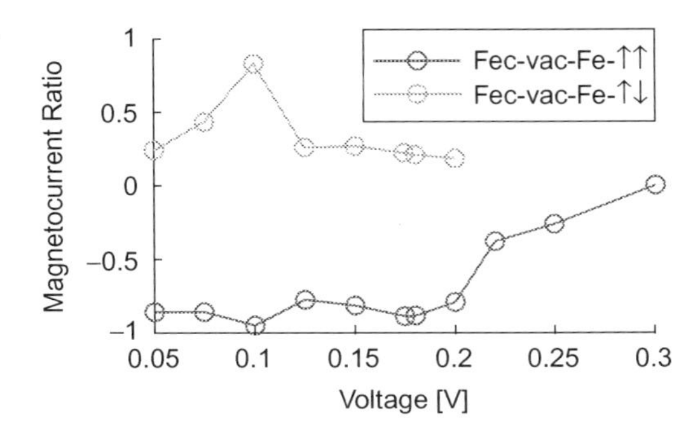

FIG. 10.14 MCR as function of the bias for *Fe-vac-Fe-cis-aGNR* in the spin-parallel and spin-antiparallel case.

much more slowly. This difference reflects strongly differing transmission profiles in the voltage regime $V > 0.2$ V. For *2Fe-cis-aGNR*↑↑, the pronounced spin minority transmission at $E = E_F$ at these voltages accounts for continued high current spin polarization. For *Fe-vac-Fe-cis-aGNR*↑↑, in contrast, no such behavior is found, as the spin majority and minority currents are equally disrupted by the presence of the vacancy between the two Fe centers.

4 Summary and outlook

To conclude this chapter, we present a summary of the findings presented here, followed by an outline of further work. Selected zGNR as well as aGNR prototypes were studied with respect to their spin-filtering capabilities. In particular, we distinguished between two spin-filter mechanisms, involving the interaction of unpolarized current with unpaired spins in the GNR fabric, or with spins in the shells of transition metal atom impurities substituting for C atoms. Both cases were explored by DFT modeling in conjunction with a Green's function approach.

Investigating the first alternative, we considered both zGNR and aGNR structures. As representatives of the zGNR class, (12,0) zGNRs with ferromagnetic coordination of the ribbon edges, and with symmetrically arranged single-C-atom vacancies in the bulk of the ribbon, were considered. Specifically, we included a system with two vacancies close to the ribbon edges (*VAC-I*) and an alternative structure with a vacancy pair in the bulk regime (*VAC-II*). The spin-polarizing effect of these devices was assessed by calculating MCR ratios in the bias interval [0, 200 mV], and at different levels of strain, ranging from $\varepsilon = 0$ to $\varepsilon = 0.05$. In both cases, sizable spin-filtering effects in excess of 80% were observed. The MCR of *VAC-II* was seen to be positive and to increase with increasing strain. *VAC-I*, on the other hand, displayed consistently negative MCR ratios. In the low-bias, low-strain region, the MCR of *VAC-I* was found to be 100%, i.e., the structure proves to be half-metallic.

These findings are rationalized in terms of vacancy-induced spin transmission gaps that determine the spin-selective features of both devices and are characteristically different for the two structural prototypes. For *VAC-I*, edge transport is hindered by the presence of the vacancies, giving rise to extended majority spin gaps and thus negative MCR ratios. The inverse effect is found for *VAC-II* where edge transmission channels are less obstructed, but bulk transmission is reduced. This configuration leads to positive MCR ratios that rise markedly as the strain increases. A general trend of decreasing spin polarization as the width of the bias window increases prevails for both structures.

A pronounced current spin polarization effect, albeit of somewhat lower magnitude, is detected for aGNR systems of the (0,8) family with two vacancies in the bulk of the ribbon. Here, spin transport through the ribbon edges is suppressed and net spin current polarization arises from interaction with unpaired spins localized at the vacant sites. Distinctly different spin polarization profiles with respect to the bias across the ribbon are obtained for parallel and antiparallel orientation of the spins at the vacancies.

The second class of potential GNR spin filters consists of (0,8) systems with one or two substitutional Fe atoms. For both 1Fe-aGNR and 2Fe-aGNR devices, very sizeable degrees of current spin polarization were recorded, amounting to more than 90% in selected ranges of the bias across the device. Examining the stability of alternative structures, 2Fe-cis-aGNR and 2Fe-trans-aGNR, with respect to both their geometries and their magnetic configurations yielded a preference for a system with both Fe atoms above the aGNR plane, and with parallel spins at the Fe sites.

From band structure as well as DOS calculation, minority spin polarization prevails close to the Fermi energy of both systems, 1Fe-aGNR and 2Fe-cis-aGNR. This feature is reflected by the respective MCRs as these are found to be consistently negative in the inspected bias regime of $\Delta V < 0.3$ V if the spin-parallel state is realized. The bias across the device may be viewed as a tuning parameter for the current spin polarization. Maximum MCR magnitudes close to 100% are recorded for 1Fe-aGNR and 2Fe-cis-aGNR. Qualitatively, the same observations are made if one allows for a vacancy between the two Fe-atom sites. For the systems with two Fe atoms, parallel and antiparallel spin configurations give rise to MCRs of negative and positive sign, respectively. While the MCR of maximum magnitude was found at different bias in the compared prototypes 1Fe-aGNR, 2Fe-cis-aGNR, and Fe-vac-Fe-cis-aGNR, its sign turned out to be identical, and its size very similar in these three device types, provided the spin orientations of the Fe atoms in the latter two systems are parallel.

Assessing the two classes of materials considered here in terms of their possible use in the context of nanospintronics, we note that the current spin polarization effect in (0, 8) aGNRs with substitutional Fe atoms shows a certain robustness with respect to the number of the impurity atoms as well as structural details of the aGNR host. While half-metallicity was detected in both zGNRs with vacancies and aGNRs with TM impurities, the latter appear to be more viable as nanospintronics devices. Thus, aGNR-based transmission elements do not suffer from the drawbacks associated with zGNR edge magnetism, involving AFM coordination across the ribbon and susceptibility to degrading chemical reactions.

In continuation of this work, we plan to extend the systems discussed here, replacing the Fe centers with transition metal clusters or chains. This step will be essential to assess these devices in terms of their potential use as spin-filtering elements in nanospintronic circuits. In particular, it will be important to identify a spin-polarizing agent of sufficient magnetic anisotropy to provide a well-defined polarization direction for transmission at finite temperature. To assess this effect, spin-orbit coupling will be taken into account. Adequate inclusion of this interaction is also essential for evaluating the role of spin relaxation processes.

Further, we will pursue the question of the magnetic effects exerted by aGNR adatoms on currents traversing them. Previous studies have dealt with problem of TM adsorption on a graphene substrate in terms of geometric, electronic, and magnetic features [51,52], or with current polarization induced by TM adatoms in graphene sheets [53]. The impact of randomized TM adsorbates on the quantum spin Hall effect has been assessed by recent modeling [53]. It will be interesting to investigate adsorption of TM atoms, molecules, and clusters as a way to endow current with a well-defined degree of spin polarization, and to compare the respective results with spin polarization by TM atom substitution, as considered in this work.

References

[1] R. Sbiaa, H. Meng, S.N. Piramanayagam, Materials with perpendicular magnetic anisotropy for magnetic random access memory, Phys. Status Solidi (RRL) 5 (2011) 413.
[2] https://thefutureofthings.com/3037-mram-the-super-memory/.
[3] S. Bhatti, R. Sbiaa, A. Hirohata, H. Ohno, S. Fukami, S.N. Piramanayagam, Spintronics based random access memory: a review, Mater. Today 20 (2017) 530.
[4] M.D. Stiles, J. Miltat, Spin Transfer Torque and Dynamics, in: B. Hillebrands, A. Thiaville (Eds.), Spin Dynamics in Confined Magnetic Structures III, Topics Appl. Physics, vol. 101, Springer-Verlag, Berlin Heidelberg, 2006, pp. 225–308.
[5] M.D. Stiles, A. Zangwill, Anatomy of spin-transfer torque, Phys. Rev. B 66 (2002) 014407.
[6] D.C. Elias, R.V. Gorbachev, A.S. Mayorov, S.V. Morozov, A.A. Zhukov, P. Blake, L.A. Ponomarenko, I.V. Grigorieva, K.S. Novoselov, F. Guinea, A.K. Geim, Dirac cones reshaped by interaction effects in suspended graphene, Nat. Mater. 7 (2011) 701.
[7] S. Konschuh, M. Gmitra, J. Fabian, Tight-binding theory of the spin-orbit coupling in graphene, Phys. Rev. B 82 (2010) 245412.
[8] M. Gmitra, S. Konschuh, C. Ertler, C. Ambrosch-Draxl, J. Fabian, Band structure topologies of graphene: spin-orbit coupling effects from first principles, Phys. Rev. B 80 (2009) 235431.
[9] M. Wojtaszek, I.J. Vera-Marun, E. Whiteway, M. Hilke, B.J.van wees, absence of hyperfine effects in ^{13}C-graphene spin-valve devices, Phys. Rev. B 89 (2014) 035417.
[10] S. Bhowmick, A.K. Singh, B.I. Yakobson, Quantum dots and nanorods of graphene embedded in hexagonal boron nitride, J. Phys. Chem. C 115 (2011) 9889.
[11] J.M. Pruneda, Origin of half-semimetallicity induced at interfaces of C-BN heterostructures, Phys. Rev. B 81 (2010) 161409.
[12] F. Hagelberg, Magnetism in Carbon Nanostructures, Cambridge University Press, 2017.
[13] D. Zhang, D.-B Zhang, F. Yang, H.-Q. Lin, H. Xu, K. Chang, Interface engineering of electronic properties of graphene/boron nitride lateral heterostructures, 2D Mater. 2 (2015) 041001.
[14] Y. Zhang, X. He, M. Sun, J. Wang, P. Ghosez, Switchable metal-to-half-metal transition at semi-hydrogenated graphene/ferroelectric interface, Nanoscale 12 (2020) 8.
[15] K. Wakabayashi, Electronic and Magnetic Properties of Nanographites, Chapter 12, in: T. Makarova, F. Palacio (Eds.), Carbon-Based Magnetism, Elsevier, 2006.
[16] S. Okada, K. Nakada, K. Kuwabara, K. Daigoku, T. Kawai, Ferromagnetic spin ordering on carbon nanotubes with topological line defects, Phys. Rev. B 74 (2006) 121412.
[17] F. Hagelberg, A. Kaiser, I. Sukuba, M. Probst, Spin filter properties of armchair graphene nanoribbons with substitutional Fe atoms, Mol. Phys. 115 (2017) 2231.
[18] G. Lee, K. Cho, Electronic structures of zigzag graphene nanoribbons with edge *hydrogenation and oxidation*, Phys. Rev. B 79 (2009) 165440.
[19] Z. Kan, C. Nelson, M. Khatun, Quantum conductance of zigzag graphene oxide nanoribbons, J. Appl. Phys. 115 (2014) 153704
[20] S.-L. Yan, M.-Q. Long, X.-J. Zhang, H. Xu, The effects of spin-filter and negative differential resistance on Fe-substituted zigzag graphene nanoribbons, Phys. Lett. A 378 (2014) 960.
[21] C. Cao, L.-N. Chen, M.Q. Long, W.-R. Huang, H. Xu, Electronic transport properties on transition-metal terminated zigzag graphene nanoribbons, J. Appl. Phys. 111 (2012) 113708.

[22] A. Garcia-Fuente, L.J. Gallego, A. Vega, Spin-polarized transport in hydrogen-passivated graphene and silicene nanoribbons with magnetic transition-metal substituents, Phys. Chem. Chem. Phys. 18 (2016) 22606.
[23] D.B. Zhang, S.-H. Wei, Inhomogeneous strain-induced half-metallicity in bent zigzag graphene nanoribbons, Comp. Mater. 3 (2017) 32.
[24] M. Wimmer, I. Adagideli, S. Berber, D. Tomanek, K. Richter, Spin currents in rough graphene nanoribbons: universal fluctuations and spin injection, Phys. Rev. Lett. 100 (2008) 177207.
[25] J.P.C. Baldwin, Y. Hancock, Effect of random edge-vacancy disorder in zigzag graphene nanoribbons, Phys. Rev. B 94 (2016) 165126.
[26] H. Sevinçli, M. Topsakal, E. Durgun, S. Ciraci, Electronic and magnetic properties of 3d transition-metal atom adsorbed graphene and graphene nanoribbons, Phys. Rev. B 77 (2008) 195434.
[27] A.V. Krasheninnikov, P.O. Lehtinen, A.S. Foster, P. Pyykkö, R.M. Nieminen, Embedding transition-metal atoms in graphene: structure, bonding, and magnetism, Phys. Rev. Lett. 102 (2009) 126807.
[28] E. Santos, A. Ayuela, D. Sánchez-Portal, First-principles study of substitutional metal impurities in graphene: structural, electronic and magnetic properties, New J. Phys. 12 (5) (2010) 053012.
[29] C.B. Crook, C. Constantin, T. Ahmed, J.-X. Zhu, A.V. Balatsky, J.T. Haraldsen, Proximity-induced magnetism in transition-metal substituted graphene, Sci. Rep. 5 (2015) 12322.
[30] C.B. Crook, G. Houchins, J.-X. Zhu, A.V. Balatsky, C. Constantin, J.T. Haraldsen, Spatial dependence of the super-exchange interactions for transition-metal trimers in graphene, J. Appl. Phys. 123 (2018) 013903.
[31] R.C. Longo, J. Carrete, L.J. Gallego, Magnetism of substitutional Fe impurities in graphene nanoribbons, J. Chem. Phys. 134 (2011) 024704.
[32] N.K. Jaiswal, P. Srivastava, Structural stability and electronic properties of Ni-doped armchair graphene nanoribbons, Solid State Commun. 151 (2011) 1490.
[33] N.K. Jaiswal, P. Srivastava, Fe-doped Armschair Grapehene nanoribbons for spintronic/interconnect applications, IEEE Trans. Nanotechnol. 12 (2013) 685.
[34] P. Esquinazi, A. Setzer, R. Höhne, C. Semmelhack, Y. Kopelevich, D. Spemann, T. Butz, B. Kohlstrunk, M. Lösche, Ferromagnetism in oriented graphite samples, Phys. Rev. B 66 (2002) 024429.
[35] http://www.openmx-square.org.
[36] T. Ozaki, T. Ozaki, H. Kino, Variationally optimized atomic orbitals for large-scale electronic structures, Phys. Rev. B 67 (2003) 155108. Numerical atomic basis orbitals from H to Kr, Phys. Rev. B 69 (2004) 195113.
[37] K. Lejaeghere, et al., Reproducibility in density functional theory calculations of solids, Science 351 (2016) 1415. https://molmod.ugent.be/deltacodesdft.
[38] H.J. Monkhorst, J.D. Pack, Special points for Brillouin zone integration, Phys. Rev. B 13 (1976) 5188.
[39] QuantumATK version 0-2018.06, Synopsys QuantumATK. https://www.synopsys.com/silicon/quantumatk.html.
[40] M. Brandbyge, J.-L. Mozos José-Luis, P. Pablo Ordejón, J. Taylor Jeremy, K. Stokbro, Density-functional method for nonequilibrium electron transport, Phys. Rev. B 65 (2002) 165401.
[41] J. Taylor, H. Guo, J. Wang, Ab initio modeling of quantum transport properties of molecular electronic devices, Phys. Rev. B 63 (2001) 245407.
[42] M. Büttiker, Y. Imry, R. Landauer, S. Pinhas, Generalized many-channel conductance formula with application to small rings, Phys. Rev. B 31 (1985) 6207.
[43] S. Datta, Quantum Transport—Atom to Transistor, Cambridge University Press, 2005.
[44] Y.-W. Son, M.L. Cohen, S.G. Louie, Energy gaps in graphene nanoribbons, Phys. Rev. Lett. 97 (2006) 216803.
[45] S. Datta, Quantum Transport—Atom to Transistor, Cambridge University Press, 2005.
[46] M. Magno, F. Hagelberg, Strained zigzag graphene nanoribbon devices with vacancies as perfect spin filters, J. Mol. Model. 24 (2018) 1.
[47] D.R. Cooper, B. D'Anjou, N. Ghattamaneni, B. Harack, M. Hilke, A. Horth, N. Majlis, M. Massicotte, L. Vandsburger, E. Whiteway, V. Yu, Experimental review of graphene, International Scholarly Research Network (ISRN), Condens. Matter Phys. 2012 (2012) 501686.
[48] D.A. Areshkin, C.T. White, Building blocks for integrated graphene circuits, NanoLett. 7 (2007) 3253.
[49] D. Gunlycke, D.A. Areshkin, J. Li, J.W. Mintmire, C.T. White, Graphene nanostrip digital memory device, NanoLett. 7 (2007) 3253.
[50] M. Bockstedte, A. Kley, J. Neugebauer, M. Scheffler, Density-functional theory calculations for polyatomic systems: electronic structure, static and elastic properties and ab initio molecular dynamics, Comput. Phys. Commun. 107 (1997) 187.

[51] M. Manadé, F. Viñes, F. Illas, Transition metal adatoms on graphene: a systematic density functional study, Carbon 95 (2015) 525.
[52] M.P. Lima, A.J.R. da Silva, A. Fazzio, Adatoms in graphene as a source of current polarization: role of the local magnetic moment, Phys. Rev. B 84 (2011) 245411.
[53] S. Ganguly, S. Basu, Magnetic adatoms in two and four terminal graphene nanoribbons: a comparison between their spin polarized transport, Phys. E. 98 (2018) 174.

Index

Note: Page numbers followed by "*f*" indicate figures and "*t*" indicate tables.

H

I

J

L

M

Printed in the United States
by Baker & Taylor Publisher Services